MAGILL'S ENCYCLOPEDIA OF SCIENCE

PLANT LIFE

MAGILL'S ENCYCLOPEDIA OF SCIENCE

PLANT LIFE

Volume 2
DNA Replication–Metabolites: Primary vs. Secondary

Editor
Bryan D. Ness, Ph.D.
Pacific Union College, Department of Biology

Project Editor
Christina J. Moose

SALEM PRESS, INC.
Pasadena, California
Hackensack, New Jersey

Editor in Chief: Dawn P. Dawson

Managing Editor: Christina J. Moose

Manuscript Editor: Elizabeth Ferry Slocum

Assistant Editor: Andrea E. Miller

Research Supervisor: Jeffry Jensen

Acquisitions Editor: Mark Rehn

Photograph Editor: Philip Bader

Production Editor: Joyce I. Buchea

Page Design and Graphics: James Hutson

Layout: William Zimmerman

Illustrator: Kimberly L. Dawson Kurnizki

Some of the updated and revised essays in this work originally appeared in *Magill's Survey of Science: Life Science* (1991), *Magill's Survey of Science: Life Science, Supplement* (1998), *Natural Resources* (1998), *Encyclopedia of Genetics* (1999), *Encyclopedia of Environmental Issues* (2000), *World Geography* (2001), and *Earth Science* (2001).

∞ The paper used in these volumes conforms to the American National Standard for Permanence of Paper for Printed Library Materials, Z39.48-1992 (R1997).

Library of Congress Cataloging-in-Publication Data

Magill's encyclopedia of science : plant life / edited by Bryan D. Ness.

    p. cm.
Includes bibliographical references (p.   ).
   ISBN 1-58765-084-3 (set : alk. paper) — ISBN 1-58765-085-1 (vol. 1 : alk. paper) —
ISBN 1-58765-086-X (vol. 2 : alk. paper) — ISBN 1-58765-087-8 (vol. 3 : alk. paper) —
ISBN 1-58765-088-6 (vol. 4 : alk. paper)
   1. Botany—Encyclopedias.   I. Ness, Bryan D.
QK7 .M34 2002
580'.3—dc21

2002013319

Second Printing

PRINTED IN THE UNITED STATES OF AMERICA

TABLE OF CONTENTS

LIST OF ILLUSTRATIONS, CHARTS, AND TABLES

List of Illustrations, Charts, and Tables

ALPHABETICAL LIST OF CONTENTS

Volume 1

Volume 2

Volume 3

Volume 4

MAGILL'S ENCYCLOPEDIA OF SCIENCE

PLANT LIFE

DNA REPLICATION

Categories: Cellular biology; genetics; reproduction and life cycles

DNA, deoxyribonucleic acid, is the hereditary material of most living creatures. It carries genetic information that determines all types of plant lives. DNA replication is a process by which a single DNA molecule is copied, resulting in two identical molecules prior to the cell division. The accuracy and precision in DNA replication has ensured the continuity of life from generation to generation.

Following James Watson and Francis Crick's landmark proposal for the structure of the deoxyribonucleic acid (DNA) molecule in 1953, many scientists turned their attention to how this molecule is replicated. The process of replication is the key to continuity of plant as well as other forms of life. The DNA molecule consists of two polymer chains (strands) forming a *double helix*. Each strand is made up of a five-carbon sugar (2′-deoxyribose), phosphoric acid, and four nitrogen-containing bases. Two bases are *purines*, which have a double-ring structure; the other two are *pyrimidines*, which contain a single ring. The purine bases are adenine (A) and guanine (G), while the pyrimidine bases are thymine (T) and cytosine (C). The two strands of a DNA molecule running in opposite directions are held together by hydrogen bonding between the A-T and G-C base pairs, which form the "rungs" of the ladder that makes the double helix. Such complementary *base pairing* is the foundation for the DNA double helix as well as its replication.

Semiconservative Replication

The replication mechanism first proposed by Watson and Crick was that the strands of the original (*parental*) duplex separate, and each individual strand serves as a pattern or template for the synthesis of a new strand (*daughter*). The daughter strands are synthesized by the addition of successive nucleotides in such a way that each base in the daughter strand is complementary to the base across the way in the template strand. This is called *semiconservative replication*.

Although the mechanism is simple in principle, replication is a complex process, with geometric problems requiring energy, a variety of enzymes, and other proteins. The end result, nevertheless, is that a single double-stranded DNA is replicated into two copies having identical sequences. Each of the two daughter double-stranded DNA copies is made up of one parental strand and one newly synthesized strand, hence the name semi-conservative (literally, "half conserving") replication.

DNA Synthesis

DNA replication in plant cells occurs in three basic steps, each catalyzed by at least one enzyme. First, the two original, or parental, DNA strands of the double helix unwind and separate. Then each parental strand is used as a template for the formation of a new daughter strand. Finally, one parental strand and its newly synthesized daughter strand wind together into one double helix, while the other parental strand and its daughter strand wind together into a second double helix.

The process begins with the separation and unwinding of segments of the parental double helix. To accomplish this, an enzyme named *DNA helicase*, powered by adenosine triphosphate (ATP), works its way between the two strands. As this enzyme "plows" its way through the double helix, it breaks the hydrogen bonds that hold together the "rungs" of the ladder, formed by the base pairs. DNA helicase then "walks" along one strand, nudging the other strand out of its way as it goes. The result is that the two DNA strands separate, thus exposing their bases, as though the ladder had been split vetically, down through the rungs.

The second step of DNA synthesis requires the enzyme *DNA polymerase*, which performs a dual function during the replication. First, it recognizes bases exposed in a parental strand and matches them up with free nucleotides that have complementary bases. Second, DNA polymerase bonds to-

Stages in DNA Replication

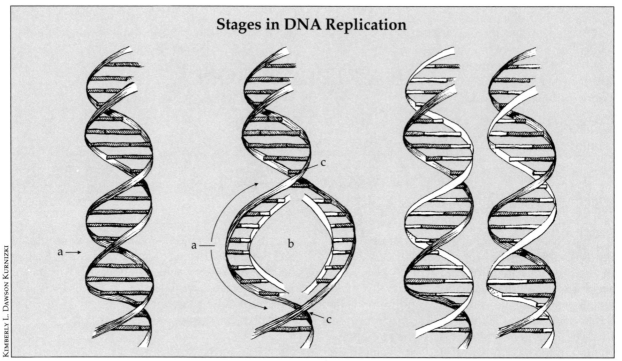

KIMBERLY L. DAWSON KURNIZKI

At left, a double-stranded DNA molecule, with the sides made of sugar-phosphate molecules and the "rungs" of base pairs. Replication begins at a point (a), with the separation of a base pair, as a result of the action of special initiator proteins (b). The molecule splits, or "unzips," in opposite directions (c) as each parental strand is used as a template for the formation of a new daughter strand (new bases pair with their appropriate "mate" bases to form new ladder "rungs"). Finally (right), one parental strand and its newly synthesized daughter strand form a new double helix, while the other parental strand and its daughter strand form a second double helix.

gether the sugars and phosphates of the complementary nucleotides to form the backbone of the daughter strand. Because the DNA polymerase can travel in only one direction on a DNA strand, from the 3′ end to 5′ end, the two DNA polymerase molecules, one on each parental strand, move in opposite directions. Only one daughter strand is synthesized continuously, while the duplication of another is done piece by piece. As the DNA helicase molecule continues to separate the parental strands, one polymerase simply follows behind it, synthesizing a long, continuous complementary daughter strand as it goes. The polymerase on the second parental strand travels away from the DNA helicase. As the helicase continues to separate the parental strands, this polymerase cannot reach the newly separated segment of the second strand. Hence a new DNA polymerase attaches to the second strand close behind the helicase to synthesize another small piece of DNA. These pieces are "sewn" together by another enzyme called *DNA ligase*. This process is repeated many times until the copying of a second parental strand is completed.

Proofreading and End-Sealing

Hydrogen bonding between complementary base pairs makes DNA replication highly accurate. However, the process is not perfect. This is a result partly of the fast pace (fifty to five hundred nucleotides added per second) and partly to the chemical flip-flops in the bases occurring spontaneously. DNA polymerase occasionally matches bases incorrectly, making on average one mistake in every ten thousand base pairs. Even this low rate of errors, if left uncorrected, would be devastating to the continuity of life. In reality, replicated DNA strands contain only one mistake in every billion base pairs. This incredible accuracy is achieved by several DNA repair enzymes, including DNA polymerase. Mismatches and errors are corrected through "proofreading" by DNA polymerase or other repair enzymes.

A separate problem in DNA replication lies at the end of a linear DNA molecule, which is not suitable for replication by polymerase. Yet another enzyme, *DNA telomerase*, is involved in solving this problem. It attaches a long stretch of repeating sequence of nucleotide 5'-TTGGGG-3' to the strand already synthesized by polymerase. This ending sequence (called the *telomere*) offers protection and provides stability for plant DNA molecules.

Plant Propagation

Overall, the faithful replication and the amazing stability of each DNA molecule ensure the continuity and survival of species from generation to generation. DNA replication precedes every cycle of mitosis, so that two daughter cells derived from cell division can inherit a full complement of genetic material. The replication is also an essential prerequisite for plant sexual reproduction, a process by which the two sexual cells (pollen and egg) produced via meiosis are united to start a new generation.

The faithful DNA replication and subsequent cell divisions (mitosis) form the basis for plant growth as well as for vegetative propagation in many plant species. Growth in size and volume of the plant results from a combination of mitosis, cell enlargement, and cell differentiation. Mitosis is also essential in wound healing through the production of a mass of cells called callus. Vegetative propagation, an asexual process through mitosis, plays an important role in agriculture. Through vegetative propagation, individual plants of the progeny population are genetic copies both of the original source plant and of one another. Such plants are known as *clones*, and the process is called cloning. Examples of cloning include grafting hardwood cuttings of grapevines and apple trees and rapid propagation of liriope by crown division. The best-known example of vegetative propagation is probably the production of Macintosh apples via grafting. More recently, micropropagation via direct cell cultures and related biotechnology has played a critical role in agriculture.

Ming Y. Zheng

See also: Biotechnology; Cell cycle; Chromatin; Chromosomes; Cloning of plants; DNA in plants; DNA: recombinant technology; Gene regulation; Genetic code; Genetics: Mendelian; Genetics: mutations; Genetics: post-Mendelian; Mitosis and meiosis; Nucleic acids; Nucleus; Plant biotechnology; RNA.

Sources for Further Study
Crick, Francis. *What Mad Pursuit: A Personal View of Scientific Discovery*. New York: Basic Books, 1998. A insider's view of the race to determine the DNA structure.
Frank-Kamenetskii, Maxim D. *Unraveling DNA: The Most Important Molecule of Life*. Reading, Mass.: Addison-Wesley, 1997. Elucidates the history of discovery related to DNA, a molecule that holds the greatest mysteries of life.
Judson, H. F. *The Eighth Day of Creation*. Cold Spring Harbor, N.Y.: Cold Spring Harbor Laboratory Press, 1993. An easy-to-understand account of the development of genetic science.
Kornberg, A., and T. A. Baker. *DNA Replication*. New York: W. H. Freeman, 1992. Provides historical accounts and clear logic for DNA synthesis and the replication process.

DORMANCY

Categories: Gardening; physiology

Dormancy is the state in which a plant or plant part exhibits little or no growth and in which most, if not all, metabolic activity ceases for a period of time.

The vast majority of plant life functions best when there is ample water and temperatures are well above freezing throughout the year. Except for those in moist, tropical regions, however, plants are exposed to dry periods and temperatures below freezing for varying lengths of time during the year. Plants, unlike animals, do not have the luxury of body insulation or locomotion. Hence, plants cannot seek shelter or use other active ways to survive water shortages and cold weather. Consequently, many plants become dormant to avoid unfavorable environmental conditions. In dormancy, their *metabolic activity* either ceases or is drastically reduced.

Dormancy evolved as a means of surviving unfavorable environmental conditions. In the temperate zones, buds normally form from spring to midsummer. While there may be a little growth in the late summer, growth virtually ceases in the fall in preparation for winter. Entering a dormant state protects the buds from freezing temperatures.

Mature seeds contain a complete embryo along with a reserve food supply—and mature seeds are formed within ripened fruit. If it were not for germination inhibitors present in the fruit or seed, the seeds would begin to germinate while still in the fruit. In addition, the seeds of plants in temperate regions most often reach the soil in late summer or early fall, when the plants are most often faced with low moisture and imminent cold weather. Were the seeds to germinate at this time, survival would be unlikely.

Patterns of Growth and Death

The type of dormant response depends on the plant's pattern of growth and death. *Perennials* are plants that live year after year, undergoing a period of dormancy during the cold season. In *herbaceous* species, the aboveground portions die, but the plants survive as specialized underground stems. Woody shrubs and trees remain alive aboveground. *Deciduous* species shed their leaves in winter, while many nondeciduous species, often called *evergreens*, keep their leaves year-round but dramatically reduce their metabolic rates.

Biennial species live for two years. The first year is devoted to vegetative growth and the formation of underground storage tissues. After the plant lies dormant through the winter months, its second year of growth results in the stored food supply being used to produce flowers and seeds. *Annuals* are plants that complete an entire life cycle in one growing season. The plants die, producing seeds, which normally remain dormant until the following growing season.

Most perennial trees and shrubs in temperate re-gions produce buds in the summer. These buds, which can eventually develop into leaves, stems, or flowers, exhibit reduced metabolic activity even before leaves begin to *senesce* (age). As temperatures decrease in the fall, complete dormancy sets in. Specialized leaves called bud scales cover the dormant tissue. These scales block the diffusion of oxygen into the bud; they also prevent the loss of water from the tissue.

Almost all flowering plants produce seeds. The seeds develop as ovules within a structural component of the flower called the *ovary*. As the ovary ripens to form the fruit, the ovules mature into seeds. Each seed is composed of a reserved food supply and a new plant with embryonic root, leaf, and stem tissue. The embryonic plant and reserve food supply are surrounded by a tough seed coat. The seeds of many species, especially trees in the temperate zones, do not germinate immediately after maturing even under ideal moisture, temperature, and nutritional conditions because there is a built-in period of dormancy.

Although seasonal dormancy is most often correlated with temperature changes, variation in precipitation is the primary factor in regions where pronounced wet and dry seasons alternate. Some deciduous trees and shrubs drop their leaves and remain dormant during the dry season and grow new leaves when the rains return. Herbaceous perennials die back and go dormant at the beginning of the dry season, then regrow their aboveground biomass. Many desert annuals have seeds that will only break dormancy when sufficient rains come, which in some regions may be only every few years. Some particularly specialized seeds germinate only after they have been tumbled in the waters of a flash flood, which scrapes their seed coats.

Bud and Seed Dormancy

A number of environmental factors appear to induce bud dormancy. In many species, bud dormancy occurs in response to low temperatures; among other species, the proper short-day *photoperiod* is responsible for initiating dormancy. In still other species, both low temperature and short days are required to trigger the onset of dormancy. Hence, dormancy is generally initiated with the onset of the short or cold days of winter. In addition, dormancy in buds has been shown to occur under situations of limited supply of nutrients or drought conditions. Dormancy can therefore be seen as a survival mechanism. When temperature, water, or nutritional conditions are no longer favorable, the buds become dormant.

Seed dormancy can also be caused by a number of different factors. For several reasons, the presence of a hard *seed coat* will very often result in dormancy of the seed. In many cases, the seed coat is impermeable to water. Because water is required for the *germination* process, the impermeable nature of the seed coat will serve as an effective inducer of dormancy. In some instances, the seed coat may be impermeable to certain gases. Both carbon dioxide and oxygen are required for germination; some seed coats prevent the diffusion of oxygen into the seed, while others are impermeable to carbon dioxide. In a few species, the seed coat physically restricts the growth of the embryo. A growing embryo must develop sufficient thrust to break through a seed coat, and in some instances the seed coat prevents this from happening.

Seed germination is also dependent on temperature. The seeds of almost all species have a minimal temperature below which they will not germinate. The exact mechanism by which low temperature causes dormancy is poorly understood, but it appears that the temperature alters membrane structure, which somehow prevents the seed from germinating.

Light is also a factor in the dormancy of many seeds. In many species, light is required for germination; in some cases, however, the exposure to light will induce dormancy. Also, some species exhibit a sensitivity to the photoperiod. Certain species are dormant during short-day cycles and germinate during long-day cycles, while others remain dormant when exposed to long-day cycles and germinate under short-day cycles. The light apparently activates a plant regulator that blocks the metabolic reactions necessary for germination.

Genetic and Chemical Control

Numerous studies show that bud dormancy is induced in some varieties of certain species but not in other varieties of the same species. This difference suggests there may be genetic variation in the control mechanism. The exact mechanisms of genetic control are not clearly understood, but there are some clues about chemical control.

The plant hormone *abscisic acid* may be associated with bud dormancy, but the evidence is inconclusive. With seed dormancy, however, the involve-

ment of abscisic acid is fairly certain. A number of studies have shown that abscisic acid, when applied to the seed, will block the activity of enzymes necessary for germination.

A number of germination-inhibiting substances are present in dormant seeds. *Respiratory inhibitors*, such as cyanide, are produced in some seeds. High concentrations of various inorganic salts prevent germination in some species, while an assortment of phenolic compounds are known to prevent the process in others. A compound known as *coumarin* is particularly widespread in seeds and is known to be an effective inhibitor of germination. A number of these germination-inhibiting substances are present in both fruit and seed and prevent the seeds from germinating within the fruit. In addition, the substances will cause the seeds to lie dormant in nature until sufficient rain has fallen to leach the substances from the seeds. This adaptation ensures that sufficient moisture will be available to support the young seedlings.

Breaking Dormancy

It is often desirable to break dormancy artificially in order to obtain faster growth or increase plant production. Treatments to release bud and seed dormancy can be categorized by their use of temperature, light, chemicals, or mechanical means. Artificial exposure to low temperatures, warmth, and altered photoperiods, either singly or in combination, mimics the natural environmental factors that break dormancy.

A number of chemicals break bud dormancy. Ethylene chlorohydrin has been used for years to release dormancy in many fruit trees, and natural hormones known as *gibberellins* will break dormancy in most cold-requiring plants when applied directly to the buds. Gibberellins, *cytokinins*, and *ethylene*, all natural plant hormones, have been shown to be involved in breaking seed dormancy, and the gibberellins and other substances, such as thiourea, are used to germinate seeds commercially. The *scarification* (mechanical breaking) of the seed coat has proven to be an effective means of overcoming dormancy in many seeds. The broken seed coats allow the seeds to take up the water and gases necessary for germination.

D. R. Gossett, updated by Bryan Ness

See also: Germination and seedling development; Hormones.

Sources for Further Study

Campbell, Neil A., and Jane B. Reece. *Biology*. 6th ed. San Francisco: Benjamin Cummings, 2002. An introductory college-level textbook for science students. The chapter on plant reproduction provides a clear, concise description of seed dormancy. The well-written text and superb graphics furnish the reader with a clear understanding of dormancy. List of suggested readings at the end of the chapter. Includes glossary.

Lang, G. A., ed. *Plant Dormancy: Physiology, Biochemistry, and Molecular Biology*. Wallingford, England: CAB International, 1996. A collection of papers from a symposium on plant dormancy. Presents a valuable review of plant dormancy for crop scientists, plant physiologists, plant molecular biologists, horticulturists, and all those involved in investigating and managing plant dormancy.

Raven, Peter H., Ray F. Evert, and Susan E. Eichhorn. *Biology of Plants*. 6th ed. New York: W. H. Freeman/Worth, 1999. An introductory college textbook for science students. Chapter 26, "External Factors and Plant Growth," provides a good discussion of dormancy. The profusely illustrated text furnishes an excellent general overview of the phenomenon in both buds and seeds. Includes glossary.

Salisbury, Frank B., and Cleon Ross. *Plant Physiology*. 4th ed. Belmont, Calif.: Wadsworth, 1999. An intermediate college-level textbook for science students. Chapter 21, "Growth Response to Temperature," gives an in-depth view of the physiological role of dormancy. An excellent explanation of the phenomenon is provided in text and graphics. Detailed bibliography at the end of the chapter.

DROUGHT

Categories: Agriculture; environmental issues

Drought is a shortage of precipitation that results in a water deficit for some activity. Droughts occur in both arid and humid regions.

One problem in analyzing and assessing the impacts of drought, as well as in delimiting drought areas, is simply defining "drought" itself. Conditions considered a drought by a farmer whose crops have withered during the summer may not be seen as a drought by a city planner. There are many types of drought: *agricultural, hydrological, economic,* and *meteorological.* The Palmer Drought Severity Index is the best known of a number of indexes that attempt to standardize the measurement of drought magnitude. Nevertheless, there still is much confusion and uncertainty on what defines a drought.

Roger G. Barry and Richard J. Chorley, in *Atmosphere, Weather, and Climate* (1998), have noted that drought conditions tend to be associated with one or more of four factors: increases in extent and persistence of subtropical high-pressure cells; changes in the summer monsoonal circulation patterns that can cause a postponement or failure of the incursion of wet maritime tropical air onto the land; lower ocean surface temperatures resulting from changes in ocean currents or increased upwelling of cold waters; and displacement of midlatitude storm tracks by drier air.

Effects of Drought

Drought can have wide-ranging impacts on the environment, communities, and farmers. Most plants and animals in arid regions are adapted to dealing with drought, either behaviorally or through specialized physical adaptations. Humans, however, are often unprepared or overwhelmed by the consequences of drought. Farmers experience decreased incomes from crop failure. Low rainfall frequently increases a crop's susceptibility to disease and pests. Drought can particularly hurt small rural communities, especially local business people who are dependent on purchases from farmers and ranchers.

Drought is a natural element of climate, and no region is immune to the drought hazard. Farmers in humid areas grow crops that are less drought-resistant than those grown in arid regions. In developing countries the effects of drought can include malnutrition and famine. A prolonged drought struck the Sahel zone of Africa from 1968 through 1974. Nearly 5 million cattle died during the drought, and more than 100,000 people died from malnutrition-related diseases during just one year of the drought.

Subsistence and traditional societies can be very resilient in the face of drought. American Indians either stored food for poor years or migrated to wetter areas. The !Kung Bushmen of southern Africa learned to change their diet, find alternate water sources, and generally adapt to the fluctuation of seasons and climate in the Kalahari Desert.

More than any other event, the Dust Bowl years of the 1930's influenced Americans' perceptions and knowledge of drought. Stories of dust storms that turned day into night, fences covered by drifting soil, and the migration of destitute farmers from the Great Plains to California captured public and government attention. The enormous topsoil loss to wind erosion, continuous crop failures, and widespread bankruptcies suggested that the United States had in some way failed to adapt to the drought hazard.

Federal Drought Response in the United States

Beginning in the 1930's, the federal government took an increasing role in drought management and relief. In 1933 the federal government created the Soil Erosion Service, known today as the Natural Resources Conservation Service. No other single federal program or organization has had a greater impact on farmers' abilities to manage the drought hazard. President Franklin D. Roosevelt's Prairie States Forestry Project (1934-1942) planted more than 230,000 acres of shelterbelts in the Plains states for wind erosion control. The federal government pur-

chased nearly 1 million acres of marginal farmland for replanting with grass. Federal agencies constructed water resource and irrigation projects.

Post-Dust Bowl droughts still caused hardships, but the brunt of the environmental, economic, and social consequences of drought were considerably lessened. Fewer dust storms ravaged the Plains. New crop varieties and better farming practices decreased crop losses during drought years. Government programs and better knowledge have enabled families and communities to cope better with drought.

Coping with Future Droughts

Numerous attempts have been made to predict droughts, especially in terms of cycles. However, attempts to predict droughts one or more years into the future have generally been unsuccessful. The

Low rainfall is one factor that can lead to drought, leaving the soil too parched for plants to grow. Drought is a natural element of climate, and no region is immune to drought hazard.

Impacts of Drought

While a farmer is aware of drought conditions in the course of his or her day-to-day activities, an urban dweller first knows about drought conditions only when he or she reads a newspaper article or views a television news item. However, droughts become important to urban inhabitants when they are affected directly.

Droughts affect urban and rural users whose water comes from wells when the water table lowers and the water supply either diminishes or becomes more expensive to pump. Redrilling wells to pump from lowered water levels can be expensive. Users hooked up to a water utility receive notices about drought from their water supplier.

In dry climates, California for example, drought is classified into five levels, depending on the severity of a drought and projected availability of water supplies. The classifications and penalties for excess use have been defined by the California State Water Resources Control Board. Drought levels increase as the projected severity increases. In the lowest level (Level 1), voluntary conservation measures are encouraged, and water conservation information is published by the supplier. As the anticipated severity increases, conservation measures become mandatory and more restrictive. The water utility imposes cutback goals. Fines for overuse may be imposed and, in egregious instances, water service can be terminated until corrective action is taken. To encourage conservation even when plentiful water supplies are projected, water rates have been implemented to include inverted rate blocks (the more one uses, the higher unit price one pays).

(usually meaning within one to twelve months). Early recognition of potential drought conditions can give policymakers and resource managers the extra time needed to adjust their management strategies. Information on soil moisture conditions aids farmers with planting and crop selection, seeding, fertilization, irrigation rates, and harvest decisions. Communities that have a few months' warning of impending drought can increase water storage, implement water conservation measures, and obtain outside sources of water.

Unfortunately, the progress made in the world's developed countries has not always been available to the developing nations. Overpopulation and overuse of agricultural lands have resulted in regional problems of *desertification* and impeded the ability of developing nations to respond. Monitoring equipment can be costly. Furthermore, drought adjustments used in the United States may not be applicable to other countries' drought situations.

shorter the prediction interval has been, the more accurate the prediction has been. Nevertheless, progress has been made in estimating drought occurrence and timing. For example, the El Niño/Southern Oscillation may be a precursor of drought in some areas. Possibly with time, the physical mechanics of climate and drought will be understood adequately for long-term predictions to have value.

Perhaps of greater value is the current capacity to detect and monitor drought in its early stages

David M. Diggs

See also: Agriculture: modern problems; Cacti and succulents; Climate and resources; Deserts; Desertification; Erosion and erosion control; Human population growth; Hydrologic cycle; Irrigation; Leaf anatomy.

Sources for Further Study

Barry, Roger G., and Richard J. Chorley. *Atmosphere, Weather, and Climate.* 7th ed. New York: Routledge, 1998. A good primer on drought and climate.

Belhassen, E., ed. *Drought Tolerance in Higher Plants: Genetical, Phyiological, and Molecular Biological Analysis.* Boston: Kluwer, 1997. Compilation of reviews on genetics and plant breeding and developments in drought tolerance in higher plants. For postgraduate students, teachers, and scientists.

Hewitt Ken, ed. *Interpretations of Calamity from the Viewpoint of Human Ecology.* Boston: Allen & Unwin, 1983. Includes a number of articles on how different societies adjust to the drought hazard.

McClure, Susan. *Water.* New York: Workman, 2000. Use and conservation of water for gardeners; includes an encyclopedia of more than one hundred drought-tolerant plants.

ECOLOGY: CONCEPT

Categories: Disciplines; ecology; ecosystems; forests and forestry

Ecology is the scientific study of ecosystems, which are generally defined as local units of nature, consisting of the aggregate of plants, animals, the physical environment, and their interactions.

Ecosystems consist of both *biotic* (living) and *abiotic* (nonliving) components. Biotic components include plants, animals, and microorganisms. The abiotic components are the physical factors of the ecosystem. The roots of ecology can be traced to the writings of early Greek philosophers such as Aristotle and Theophrastus, who were keen observers of plants and animals in their natural habitats. During the nineteenth century, German biogeographer Alexander von Humboldt and English naturalist Charles Darwin wrote detailed descriptions of their travels. They recognized that the distribution of living things is determined by such factors as rainfall, temperature, and soil.

The word "ecology" was first proposed in 1869 by the German biologist Ernst Haeckel. It soon came to be defined as "environmental biology," or the effect of environmental factors on living things. At the beginning of the twentieth century, American botanist Henry Cowles established *plant succession* as a major concept of ecology. During the next few decades, F. E. Clements helped establish plant ecology as a recognized branch of biology. Animal ecology developed separately and slightly later. Reflecting the independent development of plant and animal ecology, most ecological studies in the early twentieth century were concerned with either plant or animal communities but not both. Furthermore, most were descriptive rather than being involved with explanations of fundamental ecological processes.

However, other lines of research during this time increasingly emphasized interrelationships among all life-forms, especially those within lakes. From such beginnings emerged the concept of the ecosystem. The term, first used by British ecologist Arthur G. Tansley in the 1930's, is now considered the foundation stone of ecology.

Ecosystems at Work

Ecologists place all the organisms of an ecosystem into three categories: *producers, consumers,* and *decomposers.* Producers include algae and green plants that, because of their ability to generate biochemical energy by means of *photosynthesis,* produce all the food for the ecosystem. Consumers (herbivores, omnivores, and carnivores) are animals that feed directly on the producers. Decomposers are bacteria and fungi that break the large organic molecules of dead plants and organisms down into simpler substances. Ecosystems are dynamic. Each day, matter (nutrients) cycles through ecosystems as consumers eat producers, then moves to decomposers, which release nutrients into the soil, air, and water. Producers absorb nutrients as they photosynthesize, thus completing the cycle. Energy flows from the sun to producers, then consumers, and finally to decomposers.

Other changes in ecosystems occur over longer periods of time. Large-scale disturbances such as fire, logging, and storms initiate gradual, long-term changes known as *ecological succession.* Following a major disturbance, an ecosystem of pioneer species exists for a while but is soon replaced by a series of other temporary ecosystems. Eventually a permanent, or climax, ecosystem is formed, the nature of which is primarily determined by climate. Although generally considered to be stable, climax communities are subject to gradual changes caused by climatic fluctuations or subsequent disturbances.

Classifying Systems and Study

Ecologists attempt to name and classify ecosystems in a manner similar to the way that taxonomists name and classify species. Ecosystems may be named according to their dominant plants, such as deciduous forests, prairies, and evergreen forests. Others, such as coral reefs, are named according to their dominant animals. Physical factors are used to name deserts, ponds, tidal pools, and other ecosystems.

The science of ecology has developed a few major branches as well as several areas of specialization, with theoretical, or academic, ecology on one hand and applied, or practical, ecology on the other. Among the specialties that have developed from *theoretical ecology* are *autecology* (study at the level of individuals or species), *synecology* (study at the community level), and the *ecosystems approach*, which is largely concerned with the flow of energy and the cycling of nutrients within ecosystems. As one might expect, theoretical ecologists are generally associated with universities, where their basic research contributes to the understanding of a great diversity of ecosystems.

Applied ecologists are employed by a variety of governmental and environmental agencies as well as by universities. Their primary objective is to apply fundamental principles of ecology and related disciplines to the solution of specific problems. *Forestry*, although it developed independently from ecology, may be considered a specialty within the field of *applied ecology*. Foresters must be knowledgeable of a wide range of factors that influence the accumulation of tree biomass. Wildlife management, once concerned with only game species, has been extended to include a wide range of nongame animal species as well. Another branch of applied ecology is *conservation biology*. Concerned with biodiversity in all its aspects, conservation biologists attempt to prevent the extinction of threatened species around the globe. Disturbed ecosystems are rehabilitated by scientists working in a related field called *restoration ecology*.

The efforts of ecologists, whether theoretical or applied, represent attempts to understand and solve the many environmental challenges that humankind faces. Climate change, pollution, and other global and local problems contribute to a loss of biodiversity. All are made worse by increasing human activities.

Thomas E. Hemmerly

See also: Animal-plant interactions; Biomass related to energy; Biomes: types; Coevolution; Community-ecosystem interactions; Ecology: history; Ecosystems: overview; Ecosystems: studies; Food chain; Paleoecology; Population genetics; Succession; Trophic levels and ecological niches.

Sources for Further Study

Bush, Mark B. *Ecology of a Changing Planet*. 2d ed. Upper Saddle River, N.J.: Prentice Hall, 2000. An introductory text that outlines the fundamental ecological principles which serve as a foundation to understand environmental issues. Explores specific environmental issues including habitat fragmentation, acid deposition, and the emergence of new human diseases.

Hunter, Malcolm L., Jr. *Fundamentals of Conservation Biology*. 2d ed. Malden, Mass.: Blackwell Science, 2002. Introduces and explains the concept of conservation biology and the applied science of maintaining the earth's biological diversity. Addresses social, political, and economic issues in a manner that can be readily understood by people outside of the field who are concerned about the future of Earth and its inhabitants.

Newman, Edward I. *Applied Ecology and Environmental Management*. 2d ed. Malden, Mass.: Blackwell Science, 2000. A textbook for undergraduate students of biological or environmental sciences, integrating material usually scattered through many courses in different departments.

Smith, Robert L. *Ecology and Field Biology*. 6th ed. San Francisco: Benjamin Cummings, 2001. A comprehensive overview of all aspects of ecology, including evolution, ecosystems theory, practical applications, plants, animals, biogeochemical cycles, and global change.

ECOLOGY: HISTORY

Categories: Disciplines; ecology; ecosystems; history of plant science

Ecology is the science that studies the relationships among organisms and their biotic and abiotic environments. The term "ecology" is commonly, but mistakenly, used by people to refer to the environment or to the environmental movement.

The study of ecological topics arose in ancient Greece, but these studies were part of a catch-all science called natural history. The earliest attempt to organize an ecological science separate from natural history was made by Carolus Linnaeus in his essay *Oeconomia Naturae* (1749; *The Economy of Nature*), which focused on the balance of nature and the environments in which various natural communities existed. Although the essay was well known, the eighteenth century was dominated by biological exploration of the world, and the science of ecology did not develop.

Early Ecological Studies

The study of fossils led some naturalists to conclude that many species known only as fossils must have become extinct. However, Jean-Baptiste Lamarck argued in his *Philosophie zoologique* (1809; *Zoological Philosophy*, 1914) that fossils represented the early stages of species that evolved into different species that were still living. In order to refute this claim, geologist Charles Lyell mastered the science of biogeography and used it to argue that species do become extinct and that competition from other species seemed to be the main cause. English naturalist Charles Darwin's book *On the Origin of Species by Means of Natural Selection* (1859) blends his own researches with the influence of Linnaeus and Lyell in order to argue that some species do become extinct, but existing species have evolved from earlier ones. Lamarck had underrated and Lyell had overrated the importance of competition in nature.

Although Darwin's book was an important step toward ecological science, he and his colleagues mainly studied evolution rather than ecology. However, German evolutionist Ernst Haeckel realized the need for an ecological science and coined the name *oecologie* in 1866. It was not until the 1890's that steps were actually taken to organize this science. Virtually all of the early ecologists were specialists in the study of particular groups of organisms, and it was only in the late 1930's that some efforts were made to write textbooks covering all aspects of ecology. Since the 1890's, most ecologists have viewed themselves as *plant ecologists*, *animal ecologists*, *marine biologists*, or *limnologists*. Limnology is the study of freshwater aquatic environments.

Nevertheless, general ecological societies were established. The first was the British Ecological Society, which was founded in 1913 and began publishing the *Journal of Ecology* in the same year. Two years later ecologists in the United States and Canada founded the Ecological Society of America, which began publishing *Ecology* as a quarterly journal in 1920; in 1965 *Ecology* began appearing bimonthly. Other national societies have since been established. More specialized societies and journals also began appearing. For example, the Limnological Society of America was established in 1936 and expanded in 1948 into the American Society of Limnology and Oceanography. It publishes the journal *Limnology and Oceanography*.

Although Great Britain and Western Europe were active in establishing the study of ecological sciences, it was difficult for their trained ecologists to obtain full-time employment that utilized their expertise. European universities were mostly venerable institutions with fixed budgets; they already had as many faculty positions as they could afford, and these were all allocated to the older arts and sciences. Governments employed few, if any, ecologists. The situation was more favorable in the United States, Canada, and Australia, where universities were still growing. In the United States, the universities that became important for ecological research and the training of new ecologists were mostly in the Midwest. The reason was that eastern universities were similar to European ones in being well established, with scientists in traditional fields.

Ecology After 1950

Ecological research in the United States was not well funded until after World War II. With the advent of the Cold War, science was suddenly considered important for national welfare. In 1950 the U.S. Congress established the National Science Foundation, and ecologists were able to make the case for their research along with that of the other sciences. The Atomic Energy Commission had already begun to fund ecological researches by 1947, and under its patronage the Oak Ridge Laboratory and the University of Georgia gradually became important centers for radiation ecology research.

Another important source of research funds was the International Biological Program (IBP), which, though international in scope, depended upon national research funds. It got under way in the United States in 1968 and was still producing publications in the 1980's. Even though no new funding sources were created for the IBP, its existence meant that more research money flowed to ecologists than otherwise would have.

Ecologists learned to think big. Computers became available for ecological research shortly before the IBP got under way, and so computers and the IBP became linked in ecologists' imaginations. Earth Day, established in 1970, helped awaken Americans to the environmental crisis. The IBP encouraged a variety of studies, but in the United States, studies of biomes (large-scale environments) and ecosystems were most prominent. The biome studies were grouped under the headings of desert, eastern deciduous forest, western coniferous forest, grassland, and tundra (a proposed tropical forest program was never funded). Although the IBP has ended, a number of the biome studies continued at a reduced level.

Ecosystem studies are also large-scale, at least in comparison with many previous ecological studies, though smaller in size than a biome. The goal of ecosystem studies was to gain a total understanding of how an ecosystem—such as a lake, river valley, or forest—works. IBP funds enabled students to collect data, which computers processed. However, ecologists could not agree on what data to collect, how to compute outcomes, and how to interpret the

results. Therefore, thinking big did not always produce impressive results.

Plant Ecology

Because ecology is enormous in scope, it was bound to have growing pains. It arose at the same time as the science of genetics, but because genetics is a cohesive science, it reached maturity much sooner than ecology. Ecology can be subdivided in a wide variety of ways, and any collection of ecology textbooks will show how diversely it is organized by different ecologists. Nevertheless, self-identified professional subgroups tend to produce their own coherent findings.

Plant ecology progressed more rapidly than other subgroups and has retained its prominence. In the early nineteenth century, German naturalist Alexander von Humboldt's many publications on plant geography in relation to climate and topography were a powerful stimulus to other botanists. By the early twentieth century, however, the idea of plant communities was the main focus for plant ecologists. Henry Chandler Cowles began his studies at the University of Chicago in geology but switched to botany and studied plant communities on the Indiana dunes of Lake Michigan. He received his doctorate in 1898 and stayed at that university as a plant ecologist. He trained others in the study of community succession.

Frederic Edward Clements received his doctorate in botany in 1898 from the University of Nebraska. He carried the concept of plant community succession to an extreme by taking literally the analogy between the growth and maturation of an organism and that of a plant community. His numerous studies were funded by the Carnegie Institute in Washington, D.C., and even ecologists who disagreed with his theoretical extremes found his data useful. Henry Allan Gleason was skeptical; his studies indicated that plant species that have similar environmental needs compete with one another and do not form cohesive communities. Although Gleason first expressed his views in 1917, Clements and his disciples held the day until 1947, when Gleason's individualistic concept received the support of three leading ecologists. Debates over plant succession and the reality of communities helped increase the sophistication of plant ecologists and prepared them for later studies on biomes, ecosystems, and the degradation of vegetation by pollution, logging, and agriculture.

Marine Ecology

Marine ecology is viewed as a branch of either ecology or oceanography. Early studies were made either from the ocean shore or close to shore because of the great expense of committing oceangoing vessels to research. The first important research institute was the Stazione Zoologica at Naples, Italy, founded in 1874. Its successes soon inspired the founding of others in Europe, the United States, and other countries. Karl Möbius, a German zoologist who studied oyster beds, was an important pioneer of the community concept in ecology. Great Britain dominated the seas during the nineteenth century and made the first substantial commitment to deep-sea research by equipping the HMS *Challenger* as an oceangoing laboratory that sailed the world's seas from 1872 to 1876. Its scientists collected so many specimens and so much data that they called upon marine scientists in other countries to help them write the fifty large volumes of reports (1885-1895). The development of new technologies and the funding of new institutions and ships in the nineteenth century enabled marine ecologists to monitor the world's marine fisheries and other resources and provide advice on harvesting marine species.

Limnology is the scientific study of bodies of fresh water. The Swiss zoologist François A. Forel coined the term and also published the first textbook on the subject in 1901. He taught zoology at the Académie de Lausanne and devoted his life's researches to understanding Lake Geneva's characteristics and its plants and animals. In the United States in the early twentieth century, the University of Wisconsin became the leading center for limnological research and the training of limnologists, and it has retained that preeminence. Limnology is important for managing freshwater fisheries and water quality.

Frank N. Egerton

See also: Animal-plant interactions; Botany; Community-ecosystem interactions; Ecosystems: studies; Food chain; History of plant science; Trophic levels and ecological niches.

Sources for Further Study

Allen, Timothy F. H., and Thomas W. Hoekstra. *Toward a Unified Ecology.* New York: Columbia University Press, 1992. A theoretical perspective on ecology.

Hynes, H. Patricia. *The Recurring Silent Spring.* New York: Pergamon Press, 1989. This book remembers and commemorates Rachel Carson's classic *Silent Spring* (1962), while revealing current chemical and genetic warfare under way in agriculture and biotechnology.

Leuzzi, Linda. *Life Connections: Pioneers in Ecology.* Danbury, Conn.: Franklin Watts, 2000. Examines the lives of eight scientists who helped created the field of ecology. Includes bibliography and Internet sources.

Mills, Eric L. *Biological Oceanography: An Early History, 1870-1960.* Ithaca, N.Y.: Cornell University Press, 1989. History of marine biology includes index and references.

ECOSYSTEMS: OVERVIEW

Categories: Ecology; ecosystems

An ecosystem is made up of the complex interactions of a community of organisms of different species with one another and with their abiotic (nonliving) environment.

A *biological community* consists of a mixture of populations of individual species; a *population* consists of potentially interbreeding members of a species. Individual organisms interact with members of their own species as well as with other species. An ecosystem is formed by this web of interactions among species, along with the physical, chemical, and climatic conditions of the area.

Abiotic and Biotic Interactions

Abiotic environmental conditions include temperature, water availability, soil nutrient content, and many other factors that depend on the climate, soil, and geology of an area. Living organisms can alter their environment to some degree. A canopy formed by large forest trees, for example, will change the light, temperature, and moisture available to herbaceous plants growing near the forest floor. The environmental conditions in a particular area can also be affected by the conditions of neighboring areas. The disturbance of a stream bank can lead to erosion, which will affect aquatic habitat for a considerable distance downstream. It can be difficult to anticipate the wide-ranging affects of ecosystem disturbance.

Species and individuals within an ecosystem may interact directly with one another through the exchange of energy and material. *Predators*, for example, obtain their energy and meet their nutritional needs through consumption of prey species. Organisms also interact indirectly through modification of their surrounding environment. Earthworms physically modify soil structure, affecting aeration and the transport of water through the soil. These alterations of the physical environment, in turn, affect plant root growth and development as well as the ability of plants to secure nutrients.

Ecosystems are not closed systems: Energy and material are transferred to and from neighboring systems. The flow of energy or material among the components of an ecosystem, and exchanges with neighboring ecosystems, are governed by functions of the abiotic and biotic ecosystem components. These ecological processes operate simultaneously on many different temporal and spatial scales. At the same time that a microorganism is consuming a fallen leaf, the process of soil formation is occurring through chemical and physical weathering of parent material; plants are competing with one another for light, water, and nutrients.

Boundaries and Temporal Scales

Because of the exchange of energy and material, it is not possible to draw clear boundaries around an ecosystem. A *watershed*, for example, is formed by topographic conditions that create physical barriers guiding the gravitational flow of water, yet

Wetlands ecosystem on Bear Island, South Carolina.

wind carries seeds and pollen over these barriers, and animals can move from watershed to watershed. The strength of the interactions among neighboring systems is the basis on which humans delineate ecosystem boundaries. In truth, all ecosystems around the world interact with one another to some degree.

Ecological processes operate on many different time scales as well. Some operate over such long time scales that they are almost imperceptible to human observation. The process of soil formation occurs over many human life spans. Other processes operate over extremely short time intervals. The reproduction of soil bacteria, the response of leaves to changing temperature over the length of a day, and the time required for chemical reactions in the soil are all very short when compared to a human life span.

Ecosystem Disturbances

Ecosystems are subject to disturbance, or perturbation, when one or more ecosystem processes are interrupted. Disturbance is a natural ecological process, and the character of many ecosystems is shaped by natural disturbance patterns. There are species that require disturbance in order to regenerate themselves. These species may be present in great abundance following a disturbance. Their abundance then decreases over time, and if there is no disturbance to renew the population, they will eventually die out and no longer be present in the ecosystem.

The successful reproduction of many prairie species may be dependent on periodic fires. Suppression of fire as a means of protecting an ecosystem may lead to the local extinction of small plants, which depend on periodic fires to increase light availability by removing larger grasses and providing nutrients to the soil. The formation of sandbars in streams may be controlled by periodic flood events that remove great amounts of sediment from stream banks. Protection of existing ecosystems can depend on the protection or simulation of *natural disturbances*. This is even true of old-growth forests; the natural disturbance interval due to fire or windstorm may be centuries, and yet interruption of the natural disturbance pattern may lead to shifts in species composition or productivity.

Increasing the frequency of disturbance can also affect ecosystem structure and function. Repeated vegetation removal will favor species that take advantage of early-successional conditions at the expense of species that are more adapted to late-successional conditions. In order to ensure continued functioning of ecosystem processes and the survival of all species, it is necessary to have a mix of systems in early-successional and late-successional stages in a landscape. Human resource utilization must be managed within this context in order to ensure the long-term sustainability of all ecosystem components and to reduce the chances of extinction of some species due to human alteration of natural disturbance intervals.

Ecosystem Stability

A system is stable, or in *equilibrium*, if it can return to its previous condition at some time after disturbance. The length of time required to return to the original condition is the *recovery time*. The reestablishment of a forest following harvesting and the renewed production of forage following grazing both depend on the inherent stability of the affected ecosystem. The stability of an ecosystem is dependent on its components and their interrelationships. Disturbance may primarily affect one component of an ecosystem, as with salmon fishing in the Pacific Ocean. The ability of the entire ecosystem to adjust to this disturbance depends on the complexities of the interrelationships between the salmon, their predators and prey, and their competitors.

The length of the recovery time varies with the type of system, the natural disturbance interval, and the severity of the disturbance. A system is usually stable only within some bounds. If disturbed beyond these recovery limits the system may not return to its previous state but may settle into a new equilibrium. There are examples in the Mediterranean region of systems that were overgrazed in ancient times that have never returned to their previous species composition and productivity. Forest managers, farmers, fishermen, and others must understand the natural resiliency of the systems within which they work and stay within the bounds of stability in order to ensure sustainable resource utilization into the future.

Matter and Energy Cycles

Ecological processes work through the cycling of matter and energy within the system. *Nutrient cycling* consists of the uptake of nutrients from the soil and the transfer of these nutrients through plants, herbivores, and predators until their eventual return to the soil to begin the cycle anew. Interruption of these cycles can have far-reaching consequences in the survival of different ecosystem components.

These cycles also govern the transport of toxic substances within a system. It took many years before it was realized that persistent pesticides such as DDT would eventually be concentrated in top predators, such as raptors. The decline in populations of birds of prey because of reproductive failure caused by DDT was a consequence of the transport of the chemical through ecosystem food webs. Likewise, radionucleides from the 1986 explosion at the Chernobyl nuclear reactor have become concentrated in certain components of the ecosystems where they were deposited. This is particularly true of fungi, which take radionucleides and heavy metals from their food sources but do not shed the substances. Humans eating mushrooms from these forests can receive larger than expected doses of radiation, because the concentration in the fungi is much greater than in the surrounding system.

A basic understanding of ecosystem properties and processes is critical in designing management methods to allow continued human utilization of systems while sustaining ecosystem structure and function. With increasing human population and advancing living standards, more and more natural ecosystems are being pushed near their limits of stability. It is therefore critical for humans to understand how ecosystems are structured and function in order to ensure their sustainability in the face of continued, and often increasing, utilization.

David D. Reed

See also: Animal-plant interactions; Biomass related to energy; Biomes: types; Coevolution; Community-ecosystem interactions; Ecology: history; Ecosystems: studies; Food chain; Forest fires; Paleoecology; Population genetics; Succession; Trophic levels and ecological niches.

Sources for Further Study

Aber, J. D., and J. M. Melillo. *Terrestrial Ecosystems*. 2d ed. San Diego, Calif.: Harcourt Academic Press, 2001. Addresses the need to understand the basic units of the landscape in order to manage human impact on the environment. Integrates information from more traditional disciplines, such as the biochemistry of photosynthesis, soil chemistry, and population dynamics, in order to study the movement of energy, water, and chemical elements in the global environment.

Allen, T. F. H., and T. W. Hoekstra. *Toward a Unified Ecology*. New York: Columbia University Press, 1992. A general summary of field of ecology, with a view toward fostering common disciplinary goals. Provides a clear review of the interrelationships among ecosystem components and implications for management.

Bormann, F. H., and G. E. Likens. *Pattern and Process in a Forested Ecosystem*. New York: Springer-Verlag, 1981. Basic patterns of ecosystem structure and function are described.

Lee, K. *Compass and Gyroscope: Integrating Science and Politics for the Environment*. Washington, D.C.: Island Press, 1993. The interplay between ecosystem function and human utilization is summarized with examples from the Pacific Northwest.

ECOSYSTEMS: STUDIES

Categories: Ecology; ecosystems; history of plant science

The study of ecosystems defines a specific area of the earth and the attendant interactions among organisms and the physical-chemical environment present at the site.

Ecosystems are viewed by ecologists as basic units of the *biosphere*, much as cells are considered by biologists to be the basic units of an organism. Ecosystems are self-organized and self-regulating entities within which energy flows and resources are cycled in a coordinated, interdependent manner to sustain life. Disruptions and perturbations to, or within, the unit's organization or processes may reduce the quality of life there or cause its demise. Ecosystem boundaries are usually defined by the research or management questions being asked. An entire ocean can be viewed as an ecosystem, as can a single tree, a rotting log, or a drop of pond water. Systems with tangible boundaries—such as forests, grasslands, ponds, lakes, watersheds, seas, or oceans—are especially useful to ecosystem research.

Research Principles

The ecosystem concept was first put to use by American limnologist Raymond L. Lindeman in the classic study he conducted on Cedar Bog Lake, Minnesota, which resulted in his article "The Trophic Dynamic Aspect of Ecology" (1942). Lindeman's study, along with the publication of Eugene P. Odum's *Fundamentals of Ecology* (1953), converted the ecosystem notion into a guiding paradigm for ecological studies, thus making it a concept of theoretical and applied significance.

Ecologists study ecosystems as integrated components through which energy flows and resources cycle. Although ecosystems can be divided into many components, the four fundamental ones are *abiotic* (nonliving) *resources, producers, consumers,* and *decomposers.* The ultimate sources of energy come from outside the boundaries of the ecosystem (solar energy or chemothermo energy from deep-ocean hydrothermal vent systems). Because this energy is captured and transformed into chemical energy by producers and translocated through all biological systems via consumers and decomposers, all organisms are considered as potential sources of energy.

Abiotic resources—water, carbon dioxide, nitrogen, oxygen, and other inorganic nutrients and minerals—primarily come from within the bound-

AP/WIDE WORLD PHOTOS

The publication of Rachel Carson's classic Silent Spring *(1962) helped stimulate the environmental movement of the 1960's.*

aries of the ecosystem. From these, producers utilizing energy synthesize biomolecules, which are transformed, upgraded, and degraded as they cycle through the living systems that comprise the various components. The destiny of these *bioresources* is to be degraded to their original abiotic forms and recycled.

The ecosystem approach to environmental research is a major endeavor. It requires amassing large amounts of data relevant to the structure and function of each component. These data are then integrated among the components, in an attempt to determine linkages and relationships. This holistic ecosystem approach to research involves the use of systems information theory, predictive models, and computer application and simulations. As ecosystem ecologist Frank B. Golley stated in his book *A History of the Ecosystem Concept in Ecology* (1993), the ecosystem approach to the study of ecosystems is "machine theory applied to nature."

Research Projects

Initially, ecosystem ecologists used the principles of Arthur G. Tansley, Lindeman, and Odum to determine and describe the flow of energy and resources through organisms and their environment. Fundamental academic questions that plagued ecologists included those concerning controls on ecosystem productivity: What are the connections between animal and plant productivity? How are energy and nutrients transformed and cycled in ecosystems?

Once fundamental insights were obtained, computer-model-driven theories were constructed to provide an understanding of the biochemophysical dynamics that govern ecosystems. Responses of ecosystem components could then be examined by manipulating parameters within the simulation model. Early development of the ecosystem concept culminated, during the 1960's, in defining the approach of ecosystem studies.

Ecosystem projects were primarily funded under the umbrella of the International Biological Program (IBP). Other funding came from the Atomic Energy Commission and the National Science Foundation. The intention of the IBP was to integrate data collected by teams of scientists at research sites that were considered typical of wide regions. Although the IBP was international in scope, studies in the United States received the greatest portion of the funds—approximately $45 million during the life of IBP (1964-1974).

Five major IBP ecosystem studies, involving *grasslands, tundra, deserts, coniferous forests*, and *deciduous forests*, were undertaken. The Grasslands Project, directed by George Van Dyne, set the research stage for the other four endeavors. However, because the research effort was so extensive in scope, the objectives of the IBP were not totally realized. Because of the large number of scientists involved, little coherence in results was obtained even within the same project. A more pervasive concern, voiced by environmentalists and scientists alike, was that little of the information obtained from the ecosystem simulation models could be applied to the solution of existing environmental problems.

An unconventional project partially funded by the IBP was called the Hubbard Brook Watershed Ecosystem. Located in New Hampshire and studied by F. Herbert Bormann and Gene E. Likens, the project redirected the research approach for studying ecosystems from the IBP computer-model-driven theory to more conventional scientific methods of study. Under the Hubbard Brook approach, an ecosystem phenomenon is observed and noted. A pattern for the phenomenon's behavior is then established for observation, and questions are posed about the behavior. Hypotheses are developed to allow experimentation in an attempt to explain the observed behavior. This approach requires detailed scrutiny of the ecosystem's subsystems and their linkages. Since each ecosystem functions as a unique entity, this approach has more utility. The end results provide insights specific to the activities observed within particular ecosystems. Explanations for these observed behaviors can then be made in terms of biological, chemical, or physical principles.

Utility of the Concept

Publicity from the massive ecosystem projects and the publication of Rachel Carson's classic *Silent Spring* (1962) helped stimulate the environmental movement of the 1960's. The public began to realize that human activity was destroying the bioecological matrices that sustained life. By the end of the 1960's, the applicability of the IBP approach to ecosystem research was proving to be purely academic and provided few solutions to the problems that plagued the environment. Scientists realized that, because of the lack of fundamental knowledge about many of the systems and their links and because of the technological shortcomings that existed, ecosystems could not be divided into three to five components and analyzed by computer simulation.

The more applied approach taken in the Hubbard Brook project, however, showed that the ecosystem approach to environmental studies could be successful if the principles of the scientific method were used. The Hubbard Brook study area and the protocols used to study it were clearly defined. This ecosystem allowed hypotheses to be generated and experimentally tested. Applying the scientific method to the study of ecosystems had practical utility for the management of natural resources and for testing possible solutions to environmental problems. When perturbations such as diseases, parasites, fires, deforestation, and urban and rural development disrupt ecosystems from within, this approach helps define potential mitigation and management plans. Similarly, external causative agents within airsheds, drainage flows, or watersheds can be considered.

The principles and research approach of the ecosystem concept are being used to define and attack the impact of environmental changes caused by humans. Such problems as human population growth, apportioning of resources, toxification of biosphere, loss of *biodiversity*, global warming, acid rain, atmospheric ozone depletion, land-use changes, and eutrophication are being holistically examined. Management programs related to woodlands (the New Forestry program) and urban and rural centers (the Urban to Rural Gradient Ecology, or URGE, program), as well as other governmental agencies that are investigating water and land use, fisheries, endangered species, and exotic species introductions, have found the ecosystem perspective useful.

Ecosystems are also viewed as systems that provide the services necessary to sustain life on earth. Most people either take these services for granted or do not realize that such natural processes exist. Ecosystem research has identified seventeen naturally occurring services, including water purification, regulation, and supply, as well as atmospheric gas regulation and pollination. A 1997 article by Robert Costanza and others, "The Value of the World's Ecosystem Services and Natural Capital," placed a monetary cost to humanity should the service, for some disastrous reason, need to be maintained by human technology. The amount is staggering, averaging $33 trillion per year. Humanity could not afford this; the global gross national product is only about $20 trillion.

Academically, ecosystem science has been shown to be a tool to dissect environmental problems, but this has not been effectively demonstrated to the public and private sectors, especially decision makers and policymakers at governmental levels. The idea that healthy ecosystems provide socioeconomic benefits and services remains controversial. In order to bridge this gap between academia and the public, Scott Collins of the National Science Foundation suggested to the Association of Ecosystem Research Centers that ecosystem scientists be "bilingual"; that is, they should be able speak their scientific language and translate it so that the nonscientist can understand.

Richard F. Modlin

See also: Animal-plant interactions; Biomass related to energy; Biomes: types; Coevolution; Community-ecosystem interactions; Ecology: history; Ecosystems: overview; Food chain; Forest fires; Paleoecology; Population genetics; Succession; Trophic levels and ecological niches.

Sources for Further Study

Costanza, Robert, et al. "The Value of the World's Ecosystem Services and Natural Capital." *Nature* 387, no. 6630 (May 15, 1997): 253-260. The total economic value of seventeen ecosystems services representing sixteen biomes are estimated to determine the relative magnitude of the earth's economic value.

Daily, Gretchen C., ed. *Nature's Services: Societal Dependence on Natural Ecosystems.* Washington, D.C.: Island Press, 1997. Contains several essays examining how ongoing processes in ecosystems provide the necessary biochemophysical pathways, serving the needs that allow biological life to exist on earth.

Dodson, Stanley I., et al. *Ecology.* New York: Oxford University Press, 1998. An outstanding basic text that covers ecosystems and ecological concepts.

Golley, Frank B. *A History of the Ecosystem Concept in Ecology.* New Haven, Conn.: Yale University Press, 1993. A history of the ecosystem concept and a discussion of its impact on understanding problems and their potential solutions.

Likens, Gene E. *The Ecosystem Approach: Its Use and Abuse.* Oldendorf/Luhe, Germany: Ecology Institute, 1992. Provides critical analyses of ecosystem functions and structures.

Vogt, Kristiina A., et al. *Ecosystems: Balancing Science with Management.* New York: Springer, 1997. A complete treatment of the ecosystem concept and ecosystem management.

ELECTROPHORESIS

Category: Methods and techniques

Electrophoresis is a biochemical technique used to separate charged molecules in an electric field. Gel electrophoresis is one of the most common forms of this method, used to separate DNA, proteins, enzymes, and other molecules from the cell for laboratory investigation and manipulation.

Electrophoresis is widely used to separate, visualize, or purify charged biological molecules such as deoxyribonucleic and ribonucleic acids (DNA and RNA) and proteins, including enzymes. It is also used to estimate the size of DNA fragments and the molecular weight of proteins. Most biological molecules are electrically charged in solution; hence, when subjected to an electric field, they migrate as zones toward an electrode (a terminal source of electricity) of opposite electrical polarity. Positively charged molecules migrate to the negative electrical terminal, known as the cathode, and negatively charged molecules migrate toward the positive electrical terminal, known as the anode. The rate of migration depends on the size, shape, and charge of the molecules to be sorted as well as the strength of the electric field (voltage).

Types of Supporting Matrix

Electrophoresis is conducted in a sievelike *supporting matrix*, such as filter paper, cellulose acetate membrane, or, more commonly, a gel. Gels are made primarily of starch, agarose, or polyacrylamide. Starch and agarose are carbohydrates. *Gel electrophoresis* typically uses agarose, a purified form of agar-agar, extracted from seaweeds (marine red algae). Agarose gels have large pores, which allow large molecules to pass through.

Polyacrylamide is a synthetic polymer made of acrylamide and bisacrylamide. Acrylamide is a suspected human carcinogen (cancer-causing agent) and a potent neurotoxin (a compound that causes damage to the brain). Extreme caution is therefore required when handling polyacrylamide or its components. Polyacrylamide gels have smaller pores than agarose, therefore allowing only small molecules to pass through.

Separation of Nucleic Acids

The method of choice for separating nucleic acids, typically pieces of DNA, is agarose gel electrophoresis. The DNA fragments migrate through the agarose gel at a rate that is inversely proportional to their size. In other words, smaller fragments migrate faster through the gel than larger fragments.

In the laboratory, an agarose gel is prepared by dissolving agarose powder in a buffer solution (a salt solution) and heating to boiling. The viscous solution formed is then cooled and poured into a casting tray. A plastic-toothed comb is inserted in the melted agarose at the top. The agarose is allowed to solidify in the tray into a gelatinous slab and is then submerged into a buffer solution in a horizontal chamber. The buffer functions as a conductor of electricity through the agarose gel. After the gel is submerged, the comb is carefully removed, thereby creating a row of wells in the gel slab. The wells are then loaded with a sample consisting of a mixture of DNA fragments, sucrose or glycerol, and a blue dye. Sucrose sinks the DNA sample into the wells, while the dye marks the migration of the invisible DNA fragments through the gel. In order to establish an electric field in the chamber, a constant electric current from a power supply is generated between the electrodes at both ends of the gel. DNA is negatively charged because of its phosphate groups; therefore, the electric current drags the DNA fragments out of the wells toward the anode through a path known as a lane. Greater voltages result in faster migration of DNA fragments through the gel. The current is switched off when the blue dye moves about three-fourths of the way.

Upon completion of electrophoresis, the separated DNA fragments are made visible by staining the gel with ethidium bromide or methylene blue. Ethidium bromide is a fluorescent dye, a potent mutagen, and possible carcinogen. The stained gel is viewed with the aid of an ultraviolet box called the transilluminator. The separated DNA fragments appear as fluorescent orange bands. Each band corresponds to DNA fragments of equal length that have migrated to the same position in the gel. Methylene blue dye stains DNA bands blue under visible light.

Separation of Proteins

The principle behind the separation of proteins is similar to that of nucleic acids. Proteins can be separated by paper or cellulose acetate electrophoresis by simply placing a protein sample on a strip of filter paper or cellulose acetate saturated with a buffer, dipping the ends of the strip into chambers of buffer, and subject the strip to an electric field. The separation of most proteins, however, is performed in a polyacrylamide gel. The gel is cast and submerged in a vertical chamber of buffer.

Proteins can be separated on the basis of size (molecular weight) alone, net charge alone, or size and charge together. A common technique for separating proteins by size only is *sodium dodecyl sulfate-polyacrylamide gel electrophoresis* (SDS-PAGE). In this type of separation, a protein mixture is treated with the detergent sodium dodecyl sulfate. The detergent binds and causes the proteins to dissociate into polypeptides and become negatively charged. The proteins thereafter separate into bands according to their sizes alone. Bands are then visualized by staining with silver stain or a protein dye called coomassie blue.

Proteins can be separated on the basis of charge alone, using a method called *isoelectric focusing*. The separation is performed in a glass tube of polyacrylamide gel in which a pH gradient has been established. When a current is applied, each protein migrates until it reaches its characteristic pH (acidity or alkalinity level). At this point, the net charge on the protein becomes zero, and migration stops. The pH at which the net charge is zero is called the isoelectric focusing.

Complex mixtures of proteins of similar sizes are separated based on size and charge using the *two-dimensional gel electrophoresis*. In this technique, proteins are separated in two sequential steps. First, they are separated in a tube gel by isoelectric focusing based on their charges alone. Then the proteins migrate into a gel slab and separate by SDS-PAGE, based on their size alone. The proteins are visualized as spots in the gel slab.

Applications

Every species of organism examined by researchers has revealed immense genetic variation or polymorphism (many forms), an indication of the presence of different genotypes (genetic makeup) in the population. It is however, impossible to infer the genotypes of plants simply by observing their visible characteristics or phenotypes. In many plant science laboratories, researchers employ electrophoresis to determine the mode of reproduction of plant species, to detect genetic variation within and between plant populations, and to identify plant genotypes. Also, researchers establish genetic relatedness in plants, that is, establish the most probable paternal parent or pollen donor within and outside a study site that sired seeds collected from a known maternal plant. To accomplish these tasks, researchers rely upon protein and DNA markers generated by gel electrophoresis.

Protein markers known as *allozymes* have been used extensively in a number of genetic analyses. Allozymes are electrophoretically distinct forms of an enzyme produced by different alleles (alternate forms of a gene). (An enzyme is a protein that speeds up the rate of a chemical reaction in an organism, without being consumed in the process.) Allozymes catalyze the same chemical reactions but have slightly different sequences of amino acids, the building blocks of proteins.

To analyze allozymes, researchers extract enzymes from plants and separate them on starch or polyacrylamide gels. Gels made up of potato starch are most commonly used because of their low cost and ease of use. After electrophoresis, the gel is submerged in a solution containing a dye and a substrate appropriate for the enzyme studied. The enzyme reacts with the substrate to produce a colored band on the gel. If the gel yields one colored discrete band in a lane, then that particular plant contains just one form of the enzyme, therefore the genotype of the plant must be *homozygous* (having two identical alleles of a gene). If the gel yields two colored bands, then the plant contains two forms of the enzyme and is therefore *heterozygous* (having two different alleles of a gene). Allozyme electrophoresis enables researchers to learn about the mode of reproduction of plants. For example, plant populations with high numbers of heterozygotes indicate a high level of cross-pollination, that is, the transfer of pollen (plant male gamete) by wind, insects, birds, bats, or other animals from one flowering plant to another. High numbers of homozygotes within a plant population indicate a high level of self-pollination, the transfer of pollen within a flower or between flowers of the same plant.

In many plant laboratories today, protein markers have been superseded by DNA markers, which are fragments of DNA that are distinguished from one another because of the differences in their base sequences. To generate DNA markers, DNA is extracted from plants and cut into fragments with special enzymes known as *restriction enzymes*. The fragments are then separated by electrophoresis on an agarose gel and analyzed.

Some of the widely used DNA markers in many laboratories are different lengths of DNA fragments, known as *restriction fragment polymorphisms* (RFLPs) or short sequences of DNA bases that are repeated many times in tandem (head to tail), called variable number of tandem repeats (VNTR), a type of RFLP. RFLP analysis distinguishes between heterozygous and homozygous genotypes. Heterozygotes yield two fragments on a gel, while homozygous genotypes yield a single fragment.

Another DNA marker, known as *random amplified polymorphic DNAs* (RAPDs), have become extremely popular DNA markers with plant scientists. They are DNA fragments amplified by a technique known as polymerase chain reaction (PCR) using a short primer consisting of ten nucleotides chosen randomly and then separated by size by agarose gel electrophoresis. Polymorphisms are revealed when a DNA fragment is amplified in one plant and fails to amplify in another. On a gel, the researcher checks for the presence or absence of the marker.

Many studies involving parentage of seeds and seed pods have been done using DNA markers. The DNA profiles of seeds or seed pods are compared to maternal and paternal plants. If seeds were produced by cross-pollination, then half of the DNA bands of the seeds would be found in the maternal plant and the other half from the paternal plant.

Oluwatoyin O. Osunsanya

See also: Autoradiography; Chromatography; DNA in plants; Nucleic acids; Proteins and amino acids.

Sources for Further Study

Hames, B. D., and D. Rickwood. *Gel Electrophoresis of Proteins: A Practical Approach*. New York: IRL Press, 1990. Describes in detail major electrophoretic techniques. Emphasis is on the practical aspects involved in the techniques.

Patel, D., ed. *Gel Electrophoresis*. New York: Wiley-Liss, 1994. Text provides useful information on the electrophoresis of proteins, nucleic acids, and polysaccharides. It also provides a comprehensive reference source for reagents for electrophoresis, recipes for gels and buffers, and analysis of gels.

Westermeier, R. *Electrophoresis in Practice: A Guide to Methods and Applications of DNA and Protein Separations*. 2d ed. New York: John Wiley and Sons, 1997. Text provides a comprehensive guide to electrophoresis and a detailed troubleshooting appendix.

ENDANGERED SPECIES

Categories: Ecology; environmental issues

The International Union for the Conservation of Nature-World Conservation Union (IUCN) defines endangered species as those in immediate danger of extinction. Extinction means that the species is no longer known to exist. The IUCN defines threatened species as those at a high risk of extinction but not yet endangered. Vulnerable species are considered ones that are likely to become extinct at some point in the foreseeable future. Rare species are at risk but not yet at the vulnerable, threatened, or endangered levels.

Worldwide, the number of endangered plant species was estimated at more than 33,400 in 1999. This number is much higher than that of all of the endangered or threatened animal species combined. Although extinction is a natural process, and all species will eventually be extinct, human activities threaten the existence of plant and animal life worldwide.

Humans use plants for food; medicine; building materials; energy; to clean water, air, and soil of pollutants; to control erosion; and to convert carbon dioxide to oxygen. The process of extinction increased dramatically during the nineteenth and twentieth centuries because of habitat destruction or loss, deforestation, competition from introduced species, pollution, global warming, and plant hunting, collecting, and harvesting. Over time, pollutants and contaminates accumulate in the soil and remain in the environment, some for many decades. Pollution in the atmosphere also contributes to long-term changes in climate.

Habitat Loss

By far the most significant threat to plant species is habitat loss or destruction. Habitat loss can occur because of resource harvesting for food, medicine, and other products, deforestation, and the conversion of wilderness for agricultural, industrial, or urban uses. Wood consumption and tree clearing for agriculture and development threaten the world's forests, especially the tropical forests, which may disappear by the mid-twenty-first century if sufficient preventive action is not taken. Natural disasters, such as climatic changes, meteorites, floods, volcanic eruptions, earthquakes, drought, hurricanes, and tornados, also can be devastating to a habitat.

In Europe and Asia, the plant distribution is complex, with isolated populations of plants spread across a large area. The plants are greatly influenced by the cold climate and by humans. Plant species are disappearing, especially in Europe and the Mediterranean, because of habitat destruction and disturbances including urbanization, road construction, overgrazing, cultivation, forest plantation, fire, pollution, and overexploitation of resources, or for use in horticulture.

The mountain plants are threatened by development for industry and tourism, pollution, strip mining, walkers, and skiers. The wetlands are threatened by removal of peat for fuel, water extraction which lowers the water table, and increased drainage for building or agriculture or fear of malaria. Recreational use and susceptibility to pollution such as acid rain or fertilizer run-off present further threats.

In North America, the major causes of endangerment include loss of habitat, overexploitation of resources, introduction of invasive species, and pollution. A massive loss of wilderness has occurred through the clearing of forests, plowing of prairies, and draining of swamplands. For example, in the northeastern United States there are only 13 square miles of alpine habitat, an area in which grow thirty-three at-risk species. This area is heavily used by hikers and mountain bikers. The Florida Everglades are threatened because the water supply is diverted to supply cities, industries, and agriculture.

In California, the Channel Islands are home to seventy-six flowering plants which do not exist on the mainland. Eighteen species are located on just one island. These plants, including the San

This mandrinette shrub (Hibiscus fragilis) is one of the many endangered species on the island of Mauritius. These plants are not regenerating because of competition from introduced alien species.

Clemente broom, bush mallow (*Malacothamnus Greene*), a species of larkspur, and the San Clemente Island Indian paintbrush (*Castilleja grisea*), have been devastated by introduced grazers, browsers, and by invasive other plants. In Hawaii, more than 90 percent of native plants and almost all land birds and invertebrates are found nowhere else in world. The Hawaiian red-flowered geranium (*Geranium arboreum*) is threatened by introduced feral pigs, agricultural livestock, and competition by nonnative plants.

In developing or highly populated nations in Asia, Africa, Central and South America, the Caribbean, the Pacific Ocean islands, Australia, and New Zealand, habitat loss occurs because of population needs. Land is cleared for agriculture, develop-

A Sampling of the World's Endangered Plant Species

Common Name	Scientific Name	Habitat
African teak	*Pericopsis elata*	Semideciduous forests of Central and West Africa
Almug or red sandalwood	*Pterocarpus santalinus*	Forests of India
Atlas or Moroccan cypress	*Cupressus atlantica*	Dry woodland on steep slopes of the Atlas Mountains, Morocco
Bastard quiver tree or Basterkokerboom	*Aloe pillansii*	Hot and arid areas of Namibia and South Africa
Camphor tree	*Dryobalanops aromatica*	Peninsular Malaysia and Sumatra
Carossier palm	*Attalea crassispatha*	Haiti's southwestern peninsula in scrub forest and agricultural land
Clay's hibiscus	*Hibiscus clayi*	Dry forests in a few locations on some of the Hawaiian Islands
Commoner lignum vitae or guaiac tree	*Guaiacum officinale*	Lowland dry forests and coastland areas of Caribbean, Colombia, Venezuela
Cook's holly	*Ilex cookii*	Puerto Rican cloud forests on ridgetops of Monte Javuya in Toro Negro State Park and Cerro de Punta
Dall's pittosporum	*Pittosporum dallii*	Rocky creeks in the mountains of New Zealand
Dragon's blood tree	*Dracaena cinnabari*	Woodland areas in the center and east of Socotra Island near Saudi Arabia and the Red Sea
Ebony	*Diospyros mun*	Limestone mountains in the Lao People's Democratic Republic and Vietnam
Egyptian papyrus	*Cyperus papyrus hadidii*	Shallow, freshwater marshes in Egypt
Fiddlewood or yax-nik	*Vitex gaumeri*	Damp forests with limestone or pine ridges in Belize, Guatemala, Honduras, Mexico
Four-petal pawpaw	*Asimina tetramera*	Scrub vegetation near the Atlantic coast in Florida
Gigasiphon	*Gigasiphon macrosiphon*	Tropical forests on the coastal plains of Kenya and Tanzania
Ginkgo	*Ginkgo biloba*	Temperate forests in southern China
Hainan sonmeratia	*Sonneratia hainanensis*	Mangrove forests in Wenchang County, Hainan, China

ment, and population resettlement. In Central America and the Caribbean, the Swietenia mahogany is found only in a few protected or remote areas. The Caoba tree (*Persea theobromifolia*) was newly identified as a species as recently as 1977. The lumber is commercially important, and habitat loss has occurred as a result of the conversion of forests to banana and palm plantations. In Ecuador, only 6 percent of the original rain forest remains standing, because the rest has been converted to farmland. In Asia, including the Philippines, population pressures bring about deforestation and the clearing of land for agriculture.

In southern Africa, land is used for crops, livestock, and firewood production. Overgrazing and the introduction of agriculture have caused the Sahara Desert area to grow rapidly. The island of Madagascar has between ten thousand and twelve

Common Name	Scientific Name	Habitat
Hawaiian gardenia	*Gardenia brighamii*	Dry forest on the island of Kauai, Hawaii
Horseshoe fern	*Marattia salicina nov*	Lord Howe Island, Australia
Ley's whitebeam	*Sorbus leyana*	Carboniferous limestone cliffs in southern Breconshire, Wales
Madeira net-leaf orchid	*Goodyera macrophylla*	Cliffs and ravines in humid, maritime climate of Madeira, Portugal
Mellblom's spider orchid	*Caladenia hastata*	Coastal areas near Portland, Victoria, Australia
Millionaire's salad	*Deckenia nobilis*	Lowland forests of the Seychelles Islands off the east coast of Africa
Mongarlowe mallee	*Eucalyptus recurva*	Near Mongarlowe in New South Wales, Australia
Nubian dragon tree	*Dracaena ombet*	Tropical forests near the Red Sea Hills in Egypt, Saudi Arabia, East Africa
Saharan cypress	*Cupressus dupreziana*	Tassili N'Ajjer National Park, Algeria
Taiwan trident maple	*Acer buergerianum ssp. formosanum*	Lowland evergreen forest of Taiwan
Tennessee purple coneflower	*Echinacea tennesseensis*	Red cedar glades and dry environments in forest openings of Tennessee
Virginia round-leaf birch	*Betula uber*	Second-growth forest along Cressy Creek in Virginia
West Himalayan elm	*Ulmus walliciana*	Temperate areas near streams, rivers, and wetlands in Afghanistan, Nepal, Pakistan, India
None	*Anthoceros neesii*	Clay-loam soils in Austria, Czech Republic, Germany, Poland
None	*Diplocolea sikkimensis*	Tropical rain forests in India (Sikkim) and Nepal
None	*Distichophyllum carinatum*	Temperate mixed forests of Austria, Germany, Switzerland, China (Sichuan), Japan
None	*Andrewsianthus ferrugineus*	Coniferous forest of Bhutan, Nepal

Source: Compiled from the International Union for Conservation of Nature, World Conservation Union (IUCN) Red List of endangered species at the web site www.redlist.org (homepage www.iucn.org), and *Encyclopedia of Endangered Species*, edited by Bill Freedman, volume 2 (Detroit: Gale Research, 1999).

thousand plant species, of which 80 percent grow nowhere else in the world. Because of conversion to grassland through farming methods, only about one-fifth of the original species survive. In Australia there are 1,140 rare or threatened plants, and logging, clearing for grazing animals and crops, building developments, and mining have threatened many native species.

Plant Hunting, Collecting, and Harvesting

Habitat damage, the construction of facilities, and the opening of remote areas for human population have made many plants vulnerable to gathering and collecting. Some plants have been overharvested by gardeners, botanists, and horticulturists. One species of lady's slipper orchid (*Cypripedium calceolus*) is rare over much of its natural range except in parts of Scandinavia and the Alps because of collecting. Additionally, many mountain flowers or bulbs such as saxifrages, bellflowers, snowdrops, and cyclamen are endangered. In France and Italy, florulent saxifrage (*Saxifraga florulenta*), an alpine plant, has been overcollected by horticulturists and poachers.

Parts of the southeastern United States have poor soil that is home to the carnivorous or insectivorous plants—those that eat insects. These plants include sundews, bladderworts, Venus's flytrap, and pitcher plants. Collectors or suppliers have stripped many areas of all of these plants. In the Southwest, rare cacti are harvested for sale nationwide and worldwide. Endangered cacti include the Nellie Cory cactus (which has one remaining colony), *Epithelantha micromeres bokei*, *Ancistrocactus tubuschii*, saguaro cactus (*Carnegiea gigantea*), and *Coryphantha minima*. Near the Sierra Madre, two tree species—Guatemalan fir, or Pinabete, and the Ayuque—are endangered because of harvesting for use as Christmas trees or for the making of hand looms. Additionally, sheep eat the seedlings. In New Mexico, the gypsum wild buckwheat habitat is limited to one limestone hill, and the plants are threatened by cattle, off-road vehicles, and botanists.

In southern Mexico, there are 411 species of *epiphytes* (air plants or bromeliads in the genus *Tillandsia*), of which several are extremely rare. These plants are threatened by overcollection for the houseplant trade or conservatories. The African violet (*Saintpaulia ionantha*) of Tanzania may soon be extinct in the wild because of the horticultural trade and habitat loss due to encroaching agriculture.

Worldwide, orchids are overcollected for horticulture. Several species have been collected to extinction, are extremely rare, or have been lost because of habitat destruction. Examples include the extremely rare blue vanda (*Vanda caerulea*); *Paphiopedilum druryi*, believed extinct in its native habitat; *Dendrobium pauciflorum*, endangered and possibly extinct—only a single plant was known to exist in the wild in 1970; and the Javan phalaenopsis orchid, *Phalaenopsis javanica*. The latter was believed extinct. When it was rediscovered in 1960's, it was overcollected by commercial orchid dealers and thereby exterminated. There are no other known wild populations.

About 80 percent of the human populations in developing countries rely on traditional medicine, for which 85 percent of ingredients come from plant extracts. In Western medicine, one in four prescription medicines contain one or more plant products. Some at-risk species contain chemicals used in treating medical conditions, such as the African *Prunus africana* tree, whose bark has chemicals used to treat some prostate gland conditions, and the *Strophanthus thollonii*, a root parasite with chemicals used in heart drugs.

The Madagascan periwinkle (*Catharanthus roseus*) is commonly cultivated (its close relative *Catharanthus coriaceus* is rare and its medicinal importance unknown) and produces about seventy chemicals, some of which are useful in the treatment of cancer. The Indian podophyllum (*Podophyllum hexandrum*), a threatened species, is used to treat intestinal worms, constipation, and cancer. Rauwolfia (*Rauvolfia serpentina*), also a threatened species, is used to treat mental disorders, hypertension, and as a sedative. The lily *Amorphopahllus campanulatus* is used to treat stomachaches, and a fig, *Ficus sceptica*, is used to treat fever; both of these species are vulnerable because of habitat destruction.

The Micronesian dragon tree is believed to have magical and medicinal properties. It has been overharvested and is now extinct on several islands. In the United States' Appalachian Mountains, American ginseng is being overcollected because of an escalating demand for this plant's health benefits.

Conservation

The conservation of endangered plant species employs several compelling arguments: Plants enhance the world's beauty, have the right to exist, and are useful to people. The most persuasive argument may be that the survival of the human species depends on a healthy worldwide ecosystem. Three major goals of conservation are recovery, protection, and reintroduction.

Conservation methods depend on increasing public awareness by providing information about endangered or threatened species so that people can take action to reverse damage to the ecosystems. Other important strategies include achieving a widespread commitment to conservation and obtaining funding to protect rare or endangered species. Conservation efforts include setting aside protected areas, such as reserves, wilderness areas, and parks, and recognizing that humans must integrate and protect biodiversity where they live and work. Many countries are actively conserving species through protected areas, endangered-species acts, detailed studies of species and habitat, and information campaigns directed to the public.

Virginia L. Hodges

See also: Biological invasions; Competition; Deforestation; Ecology: history; Ecosystems: overview; Ecosystems: studies; Grazing and overgrazing; Human population growth; Invasive plants; Logging and clear-cutting.

Sources for Further Study

Burton, John A., ed. *Atlas of Endangered Species.* New York: Macmillan, 1999. Extensive discussion of endangered wildlife worldwide. Includes illustrations, lists, maps, references, and index.

Crawford, Mark. *Habitats and Ecosystems: An Encyclopedia of Endangered America.* Santa Barbara, Calif.: ABC-Clio, 1999. Brief history of conservation in the United States and a state by state overview. Includes lists, references, and index.

Freedman, Bill, ed. *Encyclopedia of Endangered Species.* Vol. 2. Detroit: Gale Research, 1999. Gives general descriptions, covers history and conservation measures, and provides concise entries on more than five hundred endangered plants and animals worldwide. Includes color photographs, maps, lists, references, and index.

Wilson, Edward O. *The Future of Life.* New York: Alfred A. Knopf, 2001. Harvard biologist Wilson, known for his work on species extinction, speaks out on the staggering impact of human activities on the many species—particularly plant species—worldwide, predicting major extinctions if the present rate continues. He also, however, identifies measures that may be taken to conserve species extinction, given the happy fact that many of them are concentrated in concentrated areas such as the rain forests.

ENDOCYTOSIS AND EXOCYTOSIS

Categories: Cellular biology; physiology; transport systems

Endocytosis is used by cells to move water, macromolecules, or larger objects, such as cell fragments or even whole cells, from outside a cell to the inside of the cell. Exocytosis is the reverse of endocytosis, that is, the movement of materials from the inside to the outside of the cell. Both types of transport move the materials using membrane-bound vesicles.

Endocytosis

In endocytosis, during which materials are moved into a cell, the cell's plasma membrane engulfs material and packs it into saclike structures called *vesicles*. The vesicles then detach from the plasma membrane and move into the cell. Once the vesicle is in the cytoplasm, it will typically fuse with some other membrane-bound organelle, such as a vacuole or the endoplasmic reticulum, and release its contents into the organelle. There are three types of endocytosis: *phagocytosis* (transport of actual particles), *pinocytosis* (transport of water, along with any solutes in the water), and *receptor-mediated endocytosis* (explained in detail below).

In receptor-mediated endocytosis, specific macromolecules outside the cell attach to the binding sites of receptor proteins. These receptors are embedded in the plasma membrane in specialized regions called *coated pits*. The macromolecules contact and bind with receptor proteins in the pit. The coated pit deepens inward to the cytosol and eventually is pinched off as a *coated vesicle* inside the cell. Coated vesicles contain the receptor proteins and whatever molecules they are importing and are coated on the outside by a protein called *clathrin*. After the contents of the vesicle are released inside the cell, the receptor proteins are reused to form new coated pits. Receptor-mediated endocytosis is especially useful for importing specific molecules, even when they are present at low concentrations.

Endocytosis is more difficult in plants than in animals because the plasma membrane of a plant cell is usually pressed against the rigid cell wall by turgor pressure, which hinders the plasma membrane from invaginating into the cytosol. Nevertheless, plant cells do have coated pits, and experiments with isolated *protoplasts* (plant cells without their cell walls) suggest that receptor-mediated endocytosis works in plant cells much as it does in animal cells.

In a special form of phagocytosis, symbiotic nitrogen-fixing bacteria (*rhizobia*) colonize root cells in legumes. First, root hairs surround rhizobia cells in a ball-like mass. Next, *infection threads*, composed of an extended portion of the cell membranes of the cell being invaded, are formed. The bacteria multiply inside the thread, which extends inwardly and penetrates through and between cortex cells. In the inner cortex cells, portions of the thread wrap around groups of bacteria in vesicles, which pinch off into the cytosol. A membrane which forms around a group of bacteria is called a *bacteroid*.

Exocytosis

Exocytosis, the transport of macromolecules and large particles outside the cell, is the reverse of endocytosis. In exocytosis, materials inside the cell are packed in a vesicle, which fuses to the plasma membrane. Some vesicles contain structural proteins and polysaccharides, whereas other

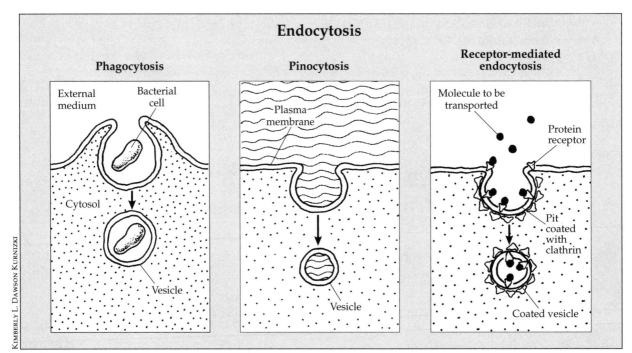

Three types of endocytosis: phagocytosis, in which the plasma membrane extends outward to engulf and draw in a particle to be taken into the cell; pinocytosis, in which the plasma membrane pouches inward, surrounding water and solutes, and then pinches off inside the cell; and receptor-mediated endocytosis, in which the molecules to be taken into the cell first bind to protein receptors in coated pits in the plasma membrane, which then form coated vesicles that pinch off and move into the cell.

vesicles, such as vacuoles, contain digestive enzymes. Each vesicle attaches to the plasma membrane. The site where it attaches opens, and the materials in the vesicle are dumped out of the cell. The vacated vesicle straightens and becomes a part of the plasma membrane. A problem arises if vesicles continually fuse with the plasma membrane, because large amounts of new membrane being added can double the amount of plasma membrane every thirty minutes, and the plasma membrane has little room to expand in cells with rigid cell walls. Nevertheless, plants cells regularly recycle excess plasma membrane via endocytosis.

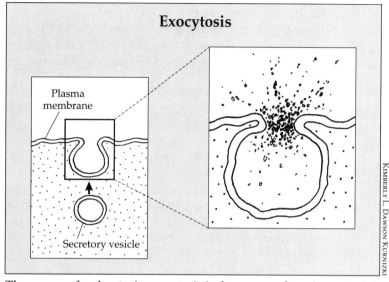

Exocytosis

The reverse of endocytosis, exocytosis is the process of moving materials outside the cell by packing them in a vesicle that fuses to the plasma membrane and then opens to dump out its contents.

Roles of Endocytosis and Exocytosis

Both exocytosis and endocytosis occur within seeds. Starch-digesting enzymes, such as alpha-amylase, move from cell to cell in cereal grains. In the cells of the *aleurone* layer, the endoplasmic reticulum manufactures these enzymes and packs them into vesicles. The enzymes are exuded from vesicles by exocytosis through the plasma membrane and must then be transported into the cells of the endosperm by endocytosis. Inside the endosperm cells, the enzyme-containing vesicle fuses with an *amyloplast*, a plastid specialized for storage of starch, where the enzymes hydrolyze starch into glucose.

Exocytosis of macromolecules plays other important roles in plants. For example, polysaccharides and proteins that are exported become structural components of cell walls. After cell division, exocytosis by secretory vesicles builds primary cell walls between newly divided nuclei. Epidermal cells of leaves extrude waxy substances onto leaf surfaces to minimize transpiration. Root-tip cells exude slimy polysaccharides to lubricate their movement as they grow and penetrate the soil.

Exocytosis of other substances by specialized cells also plays many roles in plants. For example, exocytosis is involved in the secretion of nectar by flower cells to attract pollinators. Exocytosis is also the process whereby oils are emitted by aromatic flowers, herbs, and spices. Such oils function both as attractants for pollinators and as defenses. For example, oils emitted by mustard plants irritate many animals, thus preventing many herbivores from eating them. Lignin, which is more rigid than cellulose and strengthens woody tissue, is expelled via exocytosis in woody plants and then accumulates in the middle lamella and cell walls. Enzymes that digest insects are released by the leaf cells of carnivorous plants, such as Venus's flytrap. Root exudates are released by some plants in response to environmental stress or to deter the growth of other plant species nearby.

Phytoremediation

Phytoremediation uses plant roots to clean polluted soil and water. They can remove, by endocytosis, and degrade both small organic molecular pollutants (ammunition wastes, chlorinated solvents, and herbicides) and large organic molecular pollutants (crude oil and polyaromatic hydrocarbons) in the environment. The degraded products can then be incorporated in the plant's tissues.

Domingo M. Jariel

See also: Active transport; Cell wall; Cells and diffusion; Cytosol; Endomembrane system and Golgi complex; Membrane structure; Microbodies; Osmosis, simple diffusion, and facilitated diffusion; Peroxisomes; Plasma membranes; Vacuoles; Vesicle-mediated transport.

Sources for Further Study

Gunning, Brian E. S., and Martin W. Steer. *Plant Cell Biology: Structure and Function.* Rev. ed. Sudbury, Mass.: Jones and Bartlett, 2000. Text integrates microscopy with plant cell and molecular biology. Includes more than four hundred micrographs and four pages of full-color plates.

Hay, Robert K. M., and Alastair H. Fitter. *Environmental Physiology of Plants.* 3d ed. New York: Academic Press, 1987. College-level text covers newer molecular approaches which can be used to solve problems in physiology, global change, toxicity, and more. Illustrations include color plates.

Hopkins, William G. G. *Introduction to Plant Physiology.* 2d ed. New York: John Wiley & Sons, 1998. Uses interactions between the plant and the environment as a foundation for developing plant physiology principles. Covers the role of plants in specific ecosystems, global ecology, cells, plant growth regulators, and biochemistry. Each chapter is illustrated.

Lee, A. G. *Endocytosis and Exocytosis.* Greenwich, Conn.: JAI Press, 1996. Part of the biomembranes series, covers endocytosis, exocytosis, and related processes. Includes references, index.

ENDOMEMBRANE SYSTEM AND GOLGI COMPLEX

Categories: Anatomy; cellular biology

The endomembrane system is a collective term applied to all of the membranes in a cell that are either connected with or are derived from the endoplasmic reticulum (ER), including the plasma membrane but not the membranes of chloroplasts or mitochondria. The membrane-bound organelles considered to be part of the endomembrane system are the vacuole, nuclear envelope, endoplasmic reticulum, Golgi complex, and various types of vacuoles.

Some components of the endomembrane system have direct, permanent connections with the endomembrane system (such as between the endoplasmic reticulum and the nuclear envelope), whereas other components share membrane and contents by trafficking *vesicles* (membrane-bound packages) from one component to another (for example, the ER sends numerous vesicles to the Golgi complex) across the *cytosol*. The endomembrane system is responsible for processing, sorting, and packaging membrane material, proteins embedded in membranes, and large water-soluble molecules (such as proteins or carbohydrates), either for export from the cell (called *exocytosis*) or for use within the cell. The endoplasmic reticulum is the ultimate source of all the membranes of the endomembrane system.

Golgi Complex

The Golgi complex is a major component of the endomembrane system and, in most cells, its primary role is secretion. The term "Golgi complex" refers collectively to all the *Golgi bodies* (once commonly called *dictyosomes* in plants) in a cell. It is named after Camillo Golgi (1843-1926), an Italian scientist who first described the structures in 1878. Golgi won the 1906 Nobel Prize in Physiology or Medicine for his contributions to the understanding of the structure of the nervous system.

When viewed through an electron microscope, a single Golgi body is composed of a series (of typically four to eight) round, flattened membranous sacs called *cisternae*. This "stack of sacs" has two sides; the cisterna on the *cis* side often faces the ER, while the cisterna on the *trans* side often faces away from the ER. The *medial cisternae* are in between. The *trans-Golgi network* is the collection of vesicles seen leaving the *trans* face of the Golgi body. *Intercisternal elements* are protein fibers that span the space between cisternae. They may help anchor the Golgi enzymes in the individual cisternae so they are not transported away and lost with the *shuttle vesicles*, described below. The intercisternal elements may also help stabilize the entire Golgi body.

An individual Golgi body is surrounded by a very faint, filamentous structure called the *Golgi matrix*. The matrix (along with the intercisternal elements) probably helps hold the Golgi body together as it is moved around the cell. The matrix also appears to exclude ribosomes and other cytosolic components from the immediate vicinity of the Golgi bodies and keeps the cytoplasm from interfering with the functioning of the complex. There are apparently no direct membrane connections among the individual cisternae in a Golgi body. Rather, membrane and contents move between cisternae via shuttle vesicles that pinch off of one cisterna, move in a trans direction, and fuse with an adjacent cisterna. The number of Golgi bodies in a plant cell can vary widely, from several dozen to more than ten thousand, depending on the size and function of the cell.

Entire Golgi bodies are transported around the cell via the *cytoskeleton*. They show a stop-and-go type of movement that may be associated with a need to approach a segment of the *rough ER*, or RER, to pick up proteins and then move closer to the plasma membrane or vacuole to deliver those proteins to the proper destination. Golgi bodies divide by fission, with new stacks pinching off of existing ones.

Golgi Functioning

The Golgi complex is responsible for the processing and packaging of proteins and other molecules for secretion from the cell or use within the cell. A typical pathway for the secretion of a protein to the cell exterior (that is, exocytosis) would be as follows. A ribosome bound to the RER translates a

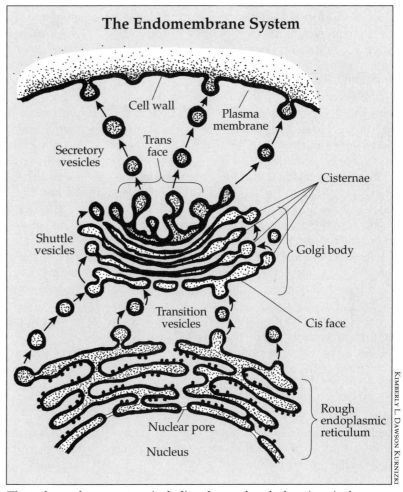

The endomembrane system, including the rough endoplasmic reticulum outside the cell's nucleus, transports materials to and from the cell's plasma membrane by means of vesicles transported through the Golgi complex (composed of flattened sacs called cisternae). The Golgi complex is responsible for the processing and packaging of proteins and other molecules for secretion from the cell or use within the cell.

piece of messenger RNA into a protein, and that protein is inserted into the lumen (the aqueous space enclosed by the RER membrane) of the RER. A vesicle forms by enclosing the protein and pinching off of the RER.

This *transition vesicle* travels across the cytosol (the traffic being directed by the *microtubules* and *microfilaments* of the cytoskeleton), fuses with the cisterna on the *cis* face of a Golgi body, and delivers the protein to the *lumen* of that cisterna. Enzymes in the lumen of the cisterna modify the protein by adding sugars (called "glycosylation") to produce a *glycoprotein*. A vesicle of Golgi membrane pinches off, forms a shuttle vesicle, and delivers the glycoprotein to the next cisterna in the stack, where the protein may be further glycosylated. The protein moves its way through the Golgi body and, eventually, leaves the *trans* face in a *secretory vesicle* that fuses with the plasma membrane. The contents of the secretory vesicle are moved out of the cell by exocytosis, while the membrane of the vesicle becomes incorporated into the plasma membrane.

In a similar fashion, the Golgi complex delivers matrix polysaccharides to the cell wall, large molecules and membrane lipids to the vacuole, or *integral* membrane proteins to the vacuolar membrane. In contrast to proteins (which are synthesized on the RER and merely modified during their transit through the Golgi body), cell wall matrix polysaccharides are synthesized from the ground up in the Golgi.

Surface-exposed proteins in the membrane of the vesicle contain the information needed to direct the vesicle to either the plasma membrane or some other place in the cell. Vesicles destined for the vacuole are visibly coated (as seen using an electron microscope) with protein, the major coat protein being *clathrin*. Vesicles destined for the plasma membrane appear smooth but undoubtedly have protein information on the exterior that directs them to the plasma membrane.

A single Golgi body can be involved in processing and packaging both glycoproteins and polysaccharides for delivery to the plasma membrane or intracellular locations at the same time. They also can be "retailored" to suit changing needs over the lifetime of a cell. That is to say, the enzymes in the cisternal lumen can be degraded and replaced with other enzymes that direct the synthesis of different molecules.

Other Endomembrane System Components

From the above, it can be seen how the ER and Golgi complex interact to deliver proteins and carbohydrates to the plasma membrane and vacuoles. In addition, the electron microscope often shows the ER to be physically connected to the nuclear envelope. Thus, the nuclear envelope is almost always included in a discussion of the endomembrane system. However, the exact functional nature of this nuclear connection remains unknown. The nuclear envelope is composed of two membranes, the outermost of which can be studded with ribosomes, much like the RER. The outer membrane of the nuclear envelope may, in some ways, be functionally similar to the RER, producing proteins which are passed across the outer membrane to the space between the two envelope membranes.

The fate of those membranes remains unclear. Whether they cross the inner membrane of the nuclear envelope as well and are delivered to the interior of the nucleus or diffuse to the lumen of a nearby section of ER and are processed through the Golgi complex is not known. The ER/nuclear envelope connection may be better understood by investigating the evolution of the eukaryotic cell.

Evolutionary Significance

According to the *endosymbiotic theory*, the ancestor of today's eukaryotic cells was a primitive prokaryote that engulfed a respiratory prokaryote (which eventually became established as mitochondria) and a photosynthetic prokaryote (which became chloroplasts). The other membrane-bound organelles of the cytoplasm were derived from infoldings of the plasma membrane. Thus the entire endomembrane system (plasma membrane, ER, Golgi complex, vacuoles, and nuclear envelope) probably has a common evolutionary background. Over time, the individual compartments became more specialized.

Robert R. Wise

See also: Active transport; Cell-to-cell communication; Chloroplasts and other plastids; Cytoplasm; Cytoskeleton; Endocytosis and exocytosis; Endoplasmic reticulum; Eukaryotic cells; Genetic code; Liquid transport systems; Microbodies; Microscopy; Mitochondria; Nucleus; Osmosis, simple diffusion, and facilitated diffusion; Plasma membranes; Vacuoles; Vesicle-mediated transport.

Sources for Further Study

Buchanan, Bob B., Wilhelm Gruissem, and Russell L. Jones. *Biochemistry and Molecular Biology of Plants*. Rockville, Md.: American Society of Plant Physiologists, 2000. Contains a thorough and detailed treatment of the endomembrane system and Golgi complex in chapter 1. Includes tables, photographs, and colored drawings.

Hopkins, William G. *Introduction to Plant Physiology.* New York: John Wiley & Sons, 1999. College-level text for upper-division plant physiology course. Chapter 1 covers the endomembrane system and Golgi complex. Includes tables, photographs.

Raven, Peter H., Ray F. Evert, and Susan E. Eichhorn. *Biology of Plants.* 6th ed. New York: W. H. Freeman/Worth, 1999. Popular introductory botany text; chapter 3 discusses the endomembrane system and Golgi complex. Includes tables, photographs.

ENDOPHYTES

Categories: Animal-plant interactions; fungi; microorganisms

Fungi that spend at least a part of their lives within the aboveground parts of living plants—in leaves, stems, and in some cases reproductive organs—but cause no outward signs of infection are called endophytes. Some endophytes protect the host plant by deterring grazing animals or pathogenic fungi.

In the 1980's scientists began to realize that a great variety of microscopic fungal species live benignly within plants, as endophytes (from the Greek words *endos*, meaning "inside," and *phyton*, for "plant"), in contrast to fungi living on the surfaces of plants, as *epiphytes* (from the Greek *epi*, meaning "upon," plus *phyton*). Most endophytic fungi are *ascomycetes*. Many appear to be close relatives of plant pathogens.

Most endophytic fungi live and feed between the host plant's cells. Those endophytes that provide a benefit to the plant in return for their keep are considered to be partners with their host, in a symbiotic relationship called *mutualism*. Endophytic mutualism is well developed in some grasses, in which the fungal partner produces alkaloid substances that deter herbivores and pathogens.

Some fungi live within a plant benignly or mutualistically for a time, and then, if environmental stress or senescence afflicts the host or conditions otherwise change, the fungi turn pathogenic. For example, in a drought-weakened tree, previously benign fungi may initiate disease symptoms. Such fungi are said to have both an endophytic and a pathogenic phase. Other fungi may have a dormant, endophytic phase, then eventually become dependent on dead organic matter for sustenance.

Two Growth Patterns

The endophytes of grasses differ in growth habit from those of woody plants (both coniferous and angiospermous). Grass endophytes have been found to grow systemically, throughout the stems and leaves, of the mature plant, producing substantial fungal biomass. Hyphae, or filaments, of the fungal body even penetrate the grass ovule, which is the reproductive structure that develops into the seed. Via the infected seed, the fungus is transmitted to the next generation and thus is perpetuated down a plant's lineage.

In contrast, in most of the woody plants that have been investigated, individual endophytes are not systemic but instead are localized within leaves or stems, where they may be confined to specific plant tissues, such as bark or xylem (wood). Woodyplant endophytes typically propagate not by invading the host's ovule but rather via spores, which are carried to other plants by air, water, or animals. Presumably, the spores are able to disperse because they are not produced inside the plant host but rather on plant parts that have dropped off or are dying. This subject has been little investigated, however.

Abundance and Diversity

Endophytic fungi are common and widespread. Although research has focused mainly on grasses and woody plants, endophytes have also been found in mosses, ferns, and herbaceous angiosperms (flowering plants). Scientists have suggested that endophytic fungi may be as widespread among plants as are *mycorrhizae* (associations between certain fungi and plant roots), which characterize the vast majority of vascular plants. In mycorrhizae, part of the fungal body is external to the plant, whereas endophytes are wholly internal.

Endophytes are diverse, especially in trees and shrubs. Individual woody species and even individual plants typically harbor scores of fungal species, as has been shown for alder, oak, beech, maple, birch, ash, pine, spruce, fir, and other plants. Nevertheless, just one or a few species or genera of fungi usually dominate the fungal community of each woody plant species. These dominants commonly do not occur in plants other than the host species or closely related species.

Tropical trees are only beginning to be investigated for endophytes, but indications are that they are particularly rich in endophytic species. Given the high diversity of tropical trees, the endophytes still awaiting discovery may represent an enormous reservoir of biodiversity. In contrast to the wide variety of endophytes in woody plants, endophytes in grasses seem to be of low diversity. Grass endophytes are all closely related species, and each grass species seems to host only one or a few of them. Grasses, however, have been little examined for the nonsystemic kinds of endophytes that woody plants harbor in such variety.

Mutualisms and Ecology

Endophytic fungi that are mutualistic, or protective of the plant, are particularly well known in the pasture grasses tall fescue (*Festuca arundinacea*) and perennial ryegrass (*Lolium perenne*). The endophytes produce alkaloid toxins that defend these plants against insect attack and also cause serious illness in grazing livestock. Some endophytes, especially in grasses, increase host tolerance of other environmental stresses, such as drought. Many of the diverse endophytic species in woody plants have no known protective function, and some seem to be latent pathogens. Among those that are protective is one in the needles of Douglas fir (*Pseudotsuga menziesii*). It produces substances toxic to midges that form galls on the needles.

The ecological effects of endophytes in natural populations seem to be very complex. Endophytes sometimes are actually antagonistic to the host plant rather than mutualistic with it. The ecological role of endophytes and their host plants in natural environments has been little studied, however.

Economic Importance

The chemical substances produced by endophytic fungi are of considerable commercial interest to the pharmaceutical and food-processing industries and as potential biocontrol agents for plant diseases and pests. Endophytic fungi may eventually help farmers grow crops with minimal use of water, fertilizers, and pesticides.

Jane F. Hill

See also: Animal-plant interactions; Ascomycetes; Biopesticides; Community-ecosystem interactions; Drought; Fungi; Mycorrhizae; Resistance to plant diseases.

Sources for Further Study

Alexopoulos, Constantine J., Charles W. Mims, and Meredith Blackwell. *Introductory Mycology*. 4th ed. New York: John Wiley and Sons, 1996. Textbook. Endophytes treated in the context of the fungal groups to which they belong. Includes references, black-and-white illustrations, glossary, and index.

Bacon, Charles W., and James F. White, Jr., eds. *Microbial Endophytes*. New York: Marcel Dekker, 2000. Papers about fungi, bacteria, and viruses that live symbiotically within plants. Includes references, black-and-white illustrations, and index.

Carroll, George C., and Donald T. Wicklow, eds. *The Fungal Community: Its Organization and Role in the Ecosystem*. New York: Marcel Dekker, 1992. A collection of papers that includes one on fungal mutualism, which discusses some aspects of endophytes. Endophytes touched on in other papers, also. Includes references, black-and-white illustrations, and systematic and subject indexes.

Redlin, Scott C., and Lori M. Carris, eds. *Endophytic Fungi in Grasses and Woody Plants: Systematics, Ecology, and Evolution*. St. Paul, Minn.: APS Press, 1996. Includes five papers presented at the 1991 American Phytopathological Society meeting, plus six additional papers. Includes black-and-white illustrations, references, and index.

ENDOPLASMIC RETICULUM

Categories: Anatomy; cellular biology

The endoplasmic reticulum is a network of sacs in the cytosol of eukaryotic cells that manufactures, processes, transports, and stores chemical compounds for use inside and outside of the cell.

The endoplasmic reticulum (ER) is an extensive, complex system of a more or less continuous distribution of convoluted membrane-bound cavities that take up a sizable portion of the cytosol. The internal space of the ER is called the *lumen*. The ER is attached to the double-layered nuclear envelope and provides a connection, or bridge, between the nucleus and the cytosol. In addition, it provides living bridges between cells by way of the *plasmodesmata*, small tubes that connect plant cells. The ER is a dynamic structure, constantly changing. It accounts for 10 percent or more of the volume of the cytosol.

In general, there are two kinds of endoplasmic reticulum, rough and smooth. *Smooth ER* is quite varied in appearance and most likely in function as well. Through a microscope, it appears as numerous nearly circular blotches, indicating that it consists of interlocking tubes of membranes. On the other hand, *rough ER* almost always appears as stacks of double membranes that are heavily dotted with *ribosomes*. Based on the consistent appearance of rough ER, it most likely consists of parallel sheets of membrane, rather than the tubular sheets that characterize smooth ER. These flattened, interconnected sacs are called *cisternae*, or cisternal cells. The cisternal cells of rough ER are also referred to as luminal cells. Rough ER and the *Golgi complex* are both composed of cisternal cells.

Smooth and Rough ER Functions

Because rough ER is covered with ribosomes, it has a bumpy appearance when viewed with an electron microscope. Rough ER is primarily involved in the production of proteins that will be exported from the cell to help with other functions of building the plant. Such proteins include antibodies, digestive enzymes, and certain hormones. Amino acid chains are assembled into proteins by the ribosomes. The protein units needed outside the manufacturing cell are transported into the rough ER for further processing. Once inside, they are shaped into the correct three-dimensional configuration that will be useful outside the cell. Necessary chemicals, such as carbohydrates or sugars, are then added to complete the proteins. The ER then transports these proteins to other areas of the ER, called *transitional ER*, where they are packed in vesicles to be sent to the Golgi complex for export, or secretion, to other parts of the plant. Occasionally, some of the completed proteins are transported to areas of the manufacturing cell where they are needed.

Because the smooth ER does not have attached ribosomes, it appears relatively smooth when viewed through an electron microscope. It also appears to bud off from vesicles that contain material from the lumen of the ER. Using the many different enzymes that are anchored to its walls, the smooth ER is involved with the synthesis, secretion, and storage of lipids as well as the manufacture of new membranes and the metabolism of carbohydrates.

Lipids are a group of fatty substances needed for building membranes and storing energy in plants. Among the more important lipids are phospholipids, which make up major components of the cell membrane. When a plant has excess energy available from photosynthesis, it sometimes stores that energy in the form of lipids known as triglycerides. When the plant is in need of more energy, the triglycerides can be broken down to produce it. Waxes are other important lipids stored in the smooth ER. They form protective coatings on the leaves of plants. Research indicates that smooth ER is also involved in the formation of cellulose for the cell wall.

Other Special ER Functions

Calcium is an essential nutrient for cellular function, growth, and development in plants. However, too much calcium can lead to cell death. To deal with high levels of calcium in the environment, many plants have developed a mechanism involving the ER to regulate bulk quantities of calcium through the formation of calcium oxalate crystals. Calcium oxalate crystals can account for more than 90 percent of the calcium found in a plant. Certain specialized ER cells, called crystal *idioblasts*, appear to participate in the formation of calcium oxalate and provide the storage locations for calcium in plants.

Plants have the capability to undergo rapid, large-scale movements when triggered by a wide variety of stimuli, such as changes in light intensity, temperature, and pressure. The ER is the plant sensor for pressure changes. For example, in Venus's flytrap and in the sensitive plant, the ER in cortex cells, referred to as *cortical ER*, provides the sense of touch. Inside the cytosol of these sensor cells, the ER aggregates at the top and the bottom of the cell.

When the cells are compressed or squeezed, the cortical ER is strained and releases accumulated calcium, producing the sense of touch. The mechanism is very similar to muscle contraction in an animal. Because the cortical ER is interconnected through the plasmodesmata, which provide communication channels among cells and end at motor cells in specialized appendages such as *pulvini* (cushionlike swellings at the bases of leaves), flowers, or specialized leaves, a pressure stimulus at one cell can trigger a response throughout the whole plant.

Alvin K. Benson

See also: Cell-to-cell communication; Cytoplasm; Cytosol; Endomembrane system and Golgi complex; Growth habits; Plant cells: molecular level; Proteins and amino acids; Ribosomes.

Sources for Further Study

Garrett, Roger A., Stephen R. Douthwaite, and Anders Liljas, eds. *The Ribosome: Structure, Function*. Washington, D.C.: ASM Press, 2000. Summarizes major advances in the understanding of ribosomes and protein synthesis, including the function of the ER. Includes color illustrations, bibliographical references, and index.

Hopkins, William G. *Introduction to Plant Physiology*. 2d ed. New York: John Wiley and Sons, 1999. Develops the foundations of plant physiology, including cellular organelles and the ER. Includes illustrations, bibliographical references, and index.

Nobel, Park S. *Physicochemical and Environmental Plant Physiology*. 2d ed. San Diego: Academic Press, 1999. Describes plant cells and tissues and the physiological processes in plants in a user-friendly, clear manner. Includes illustrations, bibliographical references, and index.

Salisbury, Frank B., and Cleon W. Ross. *Plant Physiology*. 4th ed. Pacific Grove, Calif.: Brooks/Cole, 1999. A fundamental textbook that covers plant cell biology as well as molecular biology and genetics. Includes illustrations, bibliographical references, and index.

Stern, Kingsley Rowland. *Introductory Plant Biology*. 8th ed. Boston: McGraw-Hill, 1999. A basic text on plants that provides a solid introduction to plant functions. Includes illustrations, bibliographical references, index.

ENERGY FLOW IN PLANT CELLS

Categories: Cellular biology; photosynthesis and respiration; physiology

Life on earth is dependent on the flow of energy from the sun. A small portion of the solar energy, captured in the process of photosynthesis, drives many chemical reactions associated with living systems.

In living organisms, energy flows through chemical reactions. Each chemical reaction converts one set of substances, called the reactants, into another set, the products. All chemical reactions are essentially energy transformations, in which energy stored in chemical bonds is transferred to other, newly formed chemical bonds. *Exergonic* reactions release energy, whereas *endergonic* reactions

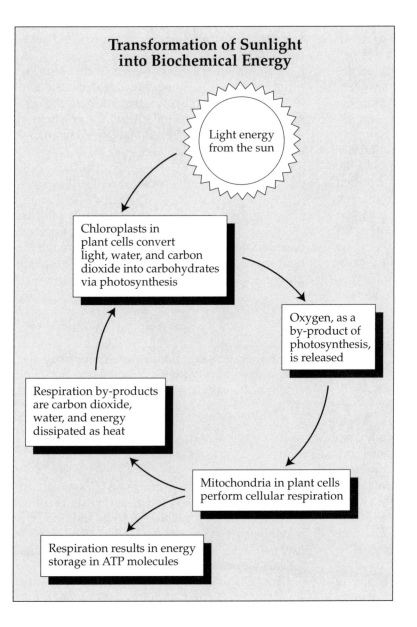

Transformation of Sunlight into Biochemical Energy

Light energy from the sun

Chloroplasts in plant cells convert light, water, and carbon dioxide into carbohydrates via photosynthesis

Oxygen, as a by-product of photosynthesis, is released

Respiration by-products are carbon dioxide, water, and energy dissipated as heat

Mitochondria in plant cells perform cellular respiration

Respiration results in energy storage in ATP molecules

plants, algae, and certain protists are the only living organisms that can produce their chemical energy using sunlight, they are called *producers*; all other life-forms are *consumers*. During seed germination, simple sugars, such as glucose, are broken down in a series of reactions called *respiration*. Energy is released to power the growth of embryo and young seedlings; hence, the reaction is exergonic. Within plant cells, both reactions occur.

In many reactions, electrons pass from one atom or molecule to another. These reactions, known as *oxidation-reduction* (or *redox*) reactions, are of great importance in living systems. The loss of an electron is known as *oxidation*, and the atom or molecule that loses the electron is said to be oxidized. Reduction involves the gain of an electron. Oxidation and reduction occur simultaneously; the electron lost by the oxidized atom or molecule is accepted by another atom or molecule, which is thus reduced.

Within plant cells, the energy-capturing reactions (photosynthesis) and the energy-releasing reactions (respiration) are redox reactions. Furthermore, all chemical reactions are orderly, linked and intertwined into sequences called *metabolic pathways*. All metabolic pathways in plant cells are finely tuned in three ways: the chemical reactions are regulated through the use of enzymes, exergonic reactions are always coupled with endergonic reactions, and energy-carrier molecules are synthesized and used for effective energy transfer.

Enzymes and Cofactors

Enzymes are biological catalysts, usually proteins, synthesized by plant cells. A number of characteristics make enzymes an essential component for energy flow in plant life. Enzymes dramatically speed up chemical reactions. Enzymes are normally very specific, catalyzing, in most cases, a single reaction that involves one or two specific mole-

require an input of energy for a reaction to occur.

In plants, such reactions occur during the process whereby plant cells convert the energy of sunlight into chemical energy that fuels plant growth and other processes. During this process, called *photosynthesis*, carbon dioxide combines with simple sugars to form more complex carbohydrates in special structures called *chloroplasts*. These chloroplasts are membrane-bound organelles that occur in the cells of plants, algae, and some protists. The energy that drives the photosynthetic reaction is derived from the photons of sunlight; hence it is an endergonic reaction (it requires energy). Because cules but leaves quite similar molecules untouched. In addition, enzyme activity is well regulated.

Many enzymes require a nonprotein component, or *cofactor*, for their optimal functions. Cofactors may be metal ions, part of or independent of the enzyme itself. Magnesium ions (Mg^{2+}), for example, are required in many important reactions in energy transfer, including photosynthesis and respiration. The two positive charges often hold the negatively charged phosphate group in position and help in moving it from one molecule to another. In other cases, ions may help enzymes maintain their proper three-dimensional conformation for optimal function. Some organic molecules can also be cofactors, including vitamins and their derivatives, and are usually called *coenzymes*. One example is the electron carrier nicotinamide adenine dinucleotide (NAD^+). NAD^+ is derived from nucleotide and vitamin-niacin. When NAD^+ accepts electrons, it is converted into $NADH + H^+$, which passes its electrons to another carrier; hence, NAD^+ is regenerated.

Plant cells regulate the amount and activity of their enzymes through various mechanisms. First, they control the synthesis of particular enzymes to meet their needs. They limit or stop the production of enzymes not needed by metabolic reactions and, hence, conserve energy. Second, plant cells may synthesize an enzyme in an inactive form and activate it only when needed. Third, plant cells can employ a feedback regulation mechanism by which an enzyme's activity is inhibited by an adequate amount of the enzyme's product. Furthermore, the activities of enzymes are affected by the environment, including temperature, pH (a measure of acidity versus alkalinity), and the presence of other chemicals. Different enzymes may require a slightly different physical environment for optimal function.

ATP: The Energy Carrier

During seed germination, stored glucose is broken down, making chemical energy available for movement, cellular repair, growth, and development. However, plant embryos cannot directly use the chemical energy derived from the breakdown of glucose. Within plant cells, most energy is transferred through a carrier—*adenosine triphosphate*, or ATP, known as the universal currency for energy transfer. Whether helping to convert light energy into chemical energy during photosynthesis or breaking down glucose in glycolysis and aerobic respiration, ATP acts as an agent to carry and transfer energy.

ATP is a nucleotide composed of the nitrogen-containing base adenine, the sugar ribose, and three phosphate groups. Energy released through glucose breakdown is used to drive the synthesis of ATP from adenosine diphosphate, or ADP, and inorganic phosphate (P_i):

$$ADP + P_i + energy \rightarrow ATP$$

The energy is largely stored in the bonds linking the phosphate groups. In reactions or processes where energy is needed, ATP releases energy through the hydrolysis and hence the removal of phosphate group:

$$ATP + H_2O \rightarrow ADP + P_i + energy \text{ (7.3 kilocalories per mole)}$$

Sometimes the second phosphate group may also be removed via hydrolysis to generate the same amount of energy and adenosine monophosphate (AMP):

$$ADP + H_2O \rightarrow AMP + P_i + energy \text{ (7.3 kilocalories per mole)}$$

The terminal phosphate group of ATP is not simply removed in most cases but is transferred to another molecule within a plant cell. This addition of a phosphate group to a molecule is defined as *phosphorylation*. The enzymes that catalyze such transfers are named *kinases*.

The following two examples of energy transfer involve ATP. The first is synthesis of sucrose by sugarcane:

$$glucose + fructose + 2\ ATP + 2\ H_2O \rightarrow sucrose + H_2O + 2\ ADP + 2\ P_i$$

The second example is the complete breakdown of glucose during cellular respiration:

$$glucose + 6\ O_2 + 36\text{-}38\ ADP + 36\text{-}38\ P_i \rightarrow 6\ CO_2 + 6\ H_2O + 36\text{-}38\ ATP$$

Either ADP or ATP can be recycled through endergonic or exergonic reactions intertwined in the metabolic pathways. In the plant kingdom, energy flow begins with photosynthesis, through which ATP and then high-energy bonds are formed as sugar by the conversion of light energy from the sun. In respiration, these bonds are broken down to carbon dioxide and water, and energy is released. Some of this energy is used to power cellular processes, but some energy is lost in each energy-conversion step. The energy flow among all other organisms also starts from photosynthesis or plants, either directly or indirectly.

Yujia Weng

Sources for Further Study

Baker, J. J. W., and G. E. Allen. *Matter, Energy, and Life*. Reading, Mass.: Addison-Wesley, 1981. Introduces basic energy-chemical principles through easy-to-understand descriptions.

Fenn, John. *Engines, Energy, and Entropy: A Thermodynacis Primer*. New York: W. H. Freeman, 1982. An excellent introduction to the laws of thermodynamics and their implications for daily life.

Hall, D. O., and K. K. Rao. *Photosynthesis*. 5th ed. New York: Cambridge University Press, 1994. An excellent, short book for nonscientists who want to understand photosynthesis.

McCarty, R. E. "H+-ATPases in Oxidative and Photosynthetic Phosphorylation." *Bioscience* 35, no. 1 (1985): 125. A detailed discussion on the structure and function of ATPases in phosphorylation and generation of ATP.

See also: Anaerobic photosynthesis; ATP and other energetic molecules; C_4 and CAM photosynthesis; Calvin cycle; Chloroplasts and other plastids; Exergonic and endergonic reactions; Glycolysis and fermentation; Krebs cycle; Mitochondria; Oxidative phosphorylation; Photorespiration; Photosynthesis; Photosynthetic light absorption; Photosynthetic light reactions; Plant cells: molecular level; Respiration.

ENVIRONMENTAL BIOTECHNOLOGY

Categories: Biotechnology; disciplines; ecology; economic botany and plant uses; environmental issues

Environmental biotechnology includes any process using biological systems (plants, microorganisms, and enzymes) to clean up and detoxify environmental contamination from hazardous and nonhazardous waste.

Bioengineering, *bioremediation*, and *biotechnological pollution control* are similar terms for the same idea: using naturally occurring plants, microorganisms, enzymes, and genetically engineered variants to clean up toxic wastes. Environmental biotechnology is becoming increasingly popular in waste treatment and remediation because it has several desirable characteristics. It is a "green" technology: It uses natural systems and naturally occurring organisms to detoxify environmental pollutants. It is not a particularly new, therefore uncertain, technology, so there are few unintended consequences of its use.

Bioremediation is inexpensive compared with other treatment technologies. If one can provide the proper environment and nutrients for the remediating organisms, there is relatively little other infrastructure involved. It can be done on-site without having to move hundreds of cubic yards of contaminated material. It can even be done in contaminated aquifers and soils that cannot be moved.

Plants and Microorganisms

Natural bioremediation of pollution is constantly occurring in the environment; without it, past pollution would still be present. Environmental biotechnology typically involves three types of organisms or biological systems: plants, microorganisms, and enzymes that may come from either group. Using plants to bioremediate an environment is referred to as *phytoremediation*. Phytoremediation is typically used when the environment is contaminated by heavy metals such as lead, mercury, or selenium. Certain plants (*Astragalus*, for example) are able to accumulate high concentrations of metals such as selenium in their tissues. The plants can be harvested, the tissue burned, and the metal-contaminated ash (now small in volume) can be stored in a hazardous waste facility.

Bioremediation most commonly refers to the use of soil microorganisms (bacteria and fungi) to degrade or immobilize pollutants. It can be used to treat a wide variety of wastes, including some nuclear wastes such as uranium. In one bioremediation process, the contaminated site is made favorable for microbial growth. Nutrients, such as nitrogen and phosphorus, are added. The area is kept moist and periodically stirred (if it is soil) to make sure it has sufficient air, or air is pumped into the system (if it is an aquifer). Microbes already present at the site start growing and use the waste as a food source.

Cometabolism and Seeding

Frequently, wastes cannot be used as a food source by microorganisms. These can still be biodegraded by a process called *cometabolism*. In cometabolism, wastes are biodegraded during the growth of microbes on some other compound. For example, trichloroethylene (TCE), one of the most common groundwater contaminants, is cometabolized during the growth of bacteria that use methane for their food source. Many other wastes, such as DDT, atrazine, and PCBs, are cometabolized by microbes in the environment.

Waiting for organisms to grow can take a long time, especially in winter. Often, environmental engineers speed the process using microorganisms grown in the laboratory on various pollutants. Therefore, when they are added to the environment in high numbers, they start bioremediating the pollutants immediately. This process is called *seeding*.

Enzymes

Sometimes a waste is so toxic or is present in such high concentration that neither plants nor microorganisms can survive in its presence. In this case, *enzymes* can be used to try to degrade the waste. Enzymes are proteins with *catalytic* activity; that is, they make chemical reactions occur faster than they normally would. Enzymes are not alive in a strict sense, but they come only from living organisms. They have an advantage over living organisms in that they can retain their catalytic activity in toxic environments. For example, horseradish peroxidase is a plant enzyme that has been used to treat chlorinated compounds. The peroxidase causes the compounds to bind together, becoming less soluble and thereby much less likely to enter the food chain of an ecosystem.

Use and Technology

Bioremediation has been used on a large scale mostly to treat oil spills. The best example of this was during the *Exxon Valdez* oil spill in Alaska in 1989. Rather than try to remove oil from beaches physically (by steam spraying or absorbing it into other materials), engineers sprayed several beaches with a nutrient solution that helped naturally occurring oil-degrading microbes in the environment to multiply and begin decomposing the pollutant. The experiment was successful, and the U.S. Environmental Protection Agency recommended that Exxon expand its bioremediation efforts to more of the affected beaches.

Environmental biotechnology is a growing industry, and numerous venture capital firms have started to supply remediation technology for various types of wastes. One application of this technology is "designer microbes" for sewage treatment facilities receiving industrial pollutants. Another involves creating unique microorganisms, using genetic engineering techniques, that have the ability to degrade new types of pollutants completely. The first living thing to be patented in the United States, a bacterium that was genetically engineered at the General Electric Company, was created specifically to degrade petroleum from oil spills.

Mark S. Coyne

See also: Biotechnology.

Sources for Further Study

Alexander, Martin. *Biodegradation and Bioremediation*. 2d ed. San Diego: Academic Press, 1999. A primer on bioremediation.

_____. "Biodegradation of Chemicals of Environmental Concern." *Science* 211, no. 9 (1980). A short, concise description of chemical reactions in bioremediation.

Congress of the United States. Office of Technology Assessment. *Bioremediation for Marine*

Oil Spills. Washington, D.C.: Government Printing Office, 1991. A fairly technical report on bioremediation in practice.

Environmental Protection Agency. *Understanding Bioremediation: A Guidebook for Citizens*. Washington, D.C.: Author, 1991. Useful pamphlet explains bioremediation at a very basic level.

Skipper, Horace, and Ron Turco, eds. *Bioremediation: Science and Applications*. Madison, Wis.: Soil Science Society of America, 1995. A collection of reports on bioremediation treatment technologies.

EROSION AND EROSION CONTROL

Categories: Agriculture; environmental issues; soil

Erosion is the loss of topsoil through several types of action of wind or water.

Erosion control is vital because soil loss from agricultural land is a major contributor to *nonpoint-source pollution* and *desertification* and represents one of the most serious threats to world food security. In the United States alone, some 2 billion tons of soil erode from cropland on an annual basis. About 60 percent, or 1.2 billion tons, is lost through water erosion, while the remainder is lost through wind erosion. This is equivalent to losing 0.3 meter (1 foot) of topsoil from 2 million acres of cropland each year. Although soil is a renewable resource, soil formation occurs at rates of just a few inches per hundred years, much too slowly to keep up with erosive forces. The loss of soil fertility is incalculable, as are the secondary effects of polluting surrounding waters and increasing *sedimentation* in rivers and streams.

Erosion removes the topsoil, the most productive soil zone for crop production, and the plant nutrients it contains. The thinning of the soil profile, which decreases a plant's rooting zone in shallow soils, can disturb the topography of cropland sufficiently to impede farm equipment operation. Erosion carries nitrates, phosphates, herbicides, pesticides, and other agricultural chemicals into surrounding waters, where they contribute to cultural eutrophication. Erosion causes sedimentation in lakes, reservoirs, and streams, which eventually require dredging.

Water Erosion

There are several types of wind and water erosion. The common steps in water erosion are *detachment*, *transport*, and *deposition*. Detachment releases soil particles from soil aggregates, transport carries the soil particles away and, in the process, scours new soil particles from aggregates. Finally, the soil particles are deposited when water flow slows. In *splash erosion*, raindrops impacting the soil can detach soil particles and hurl them considerable distances. In *sheet erosion*, a thin layer of soil is removed by tiny streams of water moving down gentle slopes. This is one of the most insidious forms of erosion because the effects of soil loss are imperceptible in the short term. *Rill erosion* is much more obvious because small channels form on a slope. These small

State	Percent of Cropland Planted Using No-Till Practices		
	1994	1996	1997
Kentucky	44	51	48
Maryland	41	46	45
Tennessee	42	44	43
West Virginia	37	39	39
Delaware	37	38	38
Ohio	35	37	36

Leading No-Tillage States, 1994-1997

Source: Data adapted from G. R. Haszler, "No Tillage Use for Crop Production in Kentucky Counties in 1997," *Soil Science News and Views* 19 (1998).

channels can be filled in by tillage. In contrast, *ephemeral gullies* are larger rills that cannot be filled by tillage. *Gully erosion* is the most dramatic type of water erosion. It leaves channels so deep that equipment operation is prevented. Gully erosion typically begins at the bottom of slopes, where the water flow is fastest, and works its way with time to the top of a slope as more erosion occurs.

Wind Erosion

Wind erosion generally accounts for less soil loss than water erosion, but in states such as Arizona, Colorado, Nevada, New Mexico, and Wyoming, it is actually the dominant type of erosion. Wind speeds 0.3 meter (1 foot) above the soil that exceed 16 to 21 kilometers (10 to 13 miles) per hour can detach soil particles. These particles, typically fine- to medium-sized sand fewer than 0.5 millimeter (0.02 inch) in diameter, begin rolling and then bouncing along the soil, progressively detaching more and more soil particles by impact. The process, called *saltation*, is responsible for 50 to 70 percent of all wind erosion. Larger soil particles are too big to become suspended and continue to roll along the soil. Their movement is called surface *creep*.

The most obvious display of wind erosion is called *suspension*, when very fine silt and clay particles detached by saltation are knocked into the air and carried for enormous distances. The Dust Bowl of the 1930's was caused by suspended silt and clay in the Great Plains of the United States. It is possible to see the effects of wind erosion on the sides of fences and similar objects. Wind passing over these obstacles deposits the soil particles it carries. Other effects of wind erosion are tattering of leaves, filling of road and drainage ditches, wearing of paint, and increasing incidence of respiratory ailments.

Soil Conservation Methods

The four most important factors affecting erosion are soil texture and structure, roughness of the soil surface, slope steepness and length, and soil cover. There are several passive and active methods of erosion control that involve these four factors. Wind erosion, for example, is controlled by creating windbreaks, rows of trees or shrubs that shorten a field and can reduce the wind velocity by

about 50 percent. Tillage perpendicular to the wind direction is also a beneficial practice, as is keeping the soil covered by plant residue as much as possible.

Highly erosive, steeply sloped land in the United States can be protected by placing it in the government-sponsored Conservation Reserve Program. The program provides incentives and assistance to farmers and ranchers for establishing conservation practices that have a beneficial effect on resources both on and off the farm. It encourages farmers to plant grass and trees to cover land that is subject to wind and water erosion.

Additional ways to prevent water erosion include planting permanent grass waterways in areas of cropland that are prone to water flow. Likewise, grass filter strips can be planted between cropland

The ecosystem of the lower coastal "terrace" of Navassa Island in the Caribbean Sea has eroded as a result of nineteenth century mining activities.

and adjacent waterways to impede the velocity of surface runoff and cause suspended soil particles to sediment and infiltrate before they can become contaminants. In this way, vegetation can improve water quality or provide food and habitat for wildlife. Tillage practices are also beneficial in combating water erosion. Tillage can be done along the contour of slopes. Long slopes can be shortened by terracing, which reduces the slope steepness.

Conservation tillage practices, such as minimal tillage and no-tillage, are being widely adapted by farmers as a simple means of erosion control. As the names imply, these are tillage practices in which as little disruption of the soil as possible occurs and in which any crop residue remaining after harvest is left on the soil surface to protect the soil from the impact of rain and wind. The surface residue also effectively impedes water flow, which causes less suspension of soil particles. Because the soil is not disturbed, practices such as no-tillage promote rapid water infiltration, which also reduces surface runoff. No-tillage is rapidly becoming the predominant tillage practice in southeastern states such as Kentucky and Tennessee, where high rainfall and erodible soils are found.

Mark S. Coyne

See also: Agriculture: modern problems; Desertification; Nutrients; Soil degradation; Soil salinization.

Sources for Further Study

Faulkner, Edward. *Plowman's Folly*. Washington, D.C.: Island Press, 1987. A classic diatribe against the effects of conventional agriculture and agricultural practices on soil erosion and environmental quality.

Laflen, John M., Junliang Tian, and Chi-Hua Huang, eds. *Soil Erosion and Dryland Farming*. Boca Raton, Fla.: CRC Press, 2000. Consists of the proceedings from an important conference on soil erosion and dryland farming, and addresses the two topics as they relate to mainland China, the most erodible place on earth. Topics such as dryland farming systems and soil water management, environmental quality and sustainability, and erosion control techniques are examined.

Plaster, Edward. *Soil Science and Management*. 3d ed. Albany, N.Y.: Delmar, 1997. Introduces the concepts behind soil science and relates these concepts to current soil management practices, such as regulatory changes and technological developments, wetlands management, and the use of Geographic Information Systems for soil mapping.

Schwab, Glen, et al. *Soil and Water Conservation Engineering*. 4th ed. New York: Wiley, 1993. Combines engineering practices for the solution of erosion and flood control, drainage and irrigational problems. Intended to make readers aware that the environment must be considered in the design of soil and water facilities.

ESTROGENS FROM PLANTS

Categories: Economic botany and plant uses; medicine and health

While the female hormones called estrogens are common in mammals, only a few plants contain estrogens. Others synthesize compounds which are chemically unrelated to estrogens but resemble them in their molecular size and shape. These compounds are called phytoestrogens (plant estrogens) and may, when ingested by animals or humans, have properties similar to those of mammalian estrogens.

The precursor of estrogens in plants and animals is the linear (straight-chain) triterpene known as squalene. Cyclization of squalene, via the intermediate cycloartenol in plants and via the intermediate lanosterol in animals, forms a group of very important compounds known as the *steroids*. Steroids include cholesterol, mammalian sex hormones (including the estrogens and androgens), corticosteroids, insects' molting hormones, and plant brassinosteroid hormones. All steroids have a tetracyclic (four-ringed) structure; the rings are named A, B, C, and D. Differences in the functional groups attached to the tetracyclic skeleton, differences in the side-chain attached to ring D, and differences in the overall shape of the molecule determine a steroid's biological activity.

Estrogens have an aromatic A ring (a ring of six carbon atoms joined by alternating single and double bonds). This constrains the junction between the A and B rings, resulting in a "flat," or planar, molecule. This shape is essential for the potent biological activity of estrogens, and chemical modifications that alter the planar nature of an estrogen molecule reduce its biological activity.

Estrogens in Plants Versus Animals

In animals, one of the most potent estrogens is *estradiol*. It triggers the production of gonadotropins leading to ovulation. It is metabolized to the less active estrogens, *estrone* and *estriol*. Estrone and estriol are produced by the placentas of pregnant mammals, and both compounds accumulate in the urine during pregnancy.

In plants, estrogens are *secondary metabolites*. Although many thousands of secondary metabolites occur in plants, the distribution of particular secondary metabolites is often limited to just a few genera. This appears to be the case for estrogens. Estrone has been isolated from the seeds of pomegranate and from the date palm, in which it is a component of the kernel oil. Estriol has been isolated from the pussy willow. It is not known what function these estrogens have in plants. It is possible that, like other secondary metabolites, they may function in plant defense.

The phytoestrogens coumestrol and daidzein are found in clover blossoms. If the content of phytoestrogens is high, reproductive cycles of grazing animals may be affected.

Isoflavonoids as Phytoestrogens

Isoflavonoids are a type of secondary metabolite and are found almost exclusively in the legume (pea) family of plants. They are known to function in plant defense. They have been shown to deter herbivores and also to facilitate a plant's defense response to pathogen attack. Interestingly, some isoflavonoids have chemical structures that, in overall size, shape, and polarity, resemble estrogens. The resemblance includes the flatness, or planarity, of the molecules and the positions and orientation of oxygen atoms. Isoflavonoids that have these molecular characteristics can mimic the biological activity of estrogens and are called *phytoestrogens*.

In terms of biosynthetic origins and chemical structure, phytoestrogens and estrogens are quite different. Phytoestrogens, being isoflavonoids, are phenolic compounds, formed from phenylalanine (an amino acid) by the shikimate pathway. In contrast, estrogens are triterpenoids, formed from acetyl coenzyme A by the isoprenoid pathway.

Isoflavonoids that are considered to be phytoestrogens exhibit only weak estrogenic activity in animals and humans. Examples of phytoestrogens are *coumestrol*, *daidzein*, and *genistein*. Coumestrol and daidzein are found in alfalfa (known as lucerne in Europe) and clover. Both of these plants belong to the legume family and are important forage crops for animals. If the content of phytoestrogens in alfalfa or clover is high, the reproductive cycles of grazing animals may be adversely affected. This can pose a problem for farmers wanting to breed livestock in the normal way. For this reason, the amount of grazing in fields of alfalfa or clover has to be restricted. Alternatively, varieties of alfalfa or clover that have been bred to contain lower levels of isoflavonoids can be grown. Unfortunately, plant varieties with lower isoflavonoid content are often more susceptible to both pathogen attack and attack by herbivorous pests.

Phytoestrogens in the Human Diet

One of the major sources of phytoestrogens in the human diet is the soybean. Genistein is the major phytoestrogen in soybeans. It is present in some soybean products such as tofu, although it is not present in soy sauce. Genistein, extracted from soybean plants, can also be obtained as a dietary supplement. Dietary supplements, which are often pills, powders, or tinctures containing plant-derived products, can be purchased over the counter. In the United States, the manufacture and sale of such products, classified as "dietary supplements," is far less closely regulated and standardized than the manufacture and sale of food and drugs.

Genistein has been promoted as a possible preventive treatment or therapy for several diseases and conditions. There are claims that it reduces hot flashes associated with menopause, that it can prevent or delay the onset of osteoporosis in postmenopausal women, and that it can lower blood cholesterol levels. In each instance the potential effectiveness of genistein would be attributable to its acting as an estrogen replacement in older women, in whom the level of estradiol is naturally low. Genistein may also be effective in the treatment of certain breast cancers that require estrogen in order to grow. In this case it is theorized that the genistein, with weak estrogen activity, acts to reduce cancer growth by competing with the more potent estradiol for the estrogen receptor.

Some of the evidence for the role of phytoestrogens in women's health is circumstantial. It is based, in part, on observations that women who live in countries such as Japan and China, where soy products are widely consumed, have a lower incidence of diseases such as osteoporosis and breast cancer. Clearly, other factors, genetic and environmental, may be contributory. Health claims attributed to phytoestrogens, including genistein, need further evaluation in well-designed clinical trials before such claims can be accepted by the scientific and medical communities or relied upon by those using dietary supplements.

Valerie M. Sponsel

See also: Medicinal plants; Metabolites: primary vs. secondary.

Sources for Further Study

Dolby, Victoria. *All About Soy Isoflavones and Women's Health.* New York: Avery, 1999. Frequently asked questions, and answers to them, written for the nonscientist.

Levetin, Estelle, and Karen McMahon. *Plants and Society.* Boston: WCB/McGraw-Hill, 1999. Chapter 13, "Legumes," and the unit on plants and human health are relevant.

ETHANOL

Category: Economic botany and plant uses

Ethanol, sometimes called grain alcohol, is an alcohol produced by fermentation of carbohydrates from a broad range of plant matter for many uses in the chemical industry. It has potentially significant use as a gasoline replacement and is also the primary alcohol component of alcoholic beverages.

Ethanol is produced by carbohydrate fermentation processes, hydration of ethylene, and, to a lesser extent, reduction of acetaldehyde obtained from acetylene. Also called ethyl alcohol, alcohol, and grain alcohol, ethanol is a colorless liquid with a mild and distinct aroma and taste. It has a boiling point of 78.3 degrees Celsius and a melting point of ⁻114.5 degrees Celsius. Ethanol is completely soluble in water and most organic solvents. It has a flash point of 8 degrees Celsius and is thus highly flammable. Ethanol undergoes numerous commercially important reactions and is thus a vital industrial chemical. It has been used as a partial replacement for gasoline (the hybrid fuel is called "gasohol"). As the major component of alcoholic beverages, ethanol has been known and recognized for thousands of years.

Primary Uses

Alcohol obtained from fermentation processes is generally included with other fermentation products and extracts from the carbohydrate-rich grains, fruits, and so on that are the raw materials for the many alcoholic beverages produced and consumed. Alcohol produced by yeast fermentation is obtained at a maximum concentration of 14 percent; therefore, alcoholic beverages other than beer and nonfortified wines require the addition of concentrated alcohol, which is obtained by distilling dilute alcohol from the fermentation of molasses and other sugar sources. In the United States and other highly industrialized countries, the alcohol added to beverages is increasingly being produced by other methods.

Ethanol is also used in large quantities for chemical synthesis in the organic chemical industry. It is used for the preparation of numerous esters vital to many polymer industries and for the production of diethyl ether (also called ether or ethyl ether), a ma-

jor solvent. Other synthetic procedures lead to the manufacture of acetaldehyde, acetic acid, ethyl halides, and acetonitrile, which are in turn employed for the preparation of drugs, explosives, adhesives, pesticides, detergents, synthetic fibers, and other substances. Ethanol itself is used in vast quantities as a solvent.

Ethanol is added to gasoline to reduce air pollution, and it is frequently considered to be a likely replacement for gasoline when petroleum resources decline or drastically increase in price. Gasohol, ethanol combined with varying amounts of gasoline, is being vigorously promoted and is already in use in Brazil and elsewhere. *Biomass conversion* (conversion of plant matter) to ethanol by cost-efficient methods will speed the entry of gasohol into large-scale use in the industrial world.

Production Processes

Until biomass conversion becomes more widely used, carbohydrate fermentation processes are destined to be a decreasingly important source of industrial ethanol. Beverage alcohol is produced from a great variety of sources, including grains, potatoes, and fruit, but fermentation-based industrial alcohol is almost entirely obtained by yeast fermentation of molasses. Molasses (50 percent sucrose residue from sugar processing) is diluted with water to approximately 15 percent and under slightly acidic conditions is fermented by yeast to give 14 percent ethanol. Fractional distillation of the solution yields the commercial product: 95 percent ethanol. Approximately two and one-half gallons of blackstrap molasses is needed to make a gallon of 190-proof ethanol. (Alcohol content is usually described in terms of its proof value, which is twice its ethanol percentage.)

Although ethylene hydration was known in the early part of the nineteenth century, it was not until 1929 that it became an industrial process. Today it is the dominant method of producing ethanol. Ethylene, obtained from the thermal cracking of petroleum fractions or from natural gas separation processes, is treated with complex phosphoric acid-based catalysts at temperatures above 300 degrees Celsius and steam at pressures of thousands of pounds per square inch. The ethanol can be fractionally distilled, and the residual ethylene can be recycled. Ethylene can also be passed into concentrated sulfuric acid, and after hydrolysis the ethanol can be distilled from the diluted sulfuric acid.

The 95 percent alcohol produced by any of these methods can be converted to nearly pure (absolute) alcohol by removal of the water by azeotropic distillation using benzene or trichloroethylene. Trace amounts of the hazardous benzene or trichloroethylene remaining make the absolute alcohol undesirable for beverage purposes but useful for industrial purposes in which the 5 percent water interferes with use requirements.

William J. Wasserman

See also: Glycolysis and fermentation; Grains; Sugars.

Sources for Further Study

Deitrich, R. A., and V. G. Erwin. *Pharmacological Effects of Ethanol on the Nervous System*. Boca Raton, Fla.: CRC Press, 1996. Examines the effects of ethanol on the major neurotransmitter systems affected by ethanol and correlates these actions with the behavioral consequences. The subject is approached first from the perspective of the neurochemical system and the behaviors resulting from ethanol's effects on that system.

Gomberg, Edith Lisansky, Helene Raskin White, and John A. Carpenter, eds. *Alcohol, Science, and Society Revisited*. Ann Arbor: University of Michigan Press, 1982. Examines the effects of ethanol on human metabolism and social interaction.

Paul, J. K., ed. *Ethyl Alcohol Production and Use as a Motor Fuel*. Park Ridge, N.J.: Noyes Data, 1979. A good general source on ethanol.

Rothman, Harry, Rod Greenshields, and Francisco Rosillo Callé. *Energy from Alcohol: The Brazilian Experience*. Lexington: University Press of Kentucky, 1983. Includes significant studies of gasohol possibilities.

Weinheim, M. Roehr, ed. *The Biotechnology of Ethanol: Classical and Future Applications*. New York: Wiley, 2001. Highlights the industrial relevance of ethanol as one of the most important products of primary metabolism. Covers the most advanced developments among classical methods as well as more unconventional techniques, before outlining various aspects of new applications and the increasing importance of ethanol as a renewable resource.

EUDICOTS

Categories: Angiosperms; *Plantae*; taxonomic groups

The eudicots, class Eudicotyledones *(literally "true dicots"), are descended from a common ancestor and comprise three-quarters of all flowering plants. It is one of the two main classes of the angiosperms, the other being the monocots, or* Monocotyledones.

Eudicots, the common name used for class *Eudicotyledones*, are the most common group of flowering plants, comprising 75 percent of all angiosperms. The other 25 percent, monocots (*Monocotyledones*), are often characterized by pollen grains that have a single aperture (or line of weakness). Eudicots have pollen grains that typically possess three apertures, referred to as triaperturate pollen. Thus, there is no monocot-dicot division among the flowering plants. Whereas "monocot" remains a useful term, "dicot" does not represent a clade (a collection of organisms which have a single common ancestor) and should no longer be used. It is more useful to refer to eudicots, which represent a well-defined clade of angiosperms.

Previously, the angiosperms were divided into two major groups, traditionally recognized as classes: the dicots (short for dicotyledons, class *Magnoliopsida*) and the monocots (monocotyledons, class *Liliopsida*). Dicots have been distinguished from monocots by several morphological (external physical) and anatomical features, but all of these were subject to exception. For example, most dicots possess two seedling leaves, or *cotyledons*, and typically

Eudicot (Dicot) Families Common in North America

Common Name	Scientific Name
Acanthus family	*Acanthaceae*
Borage family	*Boraginaceae*
Buckwheat family	*Polygonaceae*
Buttercup family	*Ranunculaceae*
Cactus family	*Cactaceae*
Daisy family	*Asteraceae*
Evening primrose family	*Onagraceae*
Gentian family	*Gentianaceae*
Ginseng family	*Araliaceae*
Madder family	*Rubiaceae*
Milkweed family	*Asclepiadaceae*
Mint family	*Lamiaceae*
Mustard family	*Brassicaceae*
Pea family	*Fabaceae*
Phlox family	*Polemoniaceae*
Pink family	*Caryophyllaceae*
Pokeweed family	*Phytolaccaceae*
Primrose family	*Primulaceae*
Purslane family	*Portulacaceae*
Rose family	*Rosaceae*
Saxifrage family	*Saxifragaceae*
Verbena family	*Verbenaceae*
Violet family	*Violaceae*
Waterleaf family	*Hydrophyllaceae*

Note: For a full list of angiosperm families, see the tables that accompany the essay "Angiosperms," in volume 1.
Source: U.S. Department of Agriculture, National Plant Data Center, *The PLANTS Database*, Version 3.1, http://plants.usda.gov. National Plant Data Center, Baton Rouge, LA 70874-4490 USA.

have net-veined leaves; monocots, in contrast, usually have one cotyledon and leaves with parallel venation.

The monocot-dicot division was recognized as early as the nineteenth century. However, later studies of *phylogeny* (the evolutionary history of a group of organisms) have demonstrated that this split does not reflect the evolutionary history of angiosperms. Phylogenetic trees (which are visual representations of the evolution of a group of organisms) showing historical relationships have been constructed based on deoxyribonucleic acid (DNA) sequences, as well as morphological, chemical, and other characteristics. These trees indicate that whereas the monocots form a *clade*, all dicots do not form a distinct group. The monocots appear among the groups of early-diverging (or early-evolving) lineages of angiosperms, all of which have traditionally been considered dicots. All early branches of the angiosperm phylogenetic tree, including the monocots, are best referred to informally as *basal angiosperms*.

Most angiosperms form a distinct clade, referred to by J. S. Doyle and C. L. Hotton as the eudicots, or true dicots. Whereas basal angiosperms are often characterized by pollen grains that have a single aperture (or line of weakness), eudicots have pollen grains that typically possess three apertures, referred to as triaperturate pollen. The eudicot clade receives strong support from analyses based on DNA sequence data. Importantly, the eudicots represent only a subset of the formerly recognized group dicots. Many basal angiosperms are traditional dicots but are not eudicots. Thus, there is no monocot-dicot split in the angiosperms. Whereas "monocot" remains a useful term, "dicot" does not represent a clade (a collection of organisms which have a single common ancestor) and should no longer be used. It is more useful to refer to eudicots, which represent a well-marked clade of flowering plants.

Classification

The eudicots contain 75 percent of all angiosperms, or about 165,000 species distributed among roughly 300 families. The eudicots include all the familiar angiosperm trees and shrubs and many herbaceous groups. Some of the larger, better-known families of eudicots include *Rosaceae* (rose family), *Fabaceae* (bean family), *Brassicaceae* (mustard family), *Ranunculaceae* (buttercup family),

Apiaceae (parsley family), Asteraceae (sunflower family), and Lamiaceae (mint family). The eudicots include many familiar trees, such as those of the Fagaceae (oak or beech family), Betulaceae (birch family), Juglandaceae (walnut or hickory family), Aceraceae (maple family), and Platanaceae (plane tree or sycamore family).

A good understanding of the major groups of eudicots has emerged from the use of DNA sequence data. The early-diverging eudicots consist of a number of ancient lineages, including Ranunculales, a group containing the Ranunculaceae and Papaveraceae (poppy family). Other early-diverging eudicots include Buxaceae (boxwood family) and Proteales; the latter includes the Platanaceae and Proteaceae (sycamore and protea families).

Following the early-diverging eudicots is a large clade, referred to as the core eudicots, that contains most euciots. The core eudicots consist of three major clades (rosids, asterids, and Caryophyllales) and several smaller ones (Santalales, Saxifragales, and Gunnerales). The rosids and asterids are very large groups, each containing roughly one-third of all angiosperms. Traditional classifications, such as presented by botanist Arthur Cronquist in 1981, do not reflect modern views of phylogenetic relationships. For comparison, the rosid clade now recognized is made up of members of the traditional subclasses Rosidae, Dilleniidae, and Asteridae (in the sense of Cronquist, 1981); the asterid clade contains members of subclasses Asteridae, Dilleniidae, and Rosidae; and Caryophyllales contains taxa previously placed in Caropyllidae and Dilleniidae.

Relationships among the core eudicots are still unclear, despite intensive study using DNA sequence data. The difficulty in clearly describing relationships among these groups appears to stem from the fact that following the origin and initial diversification of the eudicots, a rapid radiation (evolution of many organisms in a short period of time) occurred, yielding the groups of core eudicots seen today.

Evolution

The eudicots can be easily identified in the fossil record by their three-grooved pollen; they appeared in the fossil record as early as 110 million years ago. Prominent early fossil eudicots include Platanaceae. Following the origin of the eudicots, the fossil record also suggests a rapid diversification; by 90 to 80 million years ago, many of today's prominent families were established and are clearly

Horticulturally important eudicots include numerous representatives of Ranunculaceae, such as columbine.

recognizable in the fossil record. Thus, the fossil evidence suggesting a rapid radiation of eudicots agrees with the phylogeny obtained using gene sequence data.

Economic Uses

Outside the grasses, such as wheat, corn, and rice (all in the monocot family Poaceae), most plants of economic importance are eudicots. Examples of economically important eudicot families include members of the bean or legume family (Fabaceae), such as soybean, lentils, and green beans; and the sunflower family (Asteracae), which includes sunflowers, lettuce, and artichokes. The mustard family (Brassicaceae) contains numerous members of economic importance, including cabbage, kale, cauliflower, mustard, and horseradish. The rose family (Rosaceae) provides fruits such as strawberries, raspberries, apples, cherries, peaches, and plums as well as numerous ornamentals. The family Solanaceae is the source of tomatoes and potatoes.

Most familiar horticultural plants are also eudicots. They include a diverse array of trees, shrubs, and annual and perennial herbs. Horticulturally important eudicots include begonias (*Begoniaceae*), dogwoods (*Cornaceae*), rhododendrons and heaths (*Ericaceae*), numerous members of the *Asteraceae* including sunflower, asters, chrysanthemums, and marigolds; and numerous representatives of *Ranunculaceae* such as columbine, buttercups, and monkshood. Some other families that contain ornamental plants include the cactus family (*Cactacae*), the geranium family (*Geraniaceae*), and members of the mint family (*Lamiaceae*).

Because 75 percent of all flowering plants are eudicots, they are extremely diverse in ecology and morphology. The eudicots are also diverse in *habit* (pattern of growth) and include annual and perennial herbs, shrubs, and trees. In size, they range from the smallest terrestrial angiosperms, plants 1 centimeter in height (*Lepuropetalon, Parnassiaceae*), to eucalyptus trees well over 100 meters. Flowers are also highly diverse in structure, form, and size across the eudicots. The smallest eudicot flowers are those of *Lepuropetalon*, which are less than 1 millimeter in diameter; the largest eudicot flowers are more than 0.3 meter (1 foot) in length.

Douglas E. Soltis and Pamela S. Soltis

See also: Angiosperm evolution; Angiosperms; Magnoliids; Monocots vs. dicots.

Sources for Further Study

Cronquist, Arthur. *An Integrated System of Classification of Flowering Plants.* 1981. Reprint. New York: Columbia University Press, 1993. A definitive statement by a botanist with extensive knowledge of the vast range of flowering plants.

Doyle, J. A., and C. L. Hotton. "Diversification of Early Angiosperm Pollen in a Cladistic Context." In *Pollen and Spores,* edited by S. Backmore and S. H. Barnes. Oxford, England: Clarendon Press, 1991.

Heywood, V. H., ed. *Flowering Plants of the World.* Englewood Cliffs, N.J.: Prentice-Hall, 1993. A profusely illustrated guide to flowering plants. Contains information on essentially all the generally recognized families of the world, including descriptions, taxonomy, geographic distribution, and economic uses.

Kenrick, Paul, and Peter R. Crane. *The Origin and Early Diversification of Land Plants: A Cladistic Study.* Washington, D.C.: Smithsonian Institution Press, 1997. Advanced, comprehensive coverage of the evolutionary relationships among early vascular land plants.

Savolainen, V., et al. "Phylogeny of the Eudicots: A Nearly Complete Familial Analysis Based on *rbcL* Gene Sequences." *Kew Bulletin* 55 (2000): 257-309. New insights in familial relationships with some surprising results on taxonomic dispositions.

Soltis, Pamela S., Douglas E. Soltis, and Mark W. Chase. "Angiosperm Phylogeny Inferred from Multiple Genes as a Tool for Comparative Biology." *Nature* 402 (November 25, 1999): 402-404. The authors show how studies of diversification in the angiosperms in floral form, stamen organization, reproductive biology, photosynthetic pathway, nitrogen-fixing symbioses, and life histories have relied on either explicit or implied phylogenetic trees.

EUGLENOIDS

Categories: Algae; microorganisms; *Protista*; taxonomic groups; water-related life

Organisms called euglenoids, the algal phylum Euglenophyta *in the kingdom* Protista, *make up a large group of common microorganisms numbering between 750 and 900 known species.*

Euglenoids can be found in both fresh and stagnant water. Some genera of euglenoids can also be found in marine habitats. *Euglena* and *Phacus* are representative common genera. Euglenoids are unicellular, except for those of the colonial genus *Colacium*. Because euglenoids have flexible cell coverings, move about freely, and ingest their food through a structure called a gullet, many scientists have classified the euglenoids as animals. Some species of euglenoids, however, have chloroplasts and are able to supply at least some of their food needs through photosynthesis.

Structure

The cells of most euglenoids are spindle-shaped and do not have cell walls or other rigid structures covering the plasma membrane. However, one genus, *Trachelomonas*, has a covering called a lorica, which is similar to a cell wall and contains iron and magnesium minerals. Spiral strips of protein that originate in the cytoplasm support the plasma membrane, creating a structure called a *pellicle*. The pellicle may be either flexible or rigid. Euglenoids with a flexible pellicle are able to change their shape, which helps them move about in muddy habitats.

A euglenoid has two *flagella*. One, a functional flagellum, has numerous tiny hairs along one side and pulls the cell through the water. This flagellum originates in a structure called a reservoir found on the anterior end of the cell. Another, smaller flagellum is contained within the reservoir but does not protrude.

Other features of euglenoids include the presence of a *gullet*, or groove, in many species, through which food is ingested. About one-third of the euglenoids have disk-shaped *chloroplasts* and can supply at least part of their own food through photosynthesis. Even these photosynthetic euglenoids are capable of ingesting dissolved or particulate food through their gullet if necessary.

A reddish-colored structure called an *eyespot*, or *stigma*, is located in the cytoplasm near the base of the flagella. This eyespot acts as a light-sensing device. The eyespot appears to be connected to the flagellum by special strands of cytoplasm, which may serve to transmit signals from one organelle to the other.

Euglenoid cells also contain a *contractile vacuole*. The contractile vacuole functions as a pump that removes excess water from the interior of the euglenoid cell. The water is pumped out of the cell through the reservoir. A new contractile vacuole is formed after each discharge of water.

Pigments and Food Reserves

Some scientists believe that similarities between euglenoids and green algae indicate that the chloroplasts of euglenoids originated in endosymbiotic green algae. The chloroplasts of both euglenoids and green algae contain chlorophyll *a* and chlorophyll *b*. Euglenoids and green algae also have some carotenoid pigments in common, although euglenoids contain two pigments derived from carotenoids that are not found in either green algae or the higher plants.

While there are some similarities in pigmentation between euglenoids and green algae, the two groups have different food reserves. Green algae have food reserves of starch, while euglenoids have a carbohydrate food reserve called *paramylon*, which normally is present in the form of small, whitish bodies of varying shapes scattered throughout the cell.

Reproduction

Reproduction in euglenoids takes place by mitotic cell division. Even as the euglenoids are swimming about, the cell begins to divide, starting at the end of the cell, where the flagellum is located. Eventually the cell splits lengthwise, forming two complete cells. Unlike the nuclear membrane in most organisms, the membrane surrounding the euglenoid nucleus does not break down during mitosis.

Many scientists believe that euglenoids do not reproduce sexually; meiosis and gametogenesis have never been observed in euglenoids. Some scientists suspect sexual reproduction must occur, even though it has not been seen. Other scientists believe that even individuals of the same euglenoid species have different amounts of DNA in their nuclei, which would preclude meiosis. Still others believe that euglenoids branched off the main evolutionary lines of protoctists before sexual reproduction had evolved.

Some species of euglenoids have developed the capacity to make thick-walled resting cells. Inside these cells, the euglenoids can wait out unfavorable environmental conditions. When conditions are favorable, the organisms break out of their resting cells and resume their normal shapes and activities.

Carol S. Radford

See also: Algae; Green algae.

Sources for Further Study

Margulis, Lynn, and Karlene V. Schwartz. *Five Kingdoms: An Illustrated Guide to the Phyla of Life on Earth.* 3d ed. New York: W. H. Freeman, 1998. All of the major divisions of life on earth are discussed in this book.

Raven, Peter H., Ray F. Evert, and Susan E. Eichhorn. *Biology of Plants.* 6th ed. New York: W. H. Freeman/Worth, 1999. This standard college-level textbook discusses the systematics and morphology of the different divisions of the plant kingdom.

Stern, Kingsley R. *Introductory Plant Biology.* 8th ed. Boston: McGraw-Hill, 2000. Introductory botany textbook includes discussions of the different divisions of the plant kingdom.

EUKARYA

Categories: Cellular biology; evolution; taxonomic groups

The Eukarya *form one of the domains of life in the three-domain classification system.* Eukarya *consists of the advanced, complex organisms, formed by eukaryotic cells (cells with nuclei), including fungi, algae, plants, and animals. The other two domains of life,* Archaea *and* Bacteria, *consist of simpler organisms formed by prokaryotic (nucleus-free) cells.*

Two Types of Cell

The domain concept of biological organization is relatively new. As recently as the mid-twentieth century, two kingdoms—plant and animal—were widely accepted as describing the most significant split in the biological world. Every living thing was classified as either a plant or an animal. Subsequently, three additional kingdoms were recognized. Only in the late twentieth century did it become clear, based on molecular and other evidence, that distinctions at the level of the kingdom did not acknowledge the most fundamental differences among organisms. A higher category, the domain, was therefore posited. The general acceptance of the domain concept by the scientific community was an acknowledgment that, at least according to current knowledge, the differences between the prokaryotic organisms and the eukaryotic ones, and further, the split within the prokaryotic organisms, are the major dividing lines in the biological world.

Bacteria is the domain of prokaryotic organisms that are considered to be true bacteria, and *Archaea* is the domain of prokaryotic organisms able to live in extreme environments. *Eukarya* differs from the prokaryotic domains in basic characteristics of cellular organization, biochemistry, and molecular biology. Further, unlike the prokaryotic organisms, many of the *Eukarya* are truly multicellular. Eukary-

otic cells, which are structurally more complex than prokaryotic ones, have many of their cellular functions segregated into semiautonomous, membrane-bound cell regions, called *organelles*. The principal organelle is the *nucleus*, which contains the genetic material, deoxyribonucleic acid (DNA). In prokaryotic organisms, in contrast, the DNA is not segregated from the rest of the cell.

Other distinguishing organelles in eukaryotic cells include the *mitochondria*. These are the sites of *respiration*, in which energy is generated by breaking down food, in the presence of oxygen, into water and carbon dioxide. The plants and the algae have additional organelles, the *plastids*. The most common plastid is the *chloroplast*, which contains chlorophyll, the key molecule that allows algae and plants to manufacture their own food from carbon dioxide and water, by photosynthesis. In contrast, in those bacteria that are photosynthetic, chlorophyll is not confined within an organelle.

Evolutionary Origin

Prokaryotic cells are much older than eukaryotic cells and had a long reign in the primordial seas before one of them, probably a member of the domain *Archaea*, gave rise to the first eukaryotic cell, between 2.5 billion and 1 billion years ago. This was at least a billion years after life had arisen. The first eukaryotic cell lacked mitochondria and chloroplasts. Subsequently, two kinds of prokaryotic organisms belonging to the domain *Bacteria* took up residence, as symbionts, inside early eukaryotic cells and eventually became so dependent on their hosts that they could no longer live on their own. These so-called *endosymbionts* developed into mitochondria and plastids.

Mitochondria, which were acquired before chloroplasts, arose from small bacteria that were heterotrophic: They obtained their food from other organisms rather than manufacturing it themselves. Chloroplasts arose from bacteria known as *cyanobacteria*, which were autotrophic, manufacturing their food themselves by photosynthesis. Thus, the evolution of the eukaryotic cell involved three prokaryotic cells—the original archaean host cell and two kinds of endosymbiotic bacteria. The early eukaryotic organisms were single-celled and are classified in a group called protists.

Early Diversification

From their beginning as single-celled protists, eukaryotic organisms evolved rapidly. The first multicellular eukaryotic organisms appeared about 800 million years ago, during the Precambrian era, and developed into three great lineages: the fungi, plants, and animals. Scientists have accorded each of these groups "kingdom" rank within the domain *Eukarya*, as the kingdoms *Plantae*, *Animalia*, and *Fungi*. In addition, the protists, which have living representatives today, are considered to constitute a fourth kingdom, the *Protista*, within the *Eukarya*. The *Protista* consist of the predominantly unicellular phyla and some of the multicellular lines associated with them. All four of the kingdoms within the *Eukarya* arose in the sea. Transition to the land occurred later.

The three multicellular kingdoms—plants, animals, and fungi—each probably descended from a separate ancestor from among the protists, and thus each of these lineages constitutes a relatively well-defined natural kingdom. The protists, however, are something of a "catch-all" kingdom. They consist of a variety of lineages, which include both photosynthetic organisms, the algae, and nonphotosynthetic ones. Because the algae are capable of photosynthesis, older classification schemes lumped them with the plants. Scientists think that many kingdoms will ultimately be recognized among the *Protista*.

The protist that gave rise to the plant lineage was probably a now-extinct member of the family *Charophyceae*, a group of specialized, aquatic, multicellular green algae (phylum *Chlorophyta*) that includes members living today. Like the algal protists, plants are autotrophic, but plants have more complex, structurally integrated bodies than do the algae.

In contrast to the algal protists and the plants, the nonphotosynthetic protists, as well as all the fungi and the animals, obtain their food heterotrophically, from other organisms. The fungi, although lacking chloroplasts and photosynthetic pigments, were once, like the algae, classified within the plant kingdom. This was partly because, like plants, fungi are sedentary. The fungi, however, have little in common, nutritionally or structurally, with plants and are now recognized as an independent evolutionary line within the *Eukarya*. Molecular evidence indicates that the fungi are actually more closely related to animals than to plants.

Colonization of the Land

Although all of the lineages that are now recognized as kingdoms within the *Eukarya* originated from aquatic organisms, the *Eukarya* eventually achieved great success on the land. Multicellularity helped these organisms make the transition to an environment of earth and air, which was more complex and demanding than the relatively uniform conditions of the sea. With their many cells, the *Eukarya* were able to develop specialized structures for coping with this new environment. The evolution of plants shows a trend toward structures specialized for anchorage, photosynthesis, and support. This trend eventually led to the development of complex plant bodies, with roots, leaves, and stems, allowing the plants as a kingdom to be fully terrestrial, not aquatic. Had it not been for plants' pioneering of the land, animals could not have become established there, because plants form the base of terrestrial animals' food chain.

Plants may have first invaded the land sometime in the Ordovician period of the Paleozoic era, 510 million to 439 million years ago. Forms resembling modern land plants arose in the Late Silurian period of the Paleozoic, more than 408 million years ago. By the close of the Paleozoic's Devonian period, about 360 million years ago, plants had diversified into a wide variety of shapes and sizes, from small creeping forms to tall forest trees.

In fossils of early plants, fungi are often found in close association with the roots. Some scientists think that plants were able to colonize the land only because they developed symbiotic relationships with such representatives of the fungal kingdom. Scientists also think that fungi, in turn, may have been able to make the transition to land only because of their close relationship with plants. According to this view, the fungi helped the plants gain a terrestrial foothold by absorbing water and mineral nutrients from the poorly developed soils of that time and passing them on to their plant partners. The plants provided the fungi with sugars that the plants had manufactured photosyntheti-

cally. This is much the way that relationships between plant roots and certain fungi work today. These symbiotic, so-called mycorrhizal relationships are characteristic of the vast majority of plants that dominate the modern world—the plants having vascular, or conducting, tissues. Of these plants, the most important by far are the angiosperms, or flowering plants (phylum *Anthophyta*).

Jane F. Hill

See also: *Archaea*; Bacteria; Chloroplasts and other plastids; Eukaryotic cells; Evolution of cells; Evolution of plants; Mitochondria; Molecular systematics; Mycorrhizae; *Plantae*; Prokaryotes.

Sources for Further Study
Campbell, Neil A., et al. *Biology.* 6th ed. Menlo Park, Calif.: Addison Wesley Longman, 2002. Comprehensive, introductory, college-level textbook, which addresses the domain concept and the *Eukarya*. Includes color illustrations, diagrams, bibliographical references, glossary, and index.
Dyer, Betsey Dexter, and Robert Obar. *Tracing the History of Eukaryotic Cells.* New York: Columbia University Press, 1994. Includes a discussion of the role of symbiosis in the evolution of eukaryotic cells, and shows how the fossil record, genetics, and molecular evolution help illuminate cell evolution. Includes bibliographical references, appendices, and index.
Margulis, Lynn. *What Is Life?* New York: Simon and Schuster, 1995. Written for the general reader, this work includes a description of eukaryotic cells and the role of endosymbionts in their origin. Includes chapters on the plant, animal, and fungal kingdoms as well as color photos and line drawings.
Raven, Peter H., Ray F. Evert, and Susan E. Eichhorn. *Biology of Plants.* 6th ed. New York: W. H. Freeman/Worth, 1999. Comprehensive, standard college-level textbook, which addresses the domain concept and the *Eukarya* in the context of plants. Includes color photos, diagrams, glossary, index, and bibliography.
Tudge, Colin. *The Variety of Life: A Survey and a Celebration of All the Creatures That Have Ever Lived.* New York: Oxford University Press, 2000. An account, in nontechnical language, of evolution, taxonomic categories, and all the major groups of organisms. Includes bibliography, indexes.

EUKARYOTIC CELLS

Category: Cellular biology

Eukaryotic cells (as opposed to prokaryotic cells) have internal, membrane-bound organelles and a distinct nucleus that physically separates the genetic material of the cell from the all of the other parts of the cell. All protists, fungi, plants, and animals are composed of eukaryotic cells.

The cells of all organisms can be divided into two broad categories: *prokaryotic* cells and *eukaryotic*

cells. Prokaryotic cells are cells with a relatively simple structure, having no internal, membrane-

Parts of a Eukaryotic Cell

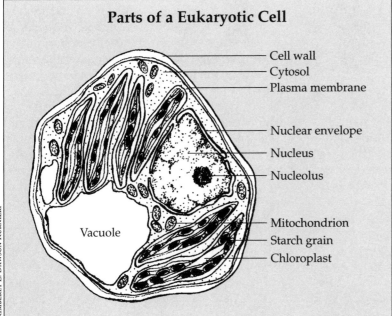

Cell wall
Cytosol
Plasma membrane

Nuclear envelope
Nucleus
Nucleolus

Vacuole

Mitochondrion
Starch grain
Chloroplast

Kimberly L. Dawson Kurnizki

Eukaryotic cells constitute all but the simplest life-forms: protists, fungi, plants, and animals. Based on an electron microscope image of a leaf cell from a corn plant, this depiction shows the basic parts of the cell. Eukaryotic cells are distinguished from more primitive prokaryotic (mainly bacterial) cells by the presence of a nucleus that contains the genetic materials as well as membrane-bound organelles such as mitochondria and, in algae and plants, plastids such as chloroplasts.

bound organelles. The most striking feature of prokaryotic cells is that they lack a distinct *nucleus*, hence the name *prokaryotic*, literally translated from its Greek roots as "before nucleus." The prokaryotic organisms comprise two domains of the three domains of life: the ancient bacteria, *Archaea*; and the modern bacteria, *Bacteria* or *Eubacteria*. The *Archaea* are single-celled organisms that often inhabit extreme environments, such as hot springs. The remainder of bacteria are classified as *Eubacteria*.

All other organisms, including fungi, plants, and animals, are composed of eukaryotic cells and belong to the domain *Eukarya*. Eukaryotic cells are more structurally complex than prokaryotic cells, having internal, membrane-bound *organelles* and a distinct nucleus that physically separates the genetic material of the cell from the all of the other parts of the cell. Based on genetic analysis, the *Archaea* and *Eukarya* are more closely related to each other than they are to the *Bacteria*, suggesting that eukaryotic cells may have arisen

from a single ancestral archaean cell.

Eukarya includes the traditional kingdoms *Plantae, Animalia, Fungi,* and *Protista*. Protists include a diverse assemblage of single-celled eukaryotic organisms including algae, amoebas, and paramecia. Because algae are photosynthetic, they have often been included in the study of plants, although they are not members of the plant kingdom.

Fungi include such organisms as smuts, rusts, molds, and mushrooms. Fungal cells have external cell walls and because of this have often been included in the study of plants. However, fungal cell walls have a completely different structure and composition from those of plant cell walls, and fungi lack plastids and photosynthetic pigments. Fungi represent a unique evolutionary line. They too, however, tend to be studied in botany courses, even though they are not plants.

Cell Parts

Eukaryotic cells are surrounded by a *cell membrane*, or *plasma membrane*, that is composed of a lipid structure in which other molecules, such as proteins and carbohydrates, are embedded. The cell membrane serves as a semipermeable, or selective, barrier between the cell and its environment. Some small, uncharged molecules can freely cross the cell membrane; others must be transported across the membrane before they can enter the cell. The cell membrane serves to protect the cell and to receive signals from the environment and other cells that help to direct cell activities.

In addition to the cell membrane, plant cells also have external *cell walls*. The presence of the external cell wall is one of the major characteristics that distinguishes plant cells from animal cells. The cell wall limits the size of the internal *protoplast* (the internal cytoplasm and nucleus) and prevents the plasma membrane from breaking when the protoplast enlarges following the uptake of water by the cell. Cell walls are not merely static support structures, however. They contain enzymes that are important in bringing essential molecules into the cell and in secreting molecules. They may also play important roles in the defense of the plant against bacterial and fungal pathogens.

Eukaryotic cells also have a prominent, membrane-bound organelle called the *nucleus*. The nucleus contains the genetic information of the cell that directs the cellular activity. A double membrane called the *nuclear envelope* surrounds the nucleus. Inside the nucleus, *deoxyribonucleic acid* (DNA) is transcribed to make molecules of *ribonucleic acid* (RNA), copies of the genetic information that can be delivered to the cytoplasm, where the RNA molecules serve to direct the manufacture of proteins. DNA in the eukaryotic nucleus exists as linear molecules that are associated with many proteins, and the DNA is packaged into a highly organized chromosomal structure by proteins called *histones*.

In addition to the nucleus, eukaryotic cells contain a number of internal membrane-bound organelles that help the cell carry out the functions necessary for life. The types of organelles found inside a eukaryotic cell reflect the function of that cell and the processes that it must carry out. Some of these organelles, such as *mitochondria* and *chloroplasts*, are important in capturing and releasing energy for cell function. Some, like the *Golgi complex* and the *endoplasmic reticulum* (ER), are involved in the manufacture, processing, and transport of proteins and other molecules within the cell. Others, such as *peroxisomes*, are involved in detoxifying chemicals and breaking down molecules.

The cell *cytoskeleton* is a highly dynamic structure that provides support and motility to cells as well as providing some of the apparatus that is used in the transduction of signals from the cell membrane to the nucleus. In plant cells, cytoskeletal elements form tracks for the movement of internal cellular organelles, such as the *cytoplasmic streaming* of chloroplasts, which can be observed by light microscopy. Work of the cytoskeleton is also necessary for the opening and closing of the stomata in plant leaves. The cytoskeleton consists of a variety of filamentlike proteins as well as proteins that serve as anchor points for filaments.

Origins of Mitochondria and Chloroplasts

The nucleus of the eukaryotic cell is not the only organelle that contains DNA and is enclosed by two membranes: The mitochondria of all cells and the chloroplasts of plant cells contain DNA and are surrounded by two membranes. The DNA of these organelles directs the synthesis of certain proteins that are necessary for the function of the organelles. This DNA is similar to DNA found in bacteria. Mitochondria and chloroplasts are thought to have evolved by a process known as *endosymbiosis*, in which bacteria were engulfed in the primitive eukaryotic cell, where they manufactured adenosine triphosphate (ATP), the nucleotide responsible for most of the chemical energy needed for metabolism, or captured energy from sunlight for the eukaryotic cell, establishing a mutually beneficial, or *symbiotic*, relationship with the eukaryotic cell.

Several lines of evidence support the endosymbiotic theory for the origin of mitochondria and chloroplasts. First, these organelles have areas of specialized cytoplasm called *nucleoids* that contain the DNA, much as bacteria do. The DNA molecules of the chloroplasts and mitochondria are circular and are associated with few proteins, like bacterial DNA, rather than linear and associated with histone proteins like most eukaryotic DNA. Chloroplasts and mitochondria also have ribosomes, structures that translate the genetic material into proteins, that are more similar to bacterial ribosomes than they are to eukaryotic ribosomes. These ribosomes are even sensitive to some of the same antibiotics, such as chloramphenicol and streptomycin, that inhibit the function of bacterial ribosomes.

Endomembrane System

The internal membranes of eukaryotic cells are dynamic, constantly changing structures. The concept of the *endomembrane system* describes all internal cytoplasmic membranes, with the exception of mitochondrial and plant plastid membranes, as a single continuum. In this model, the ER, generally the largest membrane system of eukaryotic cells, is the initial source of most other membranes. The ER is a network of interconnected, closed, membrane-bound *vesicles* that is contiguous with the nuclear envelope.

Vesicles from the ER carry proteins from the ER to the Golgi complex, fusing with its membranes. The Golgi complex can be described as a series of flattened membrane sacs, like a stack of hollow pancakes. The side closest to the nucleus receives vesicles from the ER, and the proteins inside these vesicles are processed and modified as they pass through the Golgi complex. Eventually, membrane vesicles containing the modified proteins will bud from the opposite surfaces of the Golgi complex and fuse with the cell membrane or the membranes of other organelles.

Michele Arduengo

See also: Cell theory; Cell-to-cell communication; Cell wall; Chloroplasts and other plastids; Chroma- tin; Cytoplasm; Cytoskeleton; Cytosol; DNA in plants; Endomembrane system and Golgi complex; Endoplasmic reticulum; *Eukarya*; Membrane struc- ture; Microbodies; Mitochondria; Nuclear enve- lope; Nucleolus; Nucleoplasm; Nucleus; Oil bod- ies; Peroxisomes; Plasma membranes; Proteins and amino acids; Ribosomes; RNA; Vacuoles.

Sources for Further Study
American Society for Cell Biology. *Exploring the Cell*. Available on the World Wide Web at http://www.ascb.org/pubs/exploring.pdf, this work features excellent images of cells and cellular processes.
Lodish, Harvey, et al. *Molecular Cell Biology*. 4th ed. New York: W. H. Freeman, 2000. This is one of the most authoritative cell biology reference texts available. Provides excellent graphics, CD-ROM animations of cellular processes, and primary literature references.
Raven, Peter H., Ray F. Evert, and Susan E. Eichhorn. *Biology of Plants*. 6th ed. New York: W. H. Freeman/Worth, 1999. Introductory botany text provides a chapter introducing eu- karyotic cells in general and plant cells specifically.

EUROPEAN AGRICULTURE

Categories: Agriculture; economic botany and plant uses; food; world regions

European agricultural practices are affected by the policies of the European Union, in addition to global conditions which influence farming everywhere.

Agriculture in Europe goes back to classical times. The development first of the Greek city- states, then of the Roman Empire, created urban centers that required substantial amounts of food to be imported from as far away as Egypt. In the year 2000 European agriculture was dominated by two major groups: the European Union (EU), with fif- teen member states, and those European states out- side the EU. The EU, which began with the Com- mon Market created by the Treaty of Rome, signed in 1957, initially comprised France, West Germany, Italy, Belgium, the Netherlands, and Luxembourg. By the year 2000 it had expanded to include Great Britain, Ireland, Denmark, Greece, Spain, Portugal, Finland, Sweden, and Austria.

Land and Workers
Only 11 percent of the land in the world (slightly more than 5 million square miles) is suitable for ag- riculture. Among the continents, Europe has the highest percentage of land suitable for farming: 36 percent. (In North America, the comparable figure is 22 percent.) Overall, 80 percent of the land in Eu- rope is usable in some way, either as agricultural land or as forestland.

Smaller farms are more extensive in the southern countries of the EU than in the northern countries. Some 60 percent of all farms in the EU are less than 5 hectares (12.5 acres) in size. Many of these small farms are either part-time or subsistence farms. Farms that are more than 50 hectares in size (125 acres, a small farm by U.S. standards) constitute only 6 percent of all farms but produce most of the crops.

The percentage of the labor force employed in agriculture is small where the farms are large—in Great Britain, it is a mere 2 percent. In the rest of the EU, except for some of the more recent members, such as Greece, Spain, and Portugal, the percent- ages are all in the single digits. Where the farms are

Selected Agricultural Products of Europe

small, or in non-EU countries, without the EU's agricultural policy to push production up with high prices, the percentage of the labor force employed in agriculture is much higher. In Poland, 27 percent of the labor force is employed in agriculture; in Romania, 21 percent; and in the Ukraine, 19 percent.

Crops

Europe produces about 19 percent of the world's grains eaten by humans or livestock and almost 24 percent of the world's coarse grains (barley, rye, oats). Most of all these grains are grown in Russia.

Half the world's potatoes are grown in Europe; the Russian Federation grows the largest share. Europe also grows half of the world's peas, with 40 percent produced in the Russian Federation. Three-quarters of the world's sugar beets are grown in Europe, Ukraine being the largest European producer. Rapeseed production has been increasing; Germany is its largest producer in Europe, followed closely by France. European production is a bit more than 17 percent of world production.

Europe grows 20 percent of the world's tomatoes, although the tomato is not a native European plant. Spain and Italy are the leading producers of tomatoes in Europe. Overall, Europe grows 16 percent of the world's vegetables, with Italy being the largest European producer, closely followed by the Russian Federation and Spain.

More than half of the world's grapes are grown in Europe. These grapes feed Europe's great wineries, which produce 70 percent of the world's wine, a substantial proportion of which is drunk in Europe, although it remains an important export item. Europe also produces nearly three-quarters of the world's hops, which go into the much-prized European beers. Europe grows more than half the world's olives, almost all of them in Italy, Spain, and Greece. These countries also produce about 60 percent of the world's olive oil.

Agricultural Revolution

Beginning in the 1970's, Europe underwent what has been called a new agricultural "revolution." Ownerships were consolidated, especially in Britain but also in France and Germany. As a result, owners of the larger holdings were able to invest in modern agricultural machinery. Now, 44 percent of the world's tractors are owned in Europe, mostly in France, Italy, and Poland. This has helped make European agriculture so productive that, according to the U.S. Department of Agriculture, Europe's best farms are as efficient as the best in the United States.

European farmers vastly increased their yields in the second half of the twentieth century. Britain's wheat output is up 60 percent from what it was immediately after World War II; the growth in output

Leading Agricultural Crops of European Countries with More than 20 Percent Arable Land

Country	Products	Percent Arable Land
Albania	Temperate-zone crops	21
Belarus	Grain, potatoes, vegetables	29
Belgium	Sugar beets, fresh vegetables, fruits, grain, tobacco	24
Bulgaria	Grain, oilseed, vegetables, fruits, tobacco	37
Denmark	Grain, potatoes, rapeseed, sugar beets	60
France	Wheat, cereals, sugar beets, potatoes, wine grapes	33
Germany	Potatoes, wheat, rye, barley, sugar beets, fruit, cabbage	—
Hungary	Wheat, corn, sunflower seed, potatoes, sugar beets	51
Italy	Fruits, vegetables, grapes, potatoes, sugar beets, soybeans, grain, olives	31
Liechtenstein	Wheat, barley, maize, potatoes	25
Luxembourg	Barley, oats, potatoes, wheat, fruit, wine grapes	24
Malta	Potatoes, cauliflower, grapes, wheat, barley, tomatoes, citrus, cut flowers, green peppers	38
Moldova	Vegetables, fruits, wine, grain, sugar beets, sunflower seed, tobacco	53
The Netherlands	Grains, potatoes, sugar beets, fruits, vegetables	27
Portugal	Grain, potatoes, olives, grapes	26
Romania	Corn, wheat, sugar beets, sunflower seed, potatoes, grapes	41
Russia	Grain, sugar beets, sunflower seed, vegetables, fruits	—
Spain	Grain, vegetables, olives, wine grapes, sugar beets, citrus	30
United Kingdom	Cereals, oilseed, potatoes, vegetables	25

Source: Data are from The Time Almanac 2000. Boston: Infoplease, 1999.

is nearly as great in France. In general, European agricultural productivity grew 5 percent a year between 1960 and 1999. Productivity grew much less in Eastern Europe than in Western Europe. This is partly because rainfall there varies so widely from year to year.

In 1979 the EU moved from being an importer of cereal grains to an exporter, as it did in 1975 for sugar and in 1976 for wine. Since 1960, the number of workers employed in agriculture has dropped by 50 percent, although the agricultural output remains the same or even higher. Authorities in Britain have estimated that farms there are at their most efficient when they employ no more than two or three people—a far cry from the hundreds of people who worked Europe's farms for subsistence wages in earlier centuries.

Irrigation and Drainage

Despite the generally favorable climate, Europe has 10 percent of the world's irrigated acreage. Most of that is in the Russian Federation, but Italy, Spain, and Romania also have significant amounts. A striking feature of European agriculture is the extent to which agricultural lands—some of them former wetlands—have been drained, to ensure uniform moisture for the crops being grown. In Finland, 91 percent of the agricultural land has been drained. Hungary has seen 70 percent of its land drained; the Netherlands, 65 percent; Britain, 60 percent; and Germany, 50 percent.

Environment

One factor that is assuming increasing influence over European agriculture is environmental concerns. The heavy use of fertilizers, pesticides, and herbicides has created damaging environmental conditions in some countries. The amount of cow manure generated in the Netherlands by its super-efficient dairy industry is more than the land of the entire country could absorb. The Dutch government subsidizes a company that composts some of this manure and sells it abroad as fertilizer for flowers. Sweden compensates farmers who reduce the runoff from their farms, a growing problem as the nitrogen content in water rises from fertilizer runoff. The EU has introduced a program to compensate those who set land aside for environmental protection, but more needs to be done to bring the EU's production levels closer to domestic demand, as well as to reduce the cost to consumers and taxpayers of the subsidies paid to farmers.

Organic Farming and Bioengineering

Several European countries, including the Czech Republic, France, and the United Kingdom, have introduced programs to encourage organic farming. At the beginning of the twenty-first century, however, a mere 2 percent of European crops were raised organically. Some scientists believe that environmental improvement could be generated if crops were developed that could ward off the insects that attack them, or that provide their own nitrogen, as the leguminous plants (peas and beans) do. Nevertheless, the European environmental movement has strongly opposed genetically modified foods, in part citing risks to health and the environment.

Nancy M. Gordon

See also: Biomes: types; European flora; Fruit crops; Vegetable crops; Wheat.

Sources for Further Study

Adger, W. Neil, Davide Pettenella, and Martin Whitby, eds. *Climate-Change Mitigation and European Land-Use Policies*. New York: CAB International, 1997. Examines actions, in Europe and elsewhere, to reduce emissions from forestry and agriculture through changes in land use. The book focuses on issues of efficiency, equity, and long-term impact of policy changes.

Barron, Enid M., and Ilga Nielsen, eds. *Agriculture and Sustainable Land Use in Europe: Papers from Conferences of European Environmental Advisory Councils*. Boston: Kluwer Law International, 1998. Examines problems faced by the EU concerning the future of rural communities, the maintenance of an attractive and diverse countryside, and more. This work comprises papers presented at two conferences organized by groups of European Environmental Advisory Councils. Offers the reader access to a broad range of experience and points of view.

Bowler, Ian R. *The Geography of Agriculture in Developed Market Economies*. New York: John

Wiley & Sons, 1992. Analyzes the industrialization of agriculture, factors of production in modern agriculture, and the agricultural significance of farm size and land tenure in developed economies, including those in Europe.

Brouwer, Floor, and Bob Crabtree, eds. *Environmental Indicators and Agricultural Policy.* New York: CABI, 1999. A practical review of the theory, development, and use of methods for assessing the impact of agriculture on the environment.

Huylenbroeck, Guido van. *Countryside Stewardship: Farmers, Policies, and Markets.* New York: Pergamon, 1999. Examines the economic nature of the agri-environmental measures that were intended to reduce the negative effects of agricultural production on countryside amenities and stimulate the positive contributions of farmers to countryside management. Offers a comparative analysis of stewardship policies across eight European countries.

EUROPEAN FLORA

Category: World regions

Flowering plants in Europe vary from those growing in mediterranean to alpine to Arctic regions.

Many of Europe's flowering plants are similar to those in North America, belonging to many of the same genera but to different species. Some of the most common North American flowering plants have cousins in Europe, but their location varies according to their latitude and altitude.

Climate and Soil

The most important factor determining the location of plants is climate. The continent of Europe ranges from the coastal areas on the northern shores of the Mediterranean Sea and Black Sea to the Arctic Ocean north of the Scandinavian peninsula. Although most of Europe is in the temperate climate zone, the areas that border the Mediterranean Sea are nearly all frost-free. By contrast, those parts of Europe that form the Scandinavian peninsula and northern Russia have frost-free periods each year of as little as two months. As a result, there is a south-to-north gradation of the flowering plants.

Moreover, mountains separate land that is also separated by latitude. The division is perhaps more marked than on other continents because the Alps run west to east, the highest peaks being without vegetation, whereas the mountains in North America run north to south. Since climate is strongly a function of latitude, there is, in effect, a double line separating the vegetation of the part of Europe along the Mediterranean from the part that is north of the Alps, instead of the gradual gradation that is more characteristic of North America.

Another factor in determining where flowering plants will be found is soil. The soil in the south of Europe tends to be sandy; the low annual rainfall at the Mediterranean shoreline means that what little rain there is flows rapidly through the soil, leaving relatively little for plants. In the north, much of the soil is permanently frozen, so only plants that can grow in a short period of time in the summer and survive many months of frozen life will be found there. In between these areas, European soils vary between those that make ideal growing conditions for flowering plants, the black earth soils of central Europe, and those that are thin layers over underlying rock or that trap water in the soil layers just below the surface, creating marshy conditions.

An east-to-west factor also influences which flowering plants are found where in Europe. The Atlantic coastline is warmed year-round by the Gulf Stream, so that normal temperatures in the parts of the continent touched by this current (including much of the western Baltic Sea) have warmer temperatures in winter than their latitudes would indicate. Southern Norway, for instance, is on the same latitude as Greenland, most of which is covered with ice and snow throughout the year; but many plants, including agricultural crops, grow in southern Norway because of the Gulf Stream. Because rainfall is high, these parts of Europe tend to be wetter and cooler in summer than other parts of the world at the same latitudes. Europe has been occupied by humans for such a long period of time that there are almost no parts of Europe where the vegetation has been unaffected by humans. There are virtually no "virgin forests" in Europe.

Forests

Most of Europe's trees are similar to those in North America, but these are related by genus and are not members of the same species. This is true of oaks, maples, ashes, elms, birches, beeches, chestnuts, walnuts, apples, and hornbeams as well as conifers. Some of the native tree species of Europe have been brought to North America and have become a part of the American forest. The sycamore, for example, is the same species in North America as in Europe. The sycamore is a long-lived tree; some sycamores that were planted in North America during the colonial era are still alive. The mountain ash, with its bright orange berries, is another tree from Europe (where it is known as the rowan) that has emigrated to North America. The horse chestnut is yet another tree native to Europe that has long been settled in North America.

Among the evergreens, the Norway spruce has been widely planted and now seeds itself in North America. The Scots pine has been widely planted in North America; many Christmas trees sold every year in the United States are Scots pine. The European larch has become popular with the U.S. forest industry because it is suited to *reforestation* after *clear-cutting*.

Some North American species have made the opposite journey: Much of Scotland has been reforested with the Sitka spruce, a native of the Pacific

PhotoDisc

The raspberry bush found throughout eastern North America is a European native.

Northwest. Another native American evergreen that has made the trip to Europe is the white pine, known in Europe as the Weymouth pine. Plantations of white pine have been set up all over Europe, because Europe did not have a soft pine, and the wood of the white pine is easily worked.

Apple Trees

One important European tree that has had a large impact on the United States is the apple. The apple tree that produces the familiar fruit appears to have originated in southern Russia and made its way throughout Europe. Many apple varieties were cultivated by the Romans. The English and other European settlers of North America, finding only crab apples (North America's only native apple) when they arrived, imported familiar apple varieties from Europe. Orchards were among the first things the English settlers of New England planted in Massachusetts Bay.

Shrubs

Many European shrubs have become immigrants too. The box and holly are much prized for foundation planting, and the privet makes a neat hedge. The buckthorn bush has also made the journey, although it is less widely sold at nurseries. The juniper in Europe is the same as the juniper in North America, but the raspberry is not. Currants and gooseberries are European shrubs that have been widely transplanted, although one of the European currant varieties harbors a disease that affects the North American white pine. By contrast, the cranberry is a North American shrub that has lately been transplanted to Europe. North American blueberries are quite different from the related species in Europe.

The raspberry bush found throughout eastern North America is a European native. Because birds are frequent consumers of its fruit, it has spread well beyond the beds where it was originally planted. The heather that covers many of the hills of northern Europe, where it is a native, has become popular with American florists as a filler for bouquets. One of the most common landscape shrubs in the United States, the lilac, is of European origin, but it was among the earliest to make the trip to North America. Today the foundations of old colonial homesteads can often be located because, although the house is gone, the lilacs that once surrounded it survive.

Herbs

Herbs or wildflowers have spread across the continents much as shrubs have. Many of the most common North American wildflowers, such as the dandelion, are immigrants. So is the wild strawberry as well as the plantain that infests lawns. Many wildflowers are both European and American in origin, although in most cases the species differ. Among them are the violets, some of the cinquefoils, many buttercup varieties, and the marsh marigold. The clovers that are so familiar to Americans are all imports—the white clover, the red clover, and the alsike clover—are all natives of Europe.

Several genera of grasses are found on both continents: Kentucky bluegrass is really the European smooth meadow grass. Annual rye grass, often used to green up new lawns quickly, is a European import, as is red fescue, common in hayfields. Timothy hay, cultivated in the United States as feed hay, is also an import from Europe.

The marshes and swamps of Europe are populated mostly by indigenous plants. Large numbers of sedges that are native to Europe are grouped together in the *Carex* genus and can be found in Europe's wetlands. Wetlands may have more "virgin" plant communities than anywhere else in Europe, because they were unsuitable for cultivation. Europeans have drained many of the continent's wetlands to convert the land to farmland; the most notable case is the Netherlands, where land has been reclaimed from the sea so that crops can be grown on it. Among the shrubs, Europe's wetlands, like those in North America, harbor alders, but Europe's are different species from those found in North America. The same can be said for willows, which grow well where ample moisture is available.

Commercial Plants

Ever since the first agricultural revolution, ten thousand years ago, humans have adapted plants to their needs. This is especially true for the grains, most of which originated in the Middle East. Wheat, oats, barley, rye, and others that could become food for humans were altered from their original form by careful plant breeding. This is also true for the flowers that are the staple of the florists' trade—roses, chrysanthemums, begonias, and carnations are all adaptations of wild plants.

Tulips are widely known flowers that originated in Eurasia from Austria and Italy eastward to Japan, with two-thirds of them native to the eastern Mediterranean and the southeastern parts of Russia. The Dutch cultivated tulips beginning in the 1500's and made them into a staple export. The grape hyacinth is a European native, as is the full-size hyacinth. Crocuses are natives of the Mediterranean basin.

Olive Trees

One native flowering plant of Europe deserves special mention: the olive tree. Having originated along the shores of the Mediterranean Sea, it has been cultivated and modified to increase the size of its fruit since ancient times. It remains an important agricultural resource for Mediterranean nations, especially Greece, Italy, France, and Spain. The Spanish conquerors of Central America carried the olive to the new world, and it was successfully introduced into California.

The Maquis

Uncultivated olive trees form part of the vegetation of the maquis, an area in France and Spain where the native olive grows with the carob, a small native tree like the olive, and the holm oak. Most of these trees are so stunted by the impoverished soil, heavily eroded over the centuries, that they are little more than bushes. There are also a variety of shrubs characteristics of the maquis, such as a cle-

matis vine, the Mediterranean buckthorn, and the common myrtle. A local variety of grass covers the ground between the trees and shurbs. Similar communities can also be found in Greece, where they provide grazing for goats.

Vines

The grapevine has been of commercial importance for centuries. The Greeks and Romans raised grapes and made wine from them. The wine grape appears to have originated in the Mediterranean basin, but many varieties of grape have developed. European varieties have been transplanted to North America, and the process has also worked in reverse. In the late nineteenth century, when a devastating disease known as *phylloxera* ravaged French vineyards, American grapevine rootstock was imported into France, and the French vines grafted onto it, because the American rootstock had shown itself less subject to the disease.

Another vine is the ivy. There are ivies native to almost every continent, but European ivy, sometimes called English ivy, has spread far beyond its native ground. It is popular as a wall covering and is frequently seen in gardens.

Nancy M. Gordon

See also: Arctic tundra; European agriculture; Forests; Mediterranean scrub.

Sources for Further Study

Amorosi, Thomas, et al. "Raiding the Landscape: Human Impact in the Scandinavian North Atlantic." *Human Ecology: An Interdisciplinary Journal* 25, no. 3 (September, 1997): 491-3. Tells how early European settlers who colonized the islands of the North Atlantic impacted a widely varied set of island ecosystems. Rapid degradation of flora and soil took place because of social and climatic change.

Dallman, Peter R. *Plant Life in the World's Mediterranean Climates: California, Chile, South America, Australia, and the Mediterranean Basin.* Sacramento: California Native Plant Society, 1998. Overview of the landscapes, vegetation types, and plants of the regions of the world that have a mediterranean climate.

Kirby, K., and C. Watkins, eds. *The Ecological History of European Forests.* New York: CAB International, 1998. Wide-ranging and detailed case studies of aspects of forest management, including grazing and conservation, as well as ecological history of forests and woodlands in Europe.

Weber, Ewald. "The Alien Flora of Europe: A Taxonomic and Biogeographic Analysis." *American Journal of Botany* 84, no. 6 (June, 1997): S110. Geographic and taxonomic overview of the nonindigenous plant species of Europe.

EUTROPHICATION

Categories: Algae; diseases and conditions; environmental issues; microorganisms; poisonous, toxic, and invasive plants; pollution; water-related life

The overenrichment of water by nutrients, eutrophication causes excessive plant growth and stagnation, which leads to the death of fish and other aquatic life.

The word "eutrophic" comes from the Greek *eu*, which means "good" or "well," and *trophikos*, which means "food" or "nutrition." Eutrophic waters are well-nourished and rich in nutrients; they support abundant life. Eutrophication refers to a condition in aquatic systems (ponds, lakes, and streams) in which nutrients are so abundant that plants and algae grow uncontrollably and become a problem. The plants die and decompose, and the water becomes stagnant. This ultimately causes the death of other aquatic animals, particularly fish, that cannot tolerate such conditions. Eutrophication is a major problem in watersheds and waterways such as the Great Lakes and Chesapeake Bay that are surrounded by urban populations.

The stagnation that occurs during eutrophication is attributable to the activity of microorganisms growing on the dead and dying plant material in water. As they decompose the plant material, microbes consume oxygen faster than it can be resupplied by the atmosphere. Fish, which need oxygen in the water to breathe, become starved for oxygen and suffocate. In addition, noxious gases such as hydrogen sulfide (H_2S) can be released during the decay of the plant material. The hallmark of a eutrophic environment is one that is plant-filled, littered with dead aquatic life, and smelly.

Eutrophication is actually a natural process that occurs as lakes age and fill with sediment, as deltas form, and as rivers seek new channels. The main concern with eutrophication in natural resource conservation is that human activity can accelerate the process and can cause it to occur in previously clean but nutrient-poor water. This is sometimes referred to as *cultural eutrophication*. For example, there is great concern with eutrophication in Lake Tahoe. Much of Lake Tahoe's appeal is its crystal-clear water. Unfortunately, development around Lake Tahoe is causing excess nutrients to flow into the lake and damaging the very thing that attracts people to the lake.

Roles of Nitrogen and Phosphorus

Nitrogen and phosphorus are the key nutrients involved in eutrophication, although silicon, calcium, iron, potassium, and manganese can be important. Nitrogen and phosphorus are essential in plant and animal growth. Nitrogen compounds are used in the synthesis of amino acids and proteins, whereas phosphate is found in nucleic acids and phospholipids. Nitrogen and phosphorus are usually in limited supply in lakes and rivers. Plants and animals get these nutrients through natural recycling in the water column and sediments and during seasonal variations, as algae and animals decompose, fall to the lower depths, and release their nutrients to be reused by other organisms in the ecosystem. A limited supply of nutrients—as well as variations in optimal temperature and light conditions—prevents any one species of plants or animals from dominating a water ecosystem.

Effects of Excess Nutrients

Although nutrient enrichment can have detrimental effects on a water system, an increased supply of nitrogen and phosphorus can have an initial positive effect on water productivity. Much like adding fertilizer to a lawn, increases in nutrients in a lake, river, or ocean cause it to be more productive by stimulating plant and animal growth in the entire food chain. *Phytoplankton*—microscopic algae that grow on the surface of sunlit waters—take up nutrients directly and are able to proliferate. Through photosynthesis, these primary producers synthesize organic molecules that are used by other members of the ecosystem. Increased algal growth thus stimulates the growth of zooplankton—microscopic animals that feed on algae and bacteria—as well as macroinvertebrates, fish, and other animals and plants in the food web. Indeed, many fisheries have benefited from lakes and oceans that are productive.

When enough nutrients are added to a lake or river to disrupt the natural balance of nutrient cycling, however, the excess nutrients effectively become *pollutants*. The major problem is that excess nutrients encourage profuse growth of algae and rooted aquatic weeds, species that can quickly take advantage of favorable growth conditions at the expense of slower-growing species. Algae convert carbon dioxide and water into organic molecules during photosynthesis, a process that produces oxygen. When large blooms of algae and other surface plants die, however, they sink to the bottom of the water to decompose, a process that consumes large amounts of oxygen. The net effect of increased algae production, therefore, is depletion of dissolved oxygen in the water, especially during midsummer.

Reduced oxygen levels (called *hypoxia*) can have dire consequences for lakes and rivers that support fish and bottom-dwelling animals. Oxygen depletion is greatest in the deep bottom layers of water, because gases from the oxygen-rich surface cannot readily mix with the lower layers. During summer and winter, oxygen depletion in eutrophic waters can cause massive fish kills. In extreme cases of eutrophication, the complete depletion of oxygen (*anoxia*) occurs, leading to ecosystem crashes and irreversible damage to plant and animal life. Oxygen depletion also favors the growth of anaerobic bacteria, which produce hydrogen sulfide and methane gases, leading to poor water quality and taste.

Excessive algal and plant growth has other negative effects on a water system. Algae and plants at the surface block out sunlight to plants and animals at the lower depths. Loss of aquatic plants can affect fish-spawning areas and encourage soil erosion from shores and banks.

Eutrophication often leads to loss of diversity in a water system, as high nutrient conditions favor plants and animals that are opportunistic and short-lived. Native sea grasses and delicate sea plants often are replaced by hardier weeds and rooted plants. Carp, catfish, and bluegill fish species replace more valuable coldwater species such as trout.

Thick algal growth also increases water turbidity and gives lakes and ponds an unpleasant pea-soup appearance. As algae die and decay, they wash up on shores in stinking, foamy mats.

Algal blooms of unfavorable species can produce toxins that are harmful to fish, animals, and humans. These toxins can accumulate in shellfish and have been known to cause death if eaten by humans. So-called red tides and brown tides are caused by the proliferation of unusual forms of algae, which give water a reddish or tealike appearance and in some cases produce harmful chemicals or neurotoxins.

Assessing Eutrophication

While eutrophication effects are generally caused by nutrient enrichment of a water system, not all cases of nutrient accumulation lead to increased productivity. Overall productivity is based on other factors in the water system, such as grazing pressure on phytoplankton, the presence of other chemicals or pollutants, and the physical features of a body of water. Eutrophication occurs mainly in enclosed areas such as estuaries, bays, lakes, and ponds, where water exchange and mixing are limited. Rivers and coastal areas with abundant flushing generally show less phytoplankton growth from nutrient enrichment because their waters run faster and mix more frequently. On the other hand, activities that stir up nutrient-rich sediments from the bottom, such as development along coastal waters, recreational activities, dredging, and storms, can worsen eutrophication processes.

The nutrient status of a lake or water system is often used as a measure of the extent of eutrophication. For example, lakes are often classified as *oligotrophic* (nutrient-poor), *eutrophic* (nutrient-rich), or *mesotrophic* (moderate in nutrients) based on the concentrations of nutrients and the physical appearance of the lake. Oligotrophic lakes are deep, clear, and unproductive, with little phytoplankton growth, few aquatic rooted plants, and high amounts of dissolved oxygen. In contrast, eutrophic lakes are usually shallow and highly productive, with extensive aquatic plants and sedimentation. These lakes have high nutrient levels, low amounts of dissolved oxygen, and high sediment accumulation on the lake bottom. They often show sudden blooms of green or blue-green algae (or blue-green bacteria, cyanobacteria) and support only warm-water fish species.

Mesotrophic lakes show characteristics in between those of unproductive oligotrophic waters and highly productive eutrophic waters. Mesotrophic lakes have moderate nutrient levels and phytoplankton growth and some sediment accumulation; they support primarily warm-water fish species. As a lake naturally ages over hundreds of years, it usually (but not always) gets progressively more eutrophic, as sediments fill in and eventually convert it to marsh or dryland. Nutrient enrichment from human sources can speed this process greatly.

Limiting Damage

The negative effects of eutrophication can be reduced by limiting the amount of nutrients—in most cases nitrogen and phosphorus—from entering a water system. Nutrients can enter water bodies through streams, rivers, groundwater flow, direct precipitation, and dumping and as particulate fallout from the atmosphere. While natural processes of eutrophication are virtually impossible to control, eutrophication from human activity can be reduced or reversed.

Phosphorus enrichment into water systems occurs primarily as the result of wastewater drainage into a lake, river, or ocean. Phosphate is common in industrial and domestic detergents and cleaning agents. Mining along water systems is also a major source of phosphorus. When phosphorus enters a water system, it generally accumulates in the sediments. Storms and upwelling can stir up sediments, releasing phosphorus. Treatment of wastewater to remove phosphates and the reduction of phosphates in detergents have helped to reduce phosphorus enrichment of water systems.

Nitrogen enrichment is harder to control; it is present in many forms, as ammonium, nitrates, nitrites, and nitrogen gas. The major sources of nitrogen eutrophication are synthetic fertilizers, animal wastes, and agricultural runoff. Some algae species can also fix atmospheric nitrogen directly, converting it to biologically usable forms of nitrogen. Since the atmosphere contains about 78 percent nitrogen, this can be a major source of nitrogen enrichment in waters that already have significant algal populations.

Efforts to control nitrogen and phosphorus levels have examined both point and nonpoint sources of nutrient loading. Point sources are concentrated, identifiable sites of nutrients that include municipal sewage-treatment plants, feed lots, food-processing plants, pulp mills, laundry detergents, and domestic cleaning agents. Nonpoint, or diffuse, sources of nutrients include surface runoff from rainwater, fertilizer from agricultural land and lawns, eroded soil, and roadways.

Linda Hart and Mark S. Coyne

See also: Algae; Dinoflagellates; Environmental biotechnology; Phytoplankton.

Sources for Further Study

Brönmark, Christer. *The Biology of Lakes and Ponds.* New York: Oxford University Press, 1998. A comprehensive textbook for undergraduates in biology and environmental science that integrates new approaches to the study of freshwater ecosystems with more traditional limnology. The authors combine theoretical background with empirical studies and include numerous lab experiments and a concise survey of the major organisms in lakes and ponds.

Cole, Gerald. *Textbook of Limnology.* 4th ed. Prospect Heights, Ill.: Waveland Press, 1994. Basic college-level text for the study of lakes. Includes many aspects of lake ecology, aquatic organisms, and a general discussion of eutrophication and its causes.

Harper, David. *Eutrophication of Freshwaters: Principles, Problems, and Restoration.* London: Chapman & Hall, 1992. Comprehensive book on the basics of eutrophication in fresh water, its biological effects, and methods to study and control the process. Includes a case study of lake restoration in a series of lakes in England. Many useful references for further study.

Hinga, Kenneth, Heeseon Jeon, and Noelle F. Lewis. *Marine Eutrophication Review.* Silver Springs, Md.: U.S. Department of Commerce, National Oceanic and Atmospheric Administration, Coastal Ocean Office, 1995. Report on marine eutrophication in clear, nontechnical language. Includes a good discussion on many of the methods used to study eutrophication and evaluate its impact.

Horne, A. J., and C. R. Goldman. *Limnology*. New York: McGraw-Hill, 1994. College-level book for understanding lake cycles and the classification of oligotrophic and eutrophic lakes.

Schramm, Winfrid, and Pieter H. Nienhuis, eds. *Marine Benthic Vegetation: Recent Changes and the Effects of Eutrophication*. Berlin: Springer, 1996. Mainly discusses the effects of eutrophication on marine plants but does include a good review of green tides and their impact and management in coastal waters. Includes extensive references.

Wetzel, Robert G. *Limnological Analyses*. 3d ed. New York: Springer, 2000. A research-oriented textbook consisting of field and laboratory exercises that invite active participation in the study of the physical, chemical, and biological characteristics of standing and running waters. Consists of a series of carefully designed and tested field and laboratory exercises representing the full scope of limnology.

EVOLUTION: CONVERGENT AND DIVERGENT

Categories: Ecosystems; evolution; paleobotany

Some of the most dramatic examples of natural selection are the result of adaptation in response to stressful climatic conditions. Such selection may cause unrelated species to resemble one another in appearance and function, a phenomenon known as convergence. In other situations, subpopulations of a single species may split into separate species as the result of natural selection. Such divergence is best seen on isolated islands.

Convergent Evolution

Convergent evolution occurs when organisms from different evolutionary lineages evolve similar adaptations to similar environmental conditions. This can happen even when the organisms are widely separated geographically. A classic example of convergent evolution occurred with *Cactaceae*, the cactus family, of the Americas and with the euphorbs, or *Euphorbiaceae*, the spurge family of South Africa, both of which have evolved *succulent* (water-storing) stems in response to desert conditions.

The most primitive cacti are vinelike, tropical plants of the genus *Pereskia*. These cacti, which grow on the islands of the West Indies and in tropical Central and South America, have somewhat woody stems and broad, flat leaves. As deserts developed in North and South America, members of the cactus family began to undergo selection for features that were adaptive to hotter, dryer conditions.

The stems became greatly enlarged and succulent as extensive water-storage tissues formed in the pith or cortex. The leaves became much reduced. In some cactus species, such as the common prickly pear (*Opuntia*), the leaves are small, cylindrical pegs that shrivel and fall off after a month or so of growth. In most cacti, only the leaf base forms and remains as a small hump of tissue associated with an axillary bud. In some cacti this hump is enlarged and is known as a tubercle. Axillary buds in cacti are highly specialized and are known as areoles. The "leaves" of an areole are reduced to one or more spines. Particularly in columnar cacti, the areoles are arranged in longitudinal rows along a multiple-ridged stem.

With the possible exception of the genus *Rhipsalis*, which has one species reported to occur naturally in Africa, all cacti are native to the Americas. As deserts formed in Africa, Eurasia, and Australia, different plant families evolved adaptations similar to those in cacti. The most notable examples are the candelabra euphorbs of South Africa. Desert-dwelling members of the *Euphorbiaceae* frequently have succulent, ridged, cylindrical stems resembling those of cacti. The leaves are typically reduced in size and are present only during the rainy season. They are arranged in rows along each of several ridges of the stem. Associated with each leaf are one or two spines. As a result, when the leaves shrivel and fall off during the dry season, a spiny, cactuslike stem remains.

The succulent euphorbs of Africa take on all of the forms characteristic of American cacti, from pincushions and barrels to branched and unbranched columns. Other plant families that show convergence with the cacti, in having succulent stems or leaves, are the stem succulents of the milkweed family, *Asclepiadaceae*; sunflower family, *Asteridaceae*; stonecrop family, *Crassulaceae*; purslane family, *Portulacaceae*; grape family, *Vitaceae*; leaf succulents of the ice plant family, *Aizoaceae*; daffodil family, *Amaryllidaceae*; pineapple family, *Bromeliaceae*; geranium family, *Geraniaceae*; and lily family, *Liliaceae*.

Divergent Evolution

Some of the most famous examples of divergent evolution have occurred in the Galápagos Islands. The Galápagos comprise fourteen volcanic islands located about 600 miles west of South America. A total of 543 species of vascular plants are found on the islands, 231 of which are endemic, found nowhere else on earth. Seeds of various species arrived on the islands by floating in the air or on the water or being carried by birds or humans.

With few competitors and many different open habitats, variant forms of each species could adapt to specific conditions, a process known as *adaptive radiation*. Those forms of a species best suited to each particular habitat were continually selected for and produced progeny in that habitat. Over time, this natural selection resulted in multiple new species sharing the same ancestor. The best examples of divergent evolution in the Galápagos have occurred in the *Cactaceae* and *Euphorbiaceae*. Eighteen species and variety of cacti are found on the islands, and all are endemic. Of the twenty-seven species and varieties of euphorbs, twenty are endemic.

An interesting example of the outcome of divergent evolution can be seen in the artificial selection of different cultivars (cultivated varieties) in the genus *Brassica*. The scrubby Eurasian weed colewort (*Brassica oleracea*) is the ancestor of broccoli, brussels sprouts, cabbage, cauliflower, kale, and

A classic example of convergent evolution occurred with the cactus family of the Americas and with the euphorbs, or spurge family, of South Africa, both of which have evolved succulent (water-storing) stems in response to desert conditions. In some cactus species, such as the prickly pear cactus (Opuntia)*, the leaves are small, cylindrical pegs that shrivel and fall off after a month or so of growth.*

AP/WIDE WORLD PHOTOS

kohlrabi (rutabaga). All of these vegetables are considered to belong to the same species, but since the origin of agriculture, each has been selected for a specific form that is now recognized as a distinct crop.

Marshall D. Sundberg

See also: Adaptations; Adaptive radiation; Cacti and succulents; Coevolution; Deserts; Plant domestication and breeding; Selection.

Sources for Further Study

Bowman, Robert I., Margaret Berson, and Alan E. Leviton. *Patterns of Evolution in Galápagos Organisms*. San Francisco: California Academy of Sciences, 1983. A series of symposium papers, including one on the flora of the Galápagos Islands.

Darwin, Charles. "Journal and Remarks: 1832-1836." In *Narrative of the Surveying Voyages of His Majesty's Ships "Adventure" and "Beagle" Between the Years 1826 and 1836: Describing Their Examination of the Southern Shores of South America, and the "Beagle's" Circumnavigation of the Globe*, by Robert Fitzroy. Vol. 3. Reprint. New York: AMS Press, 1966.

Harris, James G., and Melinda Woolf Harris. *Plant Identification Terminology: An Illustrated Glossary*. Spring Lake, Utah: Spring Lake, 1994. Clear line drawings illustrate specific terminology used to identify and describe plants.

Uno, Gordon, Richard Storey, and Randy Moore. *Principles of Botany*. New York: McGraw-Hill, 2001. A good introduction to all aspects of plant biology.

EVOLUTION: GRADUALISM VS. PUNCTUATED EQUILIBRIUM

Categories: Evolution; paleobotany

The gradualism model of evolution proposes that a progenitor species gradually gave rise to many new species, with no special mechanisms accounting for the origins of new genera or groups of higher classifications—only the accumulation of many small changes in the frequencies of alleles in gene pools. The punctuated equilibrium model of evolutionary change supposes long periods of little or no change interspersed with short intervals of rapid change.

Charles Darwin, author of *On the Origin of Species by Means of Natural Selection* (1859), believed that morphological change was inevitable and proceeded slowly, encompassing slight, successive, and gradual changes within lineages. Speciation, therefore, was the result of the gradual accumulation of changes within ancestral populations over time, ultimately leading to the formation of recognizably new and different species. According to Darwin, sudden, large-scale changes were improbable or impossible—an idea epitomized by the phrase *Natura non facit saltum*, or "Nature never makes leaps." Darwin's concept of the slow and gradual transformation of a species' entire ancestral population into distinct descendant species over time has been termed "phyletic gradualism," or *anagenesis*.

If true, the expectation of anagenic transformation leads to the supposition that the fossil record for any lineage should contain an "inconceivably great" number of intermediate forms. Darwin, however, realized that the fossil record is, in fact, not littered with an "interminable" and "enormous" number of intermediate forms. Darwin's solution to this problem with his theory was presented in a chapter of his book *On the Origin of Species by Means of Natural Selection* titled "On the Imperfection of the Geological Record." Here Darwin persuasively argued that the paleontological record of the past is extremely imperfect because of degradation of fossiliferous deposits and differential rates of deposition and fossilization among lineages. He also noted that, on geologic time scales, persistent, long-lived and widespread species are more likely to appear in the fossil record than short-lived species or species confined to narrow geographic ranges.

The "Modern Synthesis," Microevolution, and Species Formation

In the 1930's and 1940's Darwin's theory of natural selection was melded with then-current knowledge of genetics, heritability, and mathematics to produce the *modern synthesis theory of evolution*. Under this paradigm, evolution came to be defined as a change in gene frequency over time. The micro-evolutionary processes of mutation, migration, random genetic drift, and natural selection were recognized as the primary mechanisms that alter gene frequencies.

At this time other authorities began to carefully consider the mechanism(s) by which species arise. One particular mode of species formation, and one supported by considerable empirical evidence, is termed the *allopatric model of speciation*. Allopatric species are formed as subpopulations of a more widespread ancestral species become geographically isolated and, in time, reproductively isolated from one another.

Paleospecies and Punctuated Equilibria

These ideas, particularly those of gradual, inevitable change by microevolutionary processes and the imperfection of the fossil record, and were accepted by paleontologists for more than a century. However, in 1972 Niles Eldredge and Stephen J. Gould published a paper in which paleontologists were challenged to examine fossil sequences (and gaps) more objectively. Eldredge and Gould drew three important conclusions about the fossil record. First, some gaps in the fossil record are real and cannot be attributed to other factors; gradual series of transitional forms do occur, but they are extremely rare. Second, paleospecies often persist for millions of years without substantial morphological change. Third, they recognized that paleospecies found in older strata are sometimes rapidly replaced by morphologically different taxa in younger deposits.

Thus, a literal interpretation of the fossil record requires the acknowledgment of short, rapid bursts of evolution. That is, within the paleontological history of a lineage, morphological stasis is occasionally interrupted by near-instantaneous (in geological time) formation of morphologically different species. If these gaps are real, how are they to be explained? How is one species suddenly (in geologic time) replaced by a morphologically modified form in the fossil record?

To answer these questions, Eldredge and Gould developed the theory of *punctuated equilibria* that, in essence, fused a literal interpretation of the fossil record with the mode of speciation most often observed in extant populations (such as the allopatric model of speciation).

Eldredge and Gould's theory of punctuated equilibria is based on the following postulates or observations. First, speciation typically occurs via allopatric speciation. Second, the origin of descendant species is, in geological time, rapid and occurs in a limited geographic area. Third, most adaptive change occurs at the time of speciation. Fourth, speciation is typically followed by long periods of morphological stasis, particularly within large, widespread species. Fifth, the abrupt appearance of a new species within the range of the ancestral species is a result of ecological succession, immigration, or competition. Finally, apparent adaptive trends (such as macroevolutionary trends) observed in the fossil record are the result of species selection within lineages over time.

The authors reasoned that as long as a species is capable of successfully exploiting its habitat, adaptations that originated at the time of speciation are unlikely to be altered, and morphological stasis is the result. Thus, in terms of the geologic time scale, most species seem to persist unchanged over long periods of time. The origin of a new species, its growth in numbers, and the extension of its geographic range are, therefore, determined by reproductive and ecological characteristics. In short, under the punctuated equilibrium model new species can be successful if they are sufficiently distinct in their habitat requirements and if they are able to compete with or outcompete close relatives should they come into contact with one another. This theory of speciation and diffential reproduction and survival of species produces well-documented, large-scale evolutionary patterns or "macroevolutionary" trends.

Punctuated Equlibria and Plants

The theory of punctuated equilibria was constructed based on studies of animal fossils; unfortunately, no mention is made of plants. The theory has therefore received less attention from paleobotanists. It is true that in extant plants rapid changes in physiological or morphological characteristics associated with speciation are not uncommon. However, such phenomena are rarely documented in the fossil record.

For example, interspecific hybridization or polyploidization may (almost instantaneously) form new, reproductively isolated species whose niches may differ from those of their immediate progenitors. Thus, in plants, rapid species formation is empirically known. Stasis is also known from the fossil records of some plant lineages. For example, in terms of floral structure, extant members of the *Loraceae*, *Chloranthaceae*, *Nymphaceae*, and the magnoliids scarcely differ from their early-to mid-Cretaceous ancestors. Morphological stasis is also observed in gingkos, *Metasequoia*, cycads, lycopods, sphenopsids, and ferns as well as the genus *Pinus*, which arose in the Jurassic. Rapid radiations are also documented in the plant fossil record, as occurred after the rise of angiosperms during the early Cretaceous. Thus, the plant fossil record provides clear evidence for stasis in some lineages and rapid diversification in others. Despite these observations, paleobotanical examples of lineages whose evolution fits the predictions of the punctuated equilibrium model are few.

J. Craig Bailey

See also: Adaptive radiation; Cladistics; Competition; Evolution: convergent and divergent; Evolution of plants; Fossil plants; Genetics: post-Mendelian; Paleobotany; Population genetics; Species and speciation; Succession.

Sources for Further Study

Eldredge, N. *Time Frames.* New York: Simon and Schuster, 1985. Examples of different tempos of evolution as judged from the fossil record, including an extensive bibliography.

Eldredge, N., and S. J. Gould. "Punctuated Equilibria: An Alternative to Phyletic Gradualism." In *Models in Paleobiology*, edited by T. J. M. Schopf. San Francisco: Freeman, Cooper, 1972. The original description of the theory of punctuated equilibria.

Freeman, S., and J. C. Herron. *Evolutionary Analysis.* 2d ed. Upper Saddle River, N.J.: Prentice Hall, 2001. A concise explanation of the theory of punctuated equilibria. Includes examples of studies that support the theory and others that do not.

EVOLUTION: HISTORICAL PERSPECTIVE

Categories: Classification and systematics; evolution; history of plant science; paleobotany

Evolution is the theory that biological species undergo sufficient change with time to give rise to new species.

The concept of evolution has ancient roots. Anaximander suggested in the sixth century B.C.E. that life had originated in the seas and that humans had evolved from fish. Empedocles (c. 450 B.C.E.) and Lucretius (c. 96-55 B.C.E.), in a sense, grasped the concepts of adaptation and natural selection. They taught that bodies had originally formed from the random combination of parts, but that only harmoniously functioning combinations could survive and reproduce. Lucretius even said that the mythical centaur, half horse and half human, could never have existed because the human teeth and stomach would be incapable of chewing and digesting the kind of grassy food needed to nourish the horse's body.

Early Biological Theory

For two thousand years, however, evolution was considered an impossibility. The theory of forms (also called his *theory of ideas*) proposed by Plato (c. 428-348 B.C.E.) gave rise to the notion that each species had an unchanging "essence" incapable of evolutionary change. As a result, most scientists from Aristotle (384-322 B.C.E.) to Carolus Linnaeus (1707-1778) insisted upon the *immutability of species.*

Many of these scientists tried to arrange all species in a single linear sequence known as the *scale of being* (also called the great chain of being or *scala naturae*), a concept supported well into the nineteenth century by many philosophers and theologians as well. The sequence in this scale of being was usually interpreted as a static "ladder of perfection" in God's creation, arranged from higher to lower forms. The scale had to be continuous, for any gap would detract from the perfection of God's creation. Much exploration was devoted to searching for *missing links* in the chain, but it was generally agreed that the entire system was static and incapable of evolutionary change. Pierre-Louis Moreau de Maupertuis and Jean-Baptiste Lamarck (1744-1829) were among the scientists who tried to reinterpret the scale of being as an evolutionary sequence, but this single-sequence idea was later replaced by the concept of branching evolution proposed by Charles Darwin (1809-1882). Georges Cuvier (1769-1832) finally showed that the major groups of animals had such strikingly different anatomical structures that no possible scale of being could connect them all; the idea of a scale of being lost most of its scientific support as a result.

The theory that new biological species could arise from changes in existing species was not readily accepted at first. Linnaeus and other classical biologists emphasized the immutability of species under the Platonic-Aristotelian concept of *essentialism*. Those who believed in the concept of evolution realized that no such idea could gain acceptance until a suitable mechanism of evolution could be found. Many possible mechanisms were therefore proposed. Étienne Geoffroy Saint-Hilaire (1805-1861) proposed that the environment directly induced physiological changes, which he thought would be inherited, a theory now known as *Geoffroyism*. Lamarck proposed that there was an overall linear ascent of the scale of being but that organisms could also adapt to local environments by voluntary exercise, which would strengthen the organs used; unused organs would deteriorate. He thought that the characteristics acquired by use and disuse would be passed on to later generations, but the inheritance of acquired characteristics was later disproved. Central to both these explanations was the concept of adaptation, or the possession by organisms of characteristics that suit them to their environments or to their ways of life. In eighteenth century England, the Reverend William Paley (1743-1805) and his numerous scientific supporters believed that such adaptations could be explained only by the action of an omnipotent, benevolent God. In criticizing Lamarck, the supporters of Paley pointed out that birds migrated toward warmer climates before winter set in and that the heart of the human fetus had features that anticipated the changes of function that take place at birth. No amount of use and disuse could explain these cases of anticipation, they claimed; only an omniscient God who could foretell future events could have designed things with their future utility in mind.

Darwin's Theory

The nineteenth century witnessed a number of books asserting that living species had evolved from earlier ones. Before 1859, these works were often more geological than biological in content. Most successful among them was the anonymously published *Vestiges of the Natural History of Creation* (1844), written by Robert Chambers (1802-1871). Books of this genre sold well but contained many flaws. They proposed no mechanism to account for evolutionary change. They supported the outmoded concept of a scale of being, often as a single sequence of evolutionary "progress." In geology, they supported the outmoded theory of *catastrophism*, an idea that the history of the earth had been characterized by great cataclysmic upheavals. From 1830 on, however, that theory was being replaced by the modern theory of *uniformitarianism*, championed by Charles Lyell (1797-1875). Charles Darwin read these books and knew their faults, especially their lack of a mechanism that was compatible with Lyell's geology. In his own work, Darwin carefully tried to avoid the shortcomings of these books.

Darwin brought about the greatest revolution in biological thought by proposing both a theory of branching evolution and a mechanism of natural selection to explain how it occurred. Much of Darwin's evidence was gathered during his voyage around the world aboard HMS *Beagle* between 1831 and 1836. Darwin's stop in the Galápagos Islands and his study of tortoises and finchlike birds on these islands is usually credited with convincing him that evolution was a branching process and that *adaptation* to local environments was an essential part of the evolutionary process. Adaptation, he later concluded, came about through *natural selection*, a process that killed the maladapted variations and allowed only the well-adapted ones to survive and pass on their hereditary traits. After returning to England from his voyage, Darwin raised pigeons, consulted with various animal breeders about changes in domestic breeds, and investigated other phenomena that later enabled him to demonstrate natural selection and its power to produce evolutionary change.

Darwin delayed the publication of his book for seventeen years after he wrote his first manuscript version. He might have waited even longer, except that his hand was forced. From the East Indies, another British scientist, Alfred Russel Wallace (1823-1913), had written a description of an identical the-

Charles Darwin and the *Beagle*

In 1831, a twenty-two-year-old Charles Darwin, who had been studying for the ministry at Cambridge, by luck was offered a position as naturalist on the ship HMS *Beagle*, which was about to embark on a round-the-world voyage of exploration. His domineering father was against the trip at first, but he finally relented. The expedition would turn the young man into a scientist. Over the next five years, Darwin recorded hundreds of details about plants and animals and began to notice some consistent patterns. His work led him to develop new ideas about what causes variations in different plant and animal species:

> [The] preservation of favourable individual differences and variations, and the destruction of those which are injurious, I have called Natural Selection, or the Survival of the Fittest.... slight modifications, which in any way favoured the individuals of any species, by better adapting them to their altered conditions, would tend to be preserved....
> —*On the Origin of Species by Means of Natural Selection*, 1859

Until Darwin and such colleagues as Alfred Russel Wallace, the "fixity" or unchangingness of species had been accepted as fact,

and the appearance over time of new species remained a mystery. Darwin's lucky trip laid the foundation for today's understanding of life and its diversity.

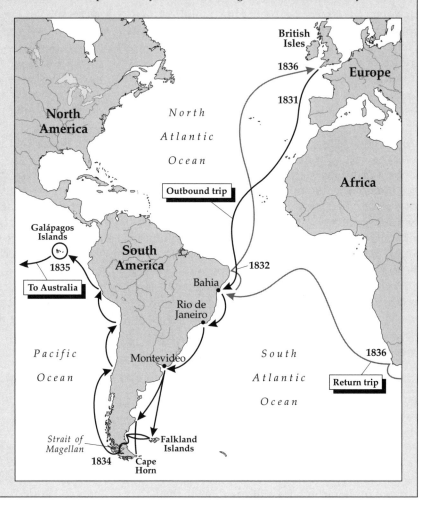

ory and submitted it to Darwin for his comments. Darwin showed Wallace's letter to Lyell, who urged that both Darwin's and Wallace's contributions be published, along with documented evidence showing that both had arrived at the same ideas independently. Darwin's great book, *On the Origin of Species by Means of Natural Selection*, was published in 1859, and it quickly won most of the scientific community to a support of the concept of branching evolution. In his later years, Darwin also published *The Descent of Man and Selection in Relation to Sex* (1871), in which he outlined his theory of *sexual selection*. According to this theory, the agent that determines the composition of the next generation may often be the opposite sex. An organism may be well adapted to live, but unless it can mate and leave offspring, it will not contribute to the next or to future generations.

After Darwin

In the early 1900's, the rise of Mendelian genetics (named for botanist Gregor Mendel, 1822-1884) initially resulted in challenges to Darwinism. Hugo de Vries (1848-1935) proposed that evolution occurred by *random mutations*, which were not necessarily adaptive. This idea was subsequently rejected, and

Mendelian genetics was reconciled with Darwinism during the period from 1930 to 1942. According to this modern synthetic theory of evolution, mutations initially occur at random, but natural selection eliminates most of them and alters the proportions among those that survive. Over many generations, the accumulation of heritable traits produces the kind of adaptive change that Darwin and others had described. The process of branching evolution through *speciation* is also an important part of the modern synthesis.

The branching of the evolutionary tree has resulted in the proliferation of species from the common ancestor of each group, a process called *adaptive radiation*. Ultimately, all species are believed to have descended from a single common ancestor. Because of the branching nature of the evolutionary process, no one evolutionary sequence can be singled out as representing any overall trend; rather, there have been different trends in different groups. Evolution is also an opportunistic process, in the sense that it follows the path of least resistance in each case. Instead of moving in straight lines toward a predetermined goal, evolving lineages often trace meandering or circuitous paths in which each change represents a momentary increase in adaptation. Species that cannot adapt to changing conditions die out and become extinct.

Evolutionary biology is itself the context into which all the other biological sciences fit. Other biologists, including physiologists and molecular biologists, study how certain processes work, but it is evolutionists who study the reasons why these processes came to work in one way and not another. Organisms and their cells are built one way and not another because their structures have evolved in a particular direction and can only be explained as the result of an evolutionary process. Not only does each biological system need to function properly, but it also must have been able to achieve its present method of functioning as the result of a long, historical, evolutionary process in which a previous method of functioning changed into the present one. If there were two or more ways of accomplishing the same result, a particular species used one of them because its ancestors were more easily capable of evolving this one method than another.

Eli C. Minkoff

See also: Genetics: Mendelian.

Sources for Further Study

Bowler, Peter J. *Evolution: The History of an Idea*. Rev. ed. Berkeley: University of California Press, 1989. A comprehensive, fascinating account of the history of evolutionary theories introduces specialist and nonspecialist alike to one of the most potent scientific ideas of modern times. This new edition is updated in its content and includes an augmented bibliography that offers an unparalleled guide to further reading.

_____. *Life's Splendid Drama: Evolutionary Biology and the Reconstruction of Life's Ancestry, 1860-1940*. Chicago: University of Chicago Press, 1996. Histories of the Darwinian revolution have paid far more attention to theoretical debates and have largely ignored the researchers who struggled to comprehend the deeper evolutionary significance of fossil bones and the structures of living animals. Bowler recovers some of this lost history in a definitive account of evolutionary morphology and its relationships with paleontology and biogeography.

Brandon, Robert N. *Concepts and Methods in Evolutionary Biology*. New York: Cambridge University Press, 1996. Robert Brandon is one of the most important and influential of contemporary philosophers of biology. This collection of his recent essays covers all the traditional topics in the philosophy of evolutionary biology and could serve as an introduction to the field.

Grant, Verne. *The Evolutionary Process: A Critical Study of Evolutionary Theory*. 2d ed. New York: Columbia University Press, 1991. A comprehensive study of the field of evolution, giving full time to critiques of the theory. Concentrates on general principals rather than specific examples.

Minkoff, Eli C. *Evolutionary Biology*. Reading, Mass.: Addison-Wesley, 1983. A comprehensive general textbook on evolutionary biology, including its historical aspects. The history of evolutionary theories and the mechanisms of the evolutionary process are described in detail. Many examples are given, and the book is profusely illustrated; it assumes no prior knowledge. Contains an excellent bibliography, arranged by topic.

Zimmer, Carl. *Evolution: The Triumph of an Idea*. New York: HarperCollins, 2001. A companion to the PBS series of the same title, this book presents both the history of the idea of evolution and the science that supports it. A thoroughly up-to-date presentation of the field at the dawn of the twenty-first century. Abundantly illustrated and written for a general audience.

EVOLUTION OF CELLS

Categories: Cellular biology; evolution; paleobotany

The earliest cells evolved sometime early in the Precambrian era, which includes the first four billion years of Earth's history. Attempts to understand life's origins are difficult, as there are very few clues left in the fossil record from those early times. The hypotheses and models of the origin of life that have been developed are based on contemporary understanding of how life works at the molecular and cellular levels and on assumptions about the conditions on Earth three billion to four billion years ago.

One assumption made about the origins of life involves the composition of the atmosphere shortly after the earth was formed. According to this assumption, the earth's atmosphere at this time contained very little free oxygen. It was an atmosphere perhaps made primarily of methane, ammonia, carbon dioxide, nitrogen, carbon monoxide, and water vapor. The first organisms are believed to have been anaerobic and did not require oxygen for *respiration*. This early atmosphere lacked a protective shield of ozone, which is derived from oxygen and which absorbs ultraviolet radiation from the sun. Intense ultraviolet radiation is lethal, and early forms of life may have evolved in water deep enough to avoid it. In some later stages of chemical evolution, however, ultraviolet radiation may actually have driven molecular interactions, producing more complex structures that were forerunners to living organisms.

Building Blocks of Life

The basic elements found in organic compounds are carbon, oxygen, hydrogen, nitrogen, phosphorus, and sulfur. These elements are also the main components of living cells and are the most plentiful elements in the solar system. The basic materials for the development of life were present on the early earth. Scientists have discovered that some of these building blocks, especially molecules of hydrogen sulfide, hydrogen cyanide, methanol, acetic acid, methyl formate, and a simple sugar called glycol aldehyde, exist in space and may have been brought to Earth by comets passing close by the plant. These compounds were, in fact, given off as Comet Hale-Bopp passed through the earth's solar system during 1995-1997. Perhaps comets "seeded" the earth with inorganic and organic compounds that triggered the chemical evolution that led to life itself.

Proteins and Amino Acids

One of the components of life is *proteins*. Proteins are made of even more basic organic molecules called *amino acids*. Proteins act as building materials for an organism's body and function in chemical reactions within an organism.

Amino acids may have been formed naturally in early earth history. In a process called photochemical dissociation, ultraviolet radiation is capable of separating the atoms in compounds such as water, ammonia, and hydrocarbons such as methane, allowing these atoms to recombine. Some of those atoms would have formed amino acids. A second form of energy that could trigger recombination of atoms is electrical discharge, occurring naturally as lightning. Lightning and ultraviolet radiation, acting separately or together, could have triggered the formation of amino acids in the atmosphere or in the oceans.

Other Building Blocks

Another of life's basic components is *nucleic acids*, such as deoxyribonucleic acid (DNA) and ribonucleic acid (RNA), which transmit genetic information from generation to generation. The building blocks of nucleic acids are nucleotides, which also participate, as nucleotide phosphates like adenosine triphosphate (ATP), in biochemical energy transactions.

Even with the first three components present on the early earth, living organisms could not develop unless those components were in close proximity to one another so they could interact. Some sort of structure was needed to contain all the components. Cell membranes would have provided just such a structure and so must be considered an essential fourth component of the first living organisms.

Wöhler's Experiment

Before 1828, many scientists believed that all organic molecules were products of living organisms. However, in 1828 the German chemist Friedrich Wöhler produced crystals of urea quite by accident after heating an inorganic compound called ammonium cyanate. Urea, a component of urine, is an organic compound. Over the next several decades, other chemists were able to duplicate Wöhler's experiment and, by using other simple inorganic compounds, succeeded in producing several other simple organic substances. This led many scientists to believe that life on Earth could have developed from inorganic materials. Between 1828 and the 1960's, however, scientists were able to produce only

simple organic molecules from inorganic substances. It was only during the 1960's that scientists could create and detect complex protein molecules.

Miller-Urey Experiments

It was not until 1953 that scientists produced amino acids and similar molecules using simulated conditions like those assumed to have been present in the early earth's atmosphere. Stanley Miller, a graduate student working for Dr. Harold Urey, cre-

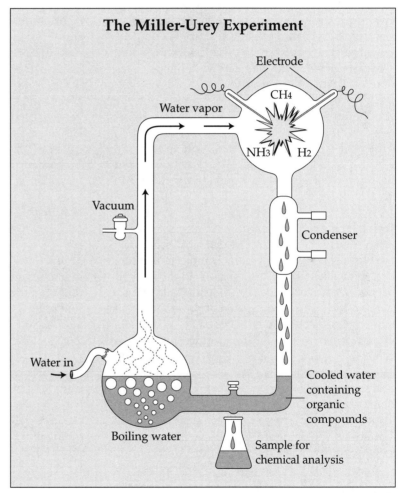

The Miller-Urey Experiment

Electrode
Water vapor
CH_4
NH_3 H_2
Vacuum
Condenser
Water in
Boiling water
Cooled water containing organic compounds
Sample for chemical analysis

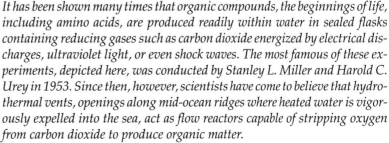

It has been shown many times that organic compounds, the beginnings of life, including amino acids, are produced readily within water in sealed flasks containing reducing gases such as carbon dioxide energized by electrical discharges, ultraviolet light, or even shock waves. The most famous of these experiments, depicted here, was conducted by Stanley L. Miller and Harold C. Urey in 1953. Since then, however, scientists have come to believe that hydrothermal vents, openings along mid-ocean ridges where heated water is vigorously expelled into the sea, act as flow reactors capable of stripping oxygen from carbon dioxide to produce organic matter.

ated an atmosphere of methane, ammonia, hydrogen, and water vapor in a bottle, thinking that this mixture would have been similar to the atmosphere of the very early earth. As the mixture was circulated through the apparatus, sparks of electricity, to simulate lightning, were discharged into the mixture. At the end of eight days, the condensed water in the apparatus had become cloudy and deep red. Analysis of the material showed that it contained a number of amino acids along with a few other, more complicated organic compounds.

Other scientists carried out similar experiments, with much the same results, leading many scientists to believe that organic compounds could be produced from a mixture of gases, including a mix of carbon dioxide, nitrogen, and water vapor, also found in Earth's early atmosphere. What was necessary for the success of these experiments was the lack of free oxygen. The experimenters felt it almost inevitable that amino acids would have developed in the earth's pre-life environment. Amino acids are relatively stable and probably became abundant in the early oceans, over time, joining together into more complex molecules.

Amino Acids to Proteinoids

In order to form proteins, amino acid molecules must lose some water from their molecular structures, which happens when amino acids are heated to temperatures of 140 degrees Celsius. Volcanic activity would have been capable of providing such temperatures. A biochemist named S. W. Fox and his coworkers were able to produce proteinlike chains from a mixture of eighteen common amino acids. Fox termed these structures *proteinoids* and thought that billions of years ago proteinoids were the transitional structures leading to true proteins. Fox actually found proteinoids similar to those he created in the laboratory in lava and cinders spewed out by Hawaiian volcanoes. Amino acids formed in the vapors emitted by the volcanoes and were combined into proteinoids by the heat of escaping gases.

Proteinoids to Microspheres

When solutions of proteinoids in hot water are cooled, they form tiny spheres with many characteristics of living cells. These *microspheres* have a filmlike outer wall, somewhat like a cell membrane; are capable of osmotic swelling and shrinking; exhibit budding, as do yeast cells; and can be ob-

served to divide into "daughter" microspheres. Sometimes these microspheres join together to form lines, or filaments, as some bacteria do, and it is possible to observe movement of internal particles within microspheres, similar to cytoplasmic streaming in living cells.

Nucleic Acids in the Laboratory

Complete long-chain nucleic acids have not yet been experimentally produced under pre-life conditions. However, short stretches of nucleic acid components were produced in 1976 by Har Gobind Khorana and his associates at the Massachusetts Institute of Technology. In yet another experiment, parts of a nucleic acid called peptide nucleic acid, or PNA, were produced by discharging electricity through a blend of methane, ammonia, nitrogen, and water. PNA is more stable than RNA and may have existed during the earth's very early days. However, the big questions remain: How did replication of these nucleic acids begin? How did the first organisms manage to pass along their genetic information to the next generation?

Life Begins in the Sea

The earliest organisms apparently originated in the sea, which contains many of the organic compounds, minerals, and other nutrients needed by living organisms. This reflects the viewpoint of many scientists, including Charles Darwin, who said in 1871 that life started in a "warm little pond," and Aleksandr I. Oparin, who proposed in the 1930's that many chemical substances would have washed out of the atmosphere and been carried into the oceans by rain, creating a "primordial soup" of nutrients for the first organisms. These organic molecules, kept in constant motion by the ocean currents, may have bumped together, recombining into larger molecules and increasing in complexity. Exactly how, or exactly when, the transition from not-quite-living to living took place is not known, but scientists assume that it could not happen under present-day conditions. Oxygen and microbial predators would destroy similar structures today.

Heterotrophs

The first living organisms were microscopic in size and unicellular. These earliest forms of life were not able to make their own food but instead assimilated small pieces of organic molecules present in the surrounding waters. They no doubt ate one another as well. Organisms with this type of nutritional mechanism are called *heterotrophs*. The food of these ancient organisms was digested externally by excreted enzymes before being ingested and metabolized by fermentation, a process that does not require free oxygen.

Autotrophs and Photoautotrophs

When nearby food resources became exhausted, the resulting food shortages might have caused selective pressures for evolutionary change. Some organisms evolved the ability to synthesize their food from simple inorganic substances. These became organisms called *autotrophs*. The autotrophs began to evolve in many different directions, using different substances, such as carbon dioxide, ammonia, and hydrogen sulfide, as food.

Another group evolved into the *photoautotrophs*, capable of carrying out *photosynthesis*, which uses solar energy to incorporate carbon dioxide into organic molecules and releases oxygen as a byproduct. Carbon combined with other elements to promote cellular growth, and the oxygen escaped into the atmosphere, which prepared the environment for the next important step in the evolution of primitive organisms.

As the photoautotrophs multiplied, photosynthesis began to gradually change Earth's original oxygen-poor atmosphere to a more oxygen-rich one. A rapid buildup of oxygen in the atmosphere was delayed, as iron in rocks exposed at the earth's surface was oxidized before oxygen could accumulate in the atmosphere. This allowed many microorganisms to evolve oxygen-mediating enzymes that permitted them to cope with the new atmosphere.

Once atmospheric oxygen concentrations reached about 10 percent, solar radiation converted part of the oxygen to ozone, forming a shield against ultraviolet radiation. Life, still primitive and vulnerable, was now protected and could expand into environments that formerly had not been able to harbor life, setting the stage for the appearance of aerobic organisms.

Aerobic Metabolism

Aerobic organisms use oxygen to convert their food into energy. Aerobic metabolism provides far more energy in relation to food consumed than does the fermentation carried out by anaerobic organisms. Aerobic metabolism provided a surplus of energy, which was an important factor in the evolution of more complex forms of life. With more energy available, organisms could move about more, colonize new niches, and engage in sexual reproduction, allowing new and innovative genetic recombinations to emerge, and may have increased the rate of evolution, which led to the evolution of complex multicellular organisms called metazoans.

Prokaryotes and Eukaryotes

The first organisms were unicellular organisms called *prokaryotes*, which today are classified in two of the three domains of life as either *Archaea* or *Bacteria*, the latter including the cyanobacteria, which are photosynthetic. Both of these domains comprise unicellular life-forms consisting of prokaryotes. Prokaryotic cells lack internal organelles and a membrane-bound nucleus. Their genetic material resides in the cytoplasm of the cell. Modern prokaryotes do have cell walls and most are able to move about. Prokaryotes are asexual, reproducing by binary fission, which limits the possibilities for variation.

Evolution proceeded from the prokaryotes to organisms with a definite nuclear wall, well-defined chromosomes, and the capacity to reproduce sexually. These more advanced life-forms are termed *eukaryotes*. Eukaryotic cells contain organelles such as mitochondria, which metabolize carbohydrates and fatty acids to carbon dioxide and water, releasing energy-rich phosphate compounds in the process. Some eukaryotes have organelles called chloroplasts, the structures in which sunlight is converted into energy in the process of photosynthesis.

The forms of life made of eukaryotic cells are now classified in the third great domain of life, *Eukarya*, which comprises most of the life-forms familiar as plants and animals—from fungi and plants to human beings—classified in several kingdoms, phyla, classes, genera, and millions of species. Biologists believe that the organelles in eukaryotic cells were once independent microorganisms that entered other cells and then established symbiotic relationships with the primary cell. This process is called the endosymbiont theory.

Ancestral Algae

The various divisions of algae make up an important group of photosynthetic eukaryotes. One group of algae, the *Chlorophyta*, or green algae, are the probable ancestors of terrestrial plants. Common characteristics of green algae and terrestrial plants include the chlorophyll pigments *a* and *b* in the chloroplasts of both green algae and land plants, a number of carotenoid pigment derivatives, starch as the carbohydrate food reserve, and cell walls made up of cellulose.

From Sea to Land

If, as scientists believe, life originated in the sea, how did it get onto the land? Did the ongoing push to find new sources of food drive organisms from the depths of the oceans to coastal areas, where environmental conditions were harsher? To cope with the intense wave action along shorelines, multicellular organisms with diversified parts, some designed to hold onto rocks, evolved. To prevent desiccation in coastal environments and to provide support against the pounding waves, these organisms also evolved rigid cell walls, which allowed them to increase in size. As plants increased in size, they had adapt to do two things: First, they needed to be able to move water and mineral nutrients from the substrate to other parts of the plant not tied directly to the ground. A second problem involved how to move the products of photosynthesis from the site of manufacture to those parts of the plant where photosynthesis could not take place.

The solution to these problems became more critical as plants began to colonize habitats farther from permanent supplies of water. Roots evolved to anchor plants in the ground and to supply the rest of the plant with water and mineral nutrients from the ground. Stems evolved to provide support for the leaves, the main organs of photosynthesis; in some cases, the stems themselves evolved the capability of photosynthesis. All of the aboveground portions of plants developed a waxy cuticle that slows down water loss and prevents desiccation. Internally, plants evolved specialized tissues, called the vascular tissues. One type of vascular tissue, the *xylem*, conducts water and mineral nutrients from the soil through the roots and stems to the leaves. The other type of vascular tissue, the *phloem*, transports the sugars manufactured in the leaves during photosynthesis to other parts of the plant body.

Many plants also evolved chemical attractants and defenses in the form of *secondary metabolites*: metabolic products, such as alkaloids, glycosides, or saponins, to deter plant-eating animals. With the appearance of flowering plants (angiosperms), plants began to evolve odors, colors, flowers, and fruits that attracted pollinators and herbivores, ensuring seed dispersal and the survival and propagation of flowering species. It was these specializations that allowed plants to move from the sea and colonize the land between 500 million and 450 million years ago.

Carol S. Radford

See also: Anaerobes and heterotrophs; *Archaea*; Bacteria; Chloroplasts and other plastids; *Eukarya*; Evolution of plants; Green algae; Prokaryotes.

Sources for Further Study

Coulter, Merle C. *The Story of the Plant Kingdom*. 3d ed. Revised by Howard J. Dittmer. Chicago: University of Chicago Press, 1964. A discussion of the evolution and biology of plants.

Margulis, Lynn, and Karlene V. Schwartz. *Five Kingdoms: An Illustrated Guide to the Phyla of Life on Earth*. 3d ed. New York: W. H. Freeman, 1998. All of the major divisions of life on Earth are discussed in this book.

Raven, Peter H., Ray F. Evert, and Susan E. Eichhorn. *Biology of Plants*. 6th ed. New York: W. H. Freeman/Worth, 1999. This standard college textbook discusses the systematics and morphology of the different divisions of the plant kingdom.

Stern, Kingsley R. *Introductory Plant Biology*. 8th ed. Boston: McGraw-Hill, 2000. An introductory botany textbook that includes discussions of the different divisions of the plant kingdom.

EVOLUTION OF PLANTS

Categories: Evolution; paleobotany; *Plantae*

As a result of prehistoric events such as the Permian-Triassic extinction event and the Cretaceous-Tertiary mass extinction event, many plant families and some ancestors of extant plant were extinct before the beginning of recorded history.

The general trend of earth's plant diversification involves four major plant groups that rose to dominance from about the Middle Silurian period to present time. The first major group providing land vegetation comprised the seedless vascular plants, represented by the phyla *Rhyniophyta*, *Zosterophyllophyta*, and *Trimerophytophyta*. The second major group appearing in the late Devonian period was made up of the ferns (*Pterophyta*). The third group, the seed plants (sometimes called the Coal Age plants), appeared at least 380 million years ago (mya). This third group includes the gymnosperms (*Gymnospermophyta*), which dominated land flora for most of the Mesozoic era until 100 mya. The last group, the flowering angiosperms (*Anthophyta*), appeared in the fossil record 130 mya. The fossil record also shows that this group of plants was abundant in most parts of the world within 30 million to 40 million years. Thus, the angiosperms have dominated land vegetation for close to 100 million years.

The Paleozoic Era

The Proterozoic and Archean eons have restricted fossil records and predate the appearance of land plants. Seedless, vascular land plants appeared in the middle of the Silurian period (437-407 mya) and are represented by the rhyniophytes or rhyniophytoids and possibly the *Lycophyta* (lycophytes or club mosses). From the primitive rhyniophytes and lycophytes, land vegetation rapidly diversified during the Devonian period (407-360 mya). Pre-fern ancestors and maybe true ferns (*Pterophyta*) were developed by the mid-Devonian. By the Late Devonian the horsetails (*Sphenophyta*) and gymnosperms (*Gymnospermophyta*) were present. By the end of the period, all major divisions of vascular plants had appeared except the angiosperms.

Development of vascular plant structures during the Devonian allowed for greater geographical diversity of plants. One such structure was flattened, planated leaves, which increased photosynthetic efficiency. Another was the development of secondary wood, allowing plants to increase significantly in structure and size, thus resulting in trees and probably forests. A gradual process was the reproductive development of the seed; the earliest structures are found in Upper Devonian deposits.

Ancestors of the conifers and cycads appeared in the Carboniferous period (360-287 mya), but their documentation is poor in the fossil record. During the early Carboniferous in the high and middle latitudes, vegetation shows a dominance of club mosses and progymnosperms (*Progymnospermophyta*). In the lower latitudes of North America and Europe, a greater diversity of club mosses and progymnosperms are found, along with a greater diversity of vegetation. Seed ferns (lagenostomaleans, calamopityaleans) are present, along with true ferns and horsetails (*Archaeocalamites*).

Late Carboniferous vegetation in the high latitudes was greatly affected by the start of the Permo-Carboniferous Ice Age. In the northern middle latitudes, the fossil record reveals a dominance of horsetails and primitive seed ferns (pteridosperms) but few other plants.

In northern low latitudes, landmasses of North America, Europe, and China were covered by shallow seas or swamps and, because they were close to the equator, experienced tropical to subtropical climatic conditions. The first tropical rain forests appeared there, known as the Coal Measure Forests or the Age of Coal. Vast amounts of peat were laid down as a result of favorable conditions of year-round growth and the giant club mosses' adaptation to the wetland tropical environments. In drier areas surrounding the lowlands, forests of horse-tails (calamites, sphenophylls), seed ferns (medul-losans, callistophytes, lagenostomaleans), cordaites, and diverse ferns (including marattialean tree ferns) existed in great abundance.

The Permian period (287-250 mya) marks a major transition of the conifers, cycads, glossopterids, gigantopterids, and the peltasperms from a poor fossil record in the Carboniferous to significantly abundant land vegetation. The two most prevalent plant assemblages of the Permian were the horse-tails, peltasperms, cycadophytes, and conifers. The second most prevalent were the gigantopterids, peltasperms, and conifers. These two plant assemblages are considered the typical paleo-equatorial lowland vegetation of the Permian. Other plants, such as the tree ferns and giant club mosses, were present in the Permian but not abundant. As a result of the Permian-Triassic extinction event, tropical swamp forests disappeared, with the extinction of the club mosses; the cordaites and glossopterids disappeared from higher latitudes; and 96 percent of all plant and animal species became extinct.

The Mesozoic Era

At the beginning of the Triassic period (248-208 mya), a meager fossil record reveals diminished land vegetation (that is, no coal formed). By the middle to late Triassic, the modern family of ferns, conifers, and a now-extinct group of plants, the bennettites (cycadeoids), inhabited most land surfaces. After the mass extinction, the bennettites moved into vacant lowland niches. They may be significant because of the similarity of their reproductive organs to the reproductive organs of the angiosperms.

Late Triassic flora in the equatorial latitudes are represented by a wide range of ferns, horsetails, pteriosperms, cycads, bennettites, leptostrobaleans, ginkgos, and conifers. The plant assemblages in the middle latitudes are similar but not as species-rich. This lack of plant variation in low and middle latitudes reflects a global frost-free climate.

In the Jurassic period (208-144 mya), land vegetation similar to modern vegetation began to appear, and the ferns of this age can be assigned to modern families: *Dipteridaceae, Matoniaceae, Gleicheni-aceae*, and *Cyatheaceae*. Conifers of this age can also be assigned to modern families: *Podocarpaceae, Araucariaceae* (Norfolk pines), *Pinaceae* (pines), and *Taxaceae* (yews). These conifers created substantial coal deposits in the Mesozoic.

During the Early to Middle Jurassic, diverse vegetation grew in the equatorial latitudes of western North America, Europe, Central Asia, and the Far East and comprised the horsetails, pteridosperms, cycads, bennettites, leptostrobaleans, ginkgos, ferns, and conifers. Warm, moist conditions also existed in the northern middle latitudes (Siberia and northwest Canada), supporting Ginkgoalean forests and leptostrobaleans. Desert conditions

By the late Triassic period, a now-extinct group of plants, the bennettites (cycadeoids, such as the fossil shown above), inhabited most land surfaces. After the mass extinction event, the bennettites moved into vacant lowland niches. They may be significant because of the similarity of their reproductive organs to the reproductive organs of the angiosperms (flowering plants).

Evolution of Plants

Origin	Geologic Period	Plant Life
420 mya	Silurian	Seedless vascular plants (earliest land plants)
400 mya	Devonian	Ferns
380 mya	Devonian	Progymnosperms
365 mya	Carboniferous	Seed plants: gymnosperms, conifers
245 mya	Permian	Cycads (earliest fossils, but may have preceded gymnosperms)
130 mya	Cretaceous	Angiosperms (flowering plants)

Notes: Dates are approximate and are often debated, but the paleobotanists generally agree on the sequence; mya = millions of years ago.

existed in central and eastern North America and North Africa, and the presence of bennettites, cycads, peltasperms and cheirolepidiacean conifers there are plant indicators of drier conditions. The southern latitudes had similar vegetation to the equatorial latitudes, but owing to drier conditions, cheirolepidiacean conifers were abundant, ginkgos scarce. This southern vegetation spread into very high latitudes, including Antarctica, because of the lack of polar ice.

In the Cretaceous period (144-66.4 mya), arid, subdesert conditions existed in South America, Central and North Africa, and central Asia. Thus, the land vegetation was dominated by cheirolepidiacean conifers and matoniacean ferns. The northern middle latitudes of Europe and North America had a more diverse vegetation comprising bennettites, cycads, ferns, peltasperms, and cheirolepidiacean conifers, with the southern middle latitudes dominated by bennettites and cheirolepidiaceans.

A major change in land vegetation took place in the late Cretaceous with the appearance and proliferation of flowering seed plants, the angiosperms. The presence of the angiosperms marked the end of the typical gymnosperm-dominated Mesozoic flora and a definite decline in the leptostrobaleans, bennettites, ginkgos, and cycads.

During the late Cretaceous in South America, central Africa, and India, arid conditions prevailed, resulting in tropical vegetation dominated by palms. The southern middle latitudes were also affected by desert conditions, and the plants that fringed these desert areas were horsetails, ferns, co-

nifers (araucarias, podocarps), and angiosperms, specifically *Nothofagus* (southern beech). The high-latitude areas were devoid of polar ice; owing to the warmer conditions, angiosperms were able to thrive. The most diverse flora was found in North America, with the presence of evergreens, angiosperms, and conifers, especially the redwood, *Sequoia*.

The Cretaceous-Tertiary (K/T) mass extinction event occurred at about 66.4 mya. This event has been hypothesized to be a meteoritic impact; whatever the cause, at this time an event took place that suddenly induced global climatic change and initiated the extinction of many species, notably the dinosaurs. The K/T had a greater effect on plants with many families than it did on plants with very few families. Those that did become extinct, such as the bennettites and caytonias, had been in decline. The greatest shock to land vegetation occurred in the middle latitudes of North America. The pollen and spore record just above the K/T boundary in the fossil record shows a dominance of ferns and evergreens. Subsequent plant colonization in North America shows a dominance of deciduous plants.

The Cenozoic Era

Increased rainfall at the beginning of the Paleogene-Neogene period (66.4-1.8 mya) supported the widespread development of rain forests in southerly areas. Rain forests are documented by larger leaf size and drip tips at leaf edge, typical characteristics of modern rain-forest floras.

Notable in this period was the polar Arcto-Tertiary forest flora found in northwest Canada at paleolatitudes of 75-80 degrees, north. Mild, moist summers alternated with continuous winter darkness, with temperatures ranging from 0 to 25 degrees Celsius. These climatic conditions supported deciduous vegetation that included *Platanaceae* (sycamore), *Judlandaceae* (walnut), *Betulaceae* (birch), *Menispermaceae, Cercidophyllaceae, Ulmaceae* (elm), *Fagaceae* (beech), *Magnoliaceae*; and gymnosperms such as *Taxodiaceae* (redwood), *Cypressaceae* (cypress), *Pinaceae* (pine), and *Ginkgoaceae* (gingko). This flora spread across North America to Europe via a land bridge between the continents.

About eleven million years ago, during the Miocene epoch, a marked change in vegetation occurred, with the appearance of grasses and their subsequent spread to grassy plains and prairies. The appearance of this widespread flora supported the development and evolution of herbivorous mammals.

The Quaternary period (1.8 mya to present) began with continental glaciation in northwest Europe, Siberia, and North America. This glaciation affected land vegetation, with plants migrating north and south in response to glacial and interglacial fluctuations. Pollen grains and spores document the presence of *Aceraceae* (maple), hazel, and *Fraxinus* (ash) during interglacial periods.

Final migrations of plant species at the close of the last ice age (about eleven thousand years ago), formed the modern geographical distribution of land plants. Some areas, such as mountain slopes or islands, have unusual distribution of plant species as a result of their isolation from the global plant migrations.

Mariana Louise Rhoades

See also: Angiosperm evolution; Angiosperms; Cycads and palms; Endangered species; Evolution of plants; Ferns; Fossil plants; Ginkgos; Gymnosperms; Horsetails; Lycophytes; Mosses; Paleobotany; *Rhyniophyta*; Seedless vascular plants; Stromatolites; *Trimerophytophyta*; *Zosterophyllophyta*.

Sources for Further Study
Cleal, Christopher J., and Barry A. Thomas. *Plant Fossils: The History of Land Vegetation*. Suffolk, England: Boydell Press, 1999. Details the record of earth's floral and land plant evolution. Includes a generous selection of fossil plant photographs, explanatory notes.

Stewart, Wilson N., and Gar A. Rothwell. *Paleobotany and the Evolution of Plants*. New York: Cambridge University Press, 1993. Comprehensive coverage on the evolution of extant and extinct plants. Includes excellent photos, drawings.

EXERGONIC AND ENDERGONIC REACTIONS

Category: Cellular biology

Exergonic reactions are spontaneous chemical reactions in which the products are at a lower energy level than the reactants; these reactions release energy. Endergonic reactions are nonspontaneous chemical reactions in which the products are at a higher energy level than the reactants; these reactions consume energy.

The primary source of energy for life on the earth is the sun, which is the energy source for photosynthesis: the biological process that transforms radiant energy into chemical energy. Chemical energy is stored in biological molecules, which can then be used as the fuel to provide an organism's energy needs. Such biological molecules include sugars (or carbohydrates), proteins, and lipids (or fats). In the reactions of metabolism, many types of molecules are synthesized (*anabolism*), and many are broken down (*catabolism*). Changes in energy content occur in all these reactions. *Bioenergetics* is the science that studies the description of the basic mechanisms that govern the transformation and use of energy by organisms. A basic tenet of bioener-

getics is that no chemical reaction can be 100 percent energy-efficient. In other words, in all reactions there is some transfer of energy, but some of it is always lost in the form of heat.

The energy (often measured in *calories*) contained in the molecular structure of a compound is called *Gibbs free energy* (after Josiah Willard Gibbs, 1839-1903, who founded the discipline of physical science) and is the energy available to perform work. The difference between the free energy of the products and the *free energy* of the reactants in a chemical reaction is called the *change in free energy* and is fundamental in determining if a reaction can occur spontaneously. If the change in free energy is negative, energy is released, and the free energy content is less in the products than in the reactants. Such reactions are considered exergonic. On the other hand, if the change in free energy is positive, the reaction is considered endergonic and is nonspontaneous (that is, endergonic reactions require a source of energy to enable them to occur).

Energy Coupling

Many cellular reactions are endergonic and cannot occur spontaneously. Nevertheless, cells can facilitate endergonic reactions using the energy released from other exergonic reactions, a process called *energy coupling*. As an example, consider a common endergonic reaction in plants in which glucose and fructose are joined together to make sucrose. To enable this reaction to take place, it is coupled with a series of other exergonic reactions as follows:

glucose + adenosine triphosphate (ATP) → glucose-p + ADP

fructose + ATP → fructose-p + adenosine diphosphate (ADP)

glucose-p + fructose-p → sucrose + 2 P_i (inorganic phosphate)

Therefore, although producing sucrose from glucose and fructose is an endergonic reaction, all three of the foregoing reactions are exergonic. This is representative of the way cells facilitate endergonic reactions.

Role of ATP

The principal molecule involved in providing the energy for endergonic cellular reactions to take place is adenosine triphosphate, or ATP, the same molecule used in the example above. ATP is typically produced by joining an inorganic phosphate to adenosine diphosphate (ADP), which is an endergonic reaction. This, too, represents a characteristic of chemical reactions: If a reaction is exergonic in one direction, it will be endergonic in the opposite direction. Thus, the breakdown of ATP is exergonic, while the production of ATP is endergonic. The energy for production of most of the ATP in plant cells comes from the *light reactions* of photosynthesis and the *electron transport system* in the mitochondria.

The enigma is why ATP, and not any other molecule, is used. Although no complete justification is available, there are several points that support its significance. First, there is the high stability of the ATP molecule at the physiological pH (around 7.4) toward hydrolysis and decomposition in the absence of an enzyme catalyst. This stability allows ATP to be stored in the cell until needed. Second, ATP is one of the molecules (a nucleotide) that is used in synthesis of DNA. Finally, the magnitude of the change in free energy involved in the ATP-ADP transformation is of an amount useful for driving many of the endergonic reactions in the cell. As a result, it can play the role of an intermediate quite easily.

Paris Svoronos, updated by Bryan Ness

See also: Anaerobic photosynthesis; ATP and other energetic molecules; C_4 and CAM photosynthesis; Calvin cycle; Chloroplasts and other plastids; Energy flow in plant cells; Glycolysis and fermentation; Krebs cycle; Mitochondria; Oxidative phosphorylation; Photorespiration; Photosynthesis; Photosynthetic light absorption; Photosynthetic light reactions; Plant cells: molecular level; Respiration.

Sources for Further Study

Hall, D. O., and K. K. Rao. *Photosynthesis*. 6th ed. New York: Cambridge University Press, 1999. Treats photosynthesis in a simple, methodical manner and explains complex concepts in an interesting and user-friendly way. Helps the student to think practically about the subject, pointing him or her toward the next stage of understanding of plant biology.

Harris, David A. *Bioenergetics at a Glance*. Cambridge, Mass.: Blackwell Science, 1995. Clear, concise introduction to the study of energy use and conversion in living organisms, with an emphasis on the biochemical aspects of plant science and physiology and cell biology.

Lawlor, David W. *Photosynthesis: Molecular, Physiological, and Environmental Processes*. 3d ed. New York: Springer-Verlag, 2001. Covers all aspects of photosynthesis, from the molecular level to plant production. Written for undergraduate or graduate students and nonspecialists who want a concise overview of the process.

Stumpf, Paul K. "ATP." *Scientific American* 188 (April, 1953): 85-90. Describes ATP's participation in biochemical reactions and cellular energy systems. Includes schemes and a picture of ATP crystals.

EXTRANUCLEAR INHERITANCE

Categories: Cellular biology; genetics; reproduction and life cycles

Extranuclear inheritance is a non-Mendelian form of heredity that involves genetic information located in cytoplasmic organelles, such as mitochondria and chloroplasts, rather than on the chromosomes found in the cell nucleus.

Extranuclear genes, also known as *cytoplasmic genes*, are located in mitochondria and chloroplasts of a cell rather than in the cell's nucleus on the chromosomes. Both egg and sperm contribute equally to the inheritance of nuclear genes, but extranuclear genes are more likely to be transmitted through the maternal line because the egg is rich in the cytoplasmic organelles where these genes are located, whereas the sperm contributes only its nucleus to the fertilized egg. Therefore, extranuclear genes do not follow genetic pioneer Gregor Mendel's statistical laws of *segregation* and *recombination*. Cytoplasmic genes are of interest in understanding evolution, genetic diseases, and the relationship between genetics and embryology.

History

Since the discovery of Mendel's principles, research in genetics has been guided by the belief that the fundamental units of inheritance are located on chromosomes in the cell nucleus. T. H. Morgan, one of the founders of modern genetics, declared that the cytoplasm could be ignored genetically. However, some biologists resisted the concept of a "nuclear monopoly" over inheritance. Embryologists, in particular, argued that nuclear genes, identical in every cell, could not explain how cells *differentiated* from one another in the course of development. They argued that differences among cells in the developing embryo must have a basis in the *cytoplasm*, the part of the cell outside the nucleus. Trying to formulate a compromise, some biologists suggested that Mendelian genes play a role in determining individual characteristics, while cytoplasmic determinants are responsible for more fundamental aspects of plants and animals. The discovery of a wide variety of cytoplasmic entities seemed to support the concept that cytoplasmic factors played a role in development and heredity.

In the 1940's, Boris Ephrussi's work on "petite" mutants in yeast suggested that inheritance of this trait depended on some factor in the cytoplasm rather than the nucleus. Yeast cells with the *petite mutation* produce abnormally small colonies when grown on a solid medium, with glucose as the energy source. Petite mutants grow slowly because they lack important membrane-bound enzymes of the respiratory system.

Similar studies have been made of slow-growing mutants of the bread mold *Neurospora*. Inheritance of the trait known as "poky" shows a non-Mendelian pattern. Microinjection of purified mitochondria from poky strains into normal strains has been used to demonstrate the cytoplasmic inheritance of this trait.

Chloroplasts

As early as 1909, geneticists were reporting examples of non-Mendelian inheritance in higher plants, usually green and white variegated patterns on leaves and stems. These patterns seemed to be related to the behavior of the *chloroplasts*, photosynthetic organelles in green plants. Because of the relatively large size of chloroplasts, scientists have been able to study their behavior in dividing cells with the light microscope since the 1880's. Like mitochondria, chloroplasts contain their own deoxyribonucleic acid (DNA) and ribonucleic acid (RNA). Although *chloroplast DNA* (cpDNA) contains many of the genes needed for chloroplast function, chloroplasts do not seem to be totally autonomous; nuclear genes are required for some chloroplast functions. Another interesting case of extranuclear inheritance in plants is that of *cytoplasmic pollen sterility*. Many species of plants seem to produce strains with cytoplasmically inherited pollen sterility.

Advances in experimental methods made in the 1960's allowed scientists to demonstrate that organelles located in the cytoplasm contain DNA. This finding came as a great surprise to most biologists. In 1966 the first vertebrate *mitochondrial DNA* (mtDNA) was isolated and characterized. Like bacterial DNA, mtDNA generally consists of a single double helix of "naked," circular DNA. The mitochondrial genome is usually smaller than that of even the simplest bacterium. Most of the proteins in the mitochondrion are encoded by nuclear genes, but mtDNA contains genes for mitochondrial ribosomal RNAs, transfer RNAs, and some of the proteins of the electron transport system of the inner membrane of the mitochondrion.

Extranuclear DNA

The DNA found in chloroplasts and mitochondria is chemically distinct from the DNA in the nucleus. Moreover, the extranuclear genetic systems behave differently from those within the nucleus. Even more surprising is the finding that mitochondria have their own, slightly different version of the genetic code, which was previously thought to be common to all organisms, from viruses to humans. In general, because of its greatly smaller size, the DNA found in cytoplasmic organelles has a limited coding capacity. Thus, by identifying the functions under the control of mitochondrial or chloroplast genes, all other functions carried on by the organelle can be assigned to the nuclear genome. Coordinating the contributions of the organelle and the nuclear genomes is undoubtedly a complex process.

In addition to the genes found in mitochondria and chloroplasts, extranuclear factors are found in various kinds of *endosymbionts* (symbiotic organisms that live within the cells of other organisms) and *bacterial plasmids*. Some biologists think that all organelles may have evolved from ancient symbiotic relationships. Endosymbionts may be bacteria, algae, fungi, protists, or viruses. Unlike the mitochondria and chloroplasts, some endosymbionts seem to have retained independent genetic systems. The "killer" particles in paramecia, discovered by T. M. Sonneborn in the 1930's, provide a his-

torically significant example. After many years of controversy, the killer particles were identified as bacterial symbionts. These cytoplasmic entities are not vital to the host cell, as the paramecia are capable of living and reproducing without them. Certain peculiar non-Mendelian conditions found in fruit flies also appear to be caused by endosymbionts.

Although bacteria lack nuclei, their circular DNA is usually referred to as bacterial chromosomes. Some bacteria also contain separate DNA circles smaller than the bacterial chromosome. In the 1950's Joshua Lederberg proposed the name "plasmid" for such extrachromosomal hereditary determinants. Some of the most interesting examples of these entities are the F (fertility) factor, the R (resistance transfer) factors, and the Col (colicin) factors. *Resistance transfer factors* can transmit resistance to antibiotics between bacteria of different species and genera. *Col factors*, toxic proteins produced by bacteria that kill other bacteria, were studied as toxins for many years before their genetic basis was discovered. Because of their simplicity, the bacterial systems are better understood and can serve as models for the kinds of studies that should be performed for extranuclear genes in higher organisms as techniques improve.

Evolutionary Advantages and Uses

The recognition of extranuclear genetic systems raises important questions about their possible evolutionary advantage. In contrast to the remarkable universality of the nuclear genetic system, extranuclear genetic systems are quite diverse in function and mechanisms of transmission. Although extranuclear genes control only a small fraction of the total hereditary material of the cell, in eukaryotic organisms the genes found in mitochondria and chloroplasts are clearly essential for maintaining life.

Although organelle DNAs clearly play an important part in cell organization, it has been difficult to pinpoint the essential roles of organelle DNA and protein-synthesizing systems. Many technical difficulties, and the traditionally low priority of this field, meant that adequate techniques for studying organelle genomes emerged slowly. Studies of cytoplasmic genetics will doubtless have significant applications in medical science and agriculture as well as an impact on understanding of the evolution of genetic control mechanisms.

For example, M. M. Rhoades's work on corn in the 1940's forced American geneticists to take note of research on cytoplasmic genes, while plant breeders began to use cytoplasmically inherited pollen sterility in the production of hybrid seed. Cytoplasmic pollen sterility is a useful trait to incorporate into commercial inbred lines because it ensures cross-pollination and thus simplifies seed production. Unfortunately, a toxin-producing fungus to which the major corn cytoplasmic gene for pollen sterility was susceptible destroyed more than 50 percent of the corn crop in certain areas of the United States in 1970. This disaster prompted a return to hand-detasseling.

Lois N. Magner

See also: Chloroplast DNA; Chloroplasts and other plastids; DNA in plants; Genetics: Mendelian; Genetics: Post-Mendelian; Mitochondria; Mitochondrial DNA; RNA.

Sources for Further Study

Attardi, Giuseppe M., and Anne Chomyn, eds. *Mitochondrial Biogenesis and Genetics*. San Diego: Academic Press, 1995. An explanation of the methods for studying the structure and function of oxidative phosphorylation complexes, import of proteins and RNA into mitochondria, ion and metabolite transport, and mitochondrial inheritance and turnover.

De Duve, Christian. *A Guided Tour of the Living Cell*. New York: Scientific American Books, 1984. Beautifully illustrated guide to the finest details of cellular architecture. The two-volume text is based on a distinguished lecture series addressed to high school students. The cytosol and its organelles are discussed in great detail with remarkable clarity and humor. Illustrations, informative and amusing diagrams, and index.

Gillham, Nicholas W. *Organelle Genes and Genomes*. Reprint. New York: Oxford University Press, 1997. Exhaustive exploration of all aspects of research on organelle genomes. Reviews the properties of chloroplast and mitochondrial genomes, describing in depth their structure, gene content, expression, evolution, and genetics. Designed for an introductory course in organelle genetics at the graduate or advanced undergraduate level.

Rochaix, J.-D., and S. Merchant, eds. *The Molecular Biology of Chloroplasts and Mitochondria in Chlamydomonas*. New York: Kluwer, 1999. A collection of essays that gives comprehensive coverage of chloroplast and mitochondrial biogenesis and function, presenting the current status of research on the molecular genetics of chloroplast gene expression. Written for advanced researchers.

Sapp, Jan. *Beyond the Gene: Cytoplasmic Inheritance and the Struggle for Authority in Genetics*. New York: Oxford University Press, 1987. History of the battle between Mendelian geneticists in the early twentieth century and advocates of cytoplasmic inheritance. Sapp argues that the "nuclear monopoly" deliberately tried to gain scientific authority by excluding competing theories. Provides valuable background information on the work and careers of early leaders in the field of extranuclear inheritance, such as Boris Ephrussi and T. M. Sonneborn. Includes bibliography and index.

FARMLAND

Categories: Agriculture; economic botany and plant uses; soil

Land used as farmland typically has good agricultural soil and is able to produce food and fiber in an efficient way.

Land suitable for agriculture is not evenly distributed throughout the world; it tends to be concentrated in limited areas. In order to be considered good farmland, land must be located at the proper elevation and slope. Because the soil supplies the mineral nutrients required for plant growth, it must also have the appropriate fertility, texture, and pH. Approximately 64 percent of the world's land has the proper topography, and about 46 percent has satisfactory soil fertility to grow crops.

Plants require large amounts of water for photosynthesis and access to soil nutrients; therefore, farmland must receive an adequate supply of moisture, either from rainfall or from irrigation water. About 46 percent of the world's land has adequate and reliable rainfall. Because plant growth is dramatically affected by temperature, farmland must be located in areas with growing seasons long enough to sustain the crop from planting to harvest.

Approximately 83 percent of the world's land has favorable temperatures. Plants also require sufficient sunlight and atmospheric carbon dioxide levels to support the photosynthetic process necessary for growth and development. Virtually all the world's land has adequate sunlight and sufficient carbon dioxide to support plant growth. Crop production requires the right combination of all these factors, and only about 7 percent of the world's land currently has the proper combination of these factors to make the production of crops feasible without additional technological advances.

Farmland in the United States

With its temperate climate, the United States devotes considerably more of its land area to agriculture than do many other parts of the world. About 45 percent of the land in the United States is used for various forms of agriculture; however, only about 20 percent of the land is actual cropland. Of the rest, approximately 4 percent is devoted to *woodlands*, and the other 21 percent is used for other purposes, such as pastures and *grasslands*. Of the farmland devoted to crop production, only 14 percent is used at any given time to produce harvestable crops. Approximately 21 percent of this harvested cropland is used to produce food grains for human consumption. Feed grains for livestock are grown on 31 percent, and the remaining 48 percent of harvested cropland is devoted to the production of other crops.

There are seven major agricultural regions in the United States. The dairy region is located in the North Atlantic states and extends westward past the Great Lakes and along the Pacific Coast. The wheat belt is centered in the central and northern Great Plains and in the Columbia Basin of the Northwest. The general and self-sufficing regions primarily made up of small, family-owned farms are found mostly in the eastern highlands region, which includes the Appalachian Mountains, a few hundred miles inland from the Atlantic Coast, and the Ozark-Ouachita mountains west of the Mississippi River. The corn and livestock belt is found throughout the Midwestern states. The range-livestock region of the western United States stretches in a band from 500 to 1,000 miles wide and extends from the Canadian border to Mexico. The western specialty-crops area is primarily composed of irrigated land in seventeen western states and produces the vast majority of the nation's vegetable crops. The cotton belt, located in the southern states (most notably Georgia, Alabama, and Mississippi), contains more farmers than any other region. While this area has been known historically for its cotton production, many other crops, including tobacco, peanuts, truck crops, and livestock are also produced in the South.

In addition to these major regions, smaller farming areas are located throughout the country. To-

PhotoDisc

Two properties of productive farmland are receipt of an adequate supply of moisture, either from rainfall or from irrigation, and location in an area with a growing season long enough to sustain crops from planting to harvest.

are produced in Maine, Minnesota, Idaho, North Dakota, and California. Citrus is grown in southern Texas, Florida, and California. Sugarcane is cultivated in southern Louisiana and Florida.

Loss of Farmland

In the United States, both the quantity of land devoted to farming and the number of farmers have been decreasing since 1965. Likewise, there has been a decrease in the amount of good farmland worldwide. Most of this decrease is attributed to a combination of *urbanization* and poor agricultural methods that have led to loss of *topsoil* through water and wind *erosion*. Historically, large tracts of farmland have been located near metropolitan areas. In recent times, these urban centers have grown outward into large suburban areas, and this sprawl has consumed many acres of farmland. Erosion destroys thousands of acres of farmland every year, and *desertification*—the conversion of productive rangeland, rain-fed cropland, or irrigated cropland into desertlike land with a resulting drop in agricultural productivity—has reduced productivity on 2 billion acres over the past fifty years. In many cases, the desertified land is no longer useful as farmland. Steps must be taken to preserve this valuable resource, or it is quite possible that the world will suffer mass food shortages in the future.

D. R. Gossett

bacco is produced throughout Kentucky, Virginia, Tennessee, and North and South Carolina. Apples and other fruits are grown in a variety of places, including the Middle Atlantic seaboard, around the Great Lakes, and the Pacific Northwest. Potatoes

See also: Agriculture: history and overview; Biomes: types; Fertilizers; Erosion and erosion control; Green Revolution; North American agriculture; North American flora; Rangeland; Soil management; Strip farming.

Sources for Further Study

Jacobson, Louis. "Planting the Seeds of Farm Preservation." *National Journal* 31, no. 44 (October 30, 1999): 3162. Examines a program to preserve farmland by local government purchase of development rights.

Janick, Jules. *Horticulture Science.* 4th ed. New York: W. H. Freeman, 1986. Contains a section on land and climatic factors associated with the horticultural industry.

Kipps, Michael S. *The Production of Field Crops: A Textbook of Agronomy.* 6th ed. New York: McGraw-Hill, 1970. An excellent discussion on the distribution of farmland.

Korfmacher, K. S. "Farmland Preservation and Sustainable Agriculture: Grassroots and Policy Connections." *American Journal of Alternative Agriculture* 15, no. 1 (Winter, 2000): 37-43. Discusses ways to integrate farmland preservation and sustainable agriculture efforts; suggests several approaches.

Staley, Samuel R., and Jefferson G. Edgens. "The Myth of Farmland Loss." *Forum for Applied Research and Public Policy* 14, no. 3 (Fall, 1999): 29-35. Questions whether farmland loss is significant enough to justify interventions in the land market through growth-management regulations.

"Urban Sprawl: Not Quite the Monster They Call It." *The Economist* 352, no. 8133 (August 21, 1999): 24. Contends that problems associated with urban sprawl are not caused by the sprawl but have other important root problems, such as overproduction driving down crop prices. Includes statistical data.

FERNS

Categories: Paleobotany; *Plantae*; seedless vascular plants; taxonomic groups

Ferns are among the most recognizable members of the phylum Pterophyta, *which are primitive, nonflowering, vascular plants that primarily reproduce by spores and occur in many variations, complicating classification.*

Approximately twelve thousand extant species of fern are classified in the phylum *Pterophyta*. These seedless plants display a diversity of physical and reproductive characteristics that separate them taxonomically. They have leaves containing branching veins known as megaphylls. Fossils from the Devonian period, about 395 million years ago, include some structures resembling *Pterophyta*. These plants are believed to have been the source for gymnosperms. Most early fernlike plants that evolved in a variety of forms during the next period, the Carboniferous (approximately 345 million to 280 million years ago), which is often referred to as the age of the ferns, became extinct afterward. Evaluating the many fossils from this period containing fernlike structures, paleobotanists have identified *Archaeopteris*, which they call the primitive fern, and *Protopteridium*, labeled the first fern, which they hypothesize are the ancestors of modern ferns. Botanists have linked two existing fern genera as possible descendants of Carboniferous ferns.

Some fossilized leaves thought to be ferns had characteristics indicating that they had seeds and were not ferns. Seed ferns, or *pteridosperms*, were trees and vines with fronds but were not directly re-lated to true ferns. Before the seed ferns became extinct in the Cretaceous period, about 136 million years ago, they were the predecessors to angiosperms. Based on anatomical comparisons, modern ferns and seed ferns may have descended from the same ancestors. Eleven modern fern families are present in fossils from the Mesozoic era, approximately 245 million years ago.

Although many modern ferns are morphologically similar to one another, they deviate in expression of specific traits. Theories differ about the evolutionary origins and development of ferns, specifically their stems and leaves. Many botanists designate the family *Schizeaceae*, or the curly grass, in phylum *Pterophyta* as the evolutionary origin of ferns. Types of curly grass also belong to family *Thelypteridaceae*. Tropical ferns are classified in family *Dicksoniaceae*. Other familiar families in phylum *Pterophyta* include *Adiantaceae*, which represents the maidenhair ferns; *Hymenophyllaceae*, which are the filmy ferns; *Blechnaceae*, the deer ferns, with reddish leaves; and *Cyathaceae*, the arborescent tree ferns. Spleenworts are in family *Aspleniaceae*. Brackens in family *Dennstaeditaceae* are further classified into the tribes *Dennstaediteae*, *Lindsaeeae*, and *Mona-*

Classification of Ferns

Class *Filicopsida*
  Order *Hydropteridales*
    Families:
      Azollaceae (azollas)
      Marsileaceae (water clovers)
      Salviniaceae (floating ferns)
  Order *Marattiales*
    Family:
      Marattiaceae (vessel ferns)
  Order *Ophioglossales*
    Family:
      Ophioglossaceae (adder's-tongues)
  Order *Polypodiales*
    Families:
      Anemiaceae (flowering ferns)
      Aspleniaceae (spleenworts)
      Blechnaceae (chain ferns)
      Cyatheaceae (tree ferns)
      Dennstaedtiaceae (bracken ferns)
      Dicksoniaceae (tree ferns)
      Dryopteridaceae (wood ferns)
      Gleicheniaceae (forking ferns)
      Grammitidaceae (kihi ferns)
      Hymenophyllaceae (filmy ferns)
      Lophosoriaceae (diamond leaf ferns)
      Lygodiaceae (climbing ferns)
      Osmundaceae (royal ferns)
      Parkeriaceae (water ferns)
      Polypodiaceae (polypodys)
      Pteridaceae (maidenhair ferns)
      Schizaeaceae (curly grasss)
      Thelypteridaceae (marsh ferns)
      Vittariaceae (shoestring ferns)

Source: Data are from U.S. Department of Agriculture, National Plant Data Center, *The PLANTS Database*, Version 3.1, http://plants.usda.gov. National Plant Data Center, Baton Rouge, LA 70874-4490 USA.

chosorae. The family *Dryopteridaceae* is divided into six tribes.

Floating *Pterophyta* plants belong to the families *Salviniaceae* and *Azollaceae*, the latter of which is considered to display the plant world's most complex reproductive system. *Marsileaceae* is the water-clover family. Variants of *Pterophyta* that grow on forest floors belong to the family *Marattiaceae*. The royal ferns (family *Osmundaceae*) exhibit some primitive characteristics such as stipular leaf bases. Family *Ophioglossaceae*, which includes adder's-tongue, are not closely related to extant ferns and express primitive traits such as high-spore-yielding sporangia, indicating an origin among the progymnosperms. The epiphytic family *Polypodiaceae*, divided by tribes, also is isolated from most *Pterophyta* ferns and is considered most closely related to the family *Gleicheniaceae*.

Pteridium, the bracken, is the most familiar fern worldwide and is among the earth's six most common and oldest plants. Classification of ferns fluctuates according to different authorities. Many known groups of ferns have not been fully examined, while other species are being discovered in previously uncharted places. Some researchers consider such plants as whiskferns (*Psilotophyta*), horsetails (*Sphenophyta*), and club mosses (*Lycophyta*) to be closely related to ferns, but most botanists stress that ferns differ morphologically from those plants, although their life cycles are alike.

Reproduction

Pterophytes such as ferns exist as two alternating forms during their sexual reproductive cycle. Initially, in the gametophyte phase, ferns are a stem known as the rhizome, from which roots and leaves, called fronds, grow as the fern matures into an adult sporophyte. The sporophyte exists separately from the gametophyte, which usually dies as soon as the sporophyte's root sinks into soil. Growing on or near the soil surface, rhizomes, the most common stem form, can be as thin as threads or as thick as ropes and sometimes have hollows that house ants that scientists believe to be living in symbiosis with the fern, protecting it from insect predators. The roots also have differing characteristics of quantity, width, length, and texture that aid in classification.

Ferns do not flower or create seeds; instead, they produce brown sacs, or sporangia, on the bottom surfaces of their fronds which contain single-cell spores. These spores vary in size and have tetrahedral or oval shapes. Spore arrangement aids fern classification. Spores are distributed by wind to germinate and form a small, heart-shaped *prothallus* which has sexual organs. The prothallus is tiny, averaging 8 millimeters (0.3 inch), and often is not visible before it dies. Water is necessary for fertilization to occur, opening the sex organs by swelling. Sperm from the antheridium fertilizes eggs, located in the archegonium on either the same or a different prothallus, which protects the resulting embryos until they mature. Embryos renew the cycle by maturing as rhizomes, which form fronds and roots to spread ferns to different areas. While spores are haploid (with one set of chromosomes), fern cells have two sets of chromosomes, one set acquired from the egg and one from sperm. Ferns create genetically identical clones in colonies.

Most fern reproduction is vegetative (asexual, without gametophytes) and occurs during the sporophytic stage, in which rhizomes produce fern clones or fragments are distributed by wind, water, or insects. The walking ferns, members of the species *Camptosorus rhizophyllus* and *Camptosorus sibiricus*, grow from sprouts emerging where parent leaves touch soil. As a result, large groups of identical ferns can be formed.

Sexual reproduction's role is to introduce ferns to new habitats and geographical areas. Genetic changes that occur because of meiotic cell division before the production of spores result in subtle variations.

Structure and Distribution

Most *Pterophyta* tissues, specifically those in ferns, are created near the top of the fronds, with the most mature cells being located at the base. Some fern genera have deviations in leaf structure. Fern leaves, known as the pteridophyll, have a fiddle-head, called the crosier, from which they unroll, instead of unfolding, as do other plants' leaves. Ferns are vascular, and most have pinnate leaves with leaflets extending from a central rib. Other types of leaves include palmate ferns, in which the leaflets emerge from one place, and staghorn ferns, with dichotomous leaves.

Leaves can be simple or compound, depending on whether they are segmented. Such appearances help identify species, especially when they are closely related. Leaves and their bases protect the stems, which have a surface consisting of one layer of flat epidermal cells with wide outer walls. Sometimes stems are covered with leaf armor, which is

Bracken ferns by a roadside in Georgia.

formed by the hardened remains of leaf bases. This armor thickens stems and can store food in addition to shielding ferns from harm. Scales and hairs are also often present on stems, guarding them and providing identifiable traits for classification. Internally, stems have vascular tissues called steles.

Fronds are of varying shapes and sizes, ranging from 1 millimeter to 30 meters (0.04 inch to 100 feet). Some leaves appear feathery, while others look solid. They exhibit different shades of green and textures, such as glossy or leathery. Fern height also ranges broadly, from a few millimeters (0.1 inch) to as tall as 10 to 25 meters (30 to 80 feet) for some tropical tree ferns.

Habitats

Pterophyta habitats are numerous. The greatest number of species live in tropical rain forests. Most ferns are terrestrial, while some ferns are vines; others are epiphytes, wrapping around trees; and some float on bodies of water. Ferns prefer damp, warm environments but can also grow in arid and cold settings. Because less moisture and lower temperature occur at higher latitudes than in tropical and subtropical zones, fewer native ferns are found at the higher latitudes, although some species can live in the polar regions of the Arctic and Antarctic. Epipteric ferns grow in rocky landscapes, with different species preferring acidic or alkaline stones and others thriving in marshes, bogs, forests, and fields. Several ferns grow at high elevations, living on volcanoes and mountains. Ferns adapt to arid, sunny, and salty conditions by developing harder tissues, waxy surfaces, hair or scale coverings, and altered life cycles.

Species are occasionally introduced into areas where they are not indigenous and then thrive to become common. Hybrids also occur between species within genera, adding to classification confusion. Rarely, hybridization between genera happens. Because they are primarily sterile, hybrids reproduce with vegetative propagation or apogamy, whereby spores with the same number of chromosomes as parent cells generate gametophytes that bud a sporophyte without undergoing fertilization.

Some ferns, including the bracken, are regarded as weeds because they cover fields and bodies of water, blocking light and oxygen necessary for other organisms to survive.

Uses

Economically, *Pterophyta* are not as significant as other plants. Nevertheless, some ferns species are edible, with crosiers being considered delicacies. Other ferns are used medicinally.

By contrast, *Osmunda* and *Pteridium* ferns are considered to be carcinogenic. The aquatic mosquito fern (*Azolla*) hosts *Anabaena azollae*, which converts nitrogen for use by plants such as rice, enhancing production in rice paddies and other fields. Animals often root for fern rhizomes, which store starches. The braken fern *Pteris vittata* absorbs arsenic, a carcinogenic heavy metal, from soil. By removing this toxin, ferns can restore contaminated areas into viable agricultural, industrial, and recreational sites.

The most important economic use of *Pterophyta* is as ornamental garden plants and houseplants. Fern fronds are used in cut-flower arrangments as ornamental greenery. Botanical collectors have identified rare ferns, for which they pay high prices; there is even an underground economy of smuggling ferns illegally from protected areas. Masses of fern roots are used to cultivate such epiphytic greenhouse plants as orchids. In the Middle Ages, some people believed that at midnight on June 24, St. John's Day, ferns would produce blossoms which contained magical seeds.

Elizabeth D. Schafer

See also: Bryophytes; Evolution of plants; Psilotophytes; Seedless vascular plants.

Sources for Further Study

Camus, Josephine M., A. Clive Jermy, and Barry A. Thomas. *A World of Ferns*. London: Natural History Museum Publications, 1991. Pictorial reference describing ferns distributed globally. Includes color illustrations, bibliography, and index.

Dyer, A. F., ed. *The Experimental Biology of Ferns*. New York: Academic Press, 1979. A thorough discussion of how botanists conduct fern investigations. Includes illustrations, bibliographies, and indexes.

Foster, F. Gordon. *Ferns to Know and Grow*. 3d ed. Portland, Oreg.: Timber Press, 1995. Information about planting and arranging ferns ornamentally. Includes an updated nomenclature prepared by George Yatskievych, illustrations, and index.

Hoshizaki, Barbara Joe, and Robbin C. Moran. *Fern Grower's Manual*. Rev. ed. Portland, Oreg.: Timber Press, 2001. Extensive information for amateur and professional gardeners regarding landscaping with ferns. Includes color illustrations, map, and bibliographical references.

Mickel, John. *Ferns for American Gardens: The Definitive Guide to Selecting and Growing More than Five Hundred Kinds of Hardy Ferns*. New York: Macmillan, 1994. Provides information about both popular and obscure ferns propagated in the United States, covering cultivation tips, growing zones, size, and how to use ferns for landscaping. Includes color photographs, illustrations, and source list.

Rickard, Martin. *The Plantfinder's Guide to Garden Ferns*. Portland, Oreg.: Timber Press, 2000. Advises how to select and nurture suitable ferns for domestic displays. Includes color illustrations, bibliographical references.

Wee, Yeow Chin. *Ferns of the Tropics*. Portland, Oreg.: Timber Press, 1998. Describes indigenous tropical ferns. Includes color illustrations, bibliography, and index.

FERTILIZERS

Categories: Agriculture; economic botany and plant uses; nutrients and nutrition; soil

Fertilizers are materials used to modify the chemical composition of soil in order to enhance plant growth. They represent an important use of natural resources because agricultural systems depend upon an ability to retain soil fertility.

Soil is a dynamic, chemically reactive medium, and agricultural soils must provide structural support for plants, contain a sufficient supply of plant nutrients, and exhibit an adequate capacity to hold and exchange minerals.

Topsoil, the 6-inch layer of soil covering the earth's landmasses, is the *root zone* for the majority of the world's food and fiber crops. As plants grow and develop, they remove the essential mineral nutrients from the soil. Because crop production normally requires the removal of plants or plant parts, nutrients are continuously being removed from the soil. Therefore, the long-term agricultural use of any soil requires periodic fertilization to replace these lost nutrients. Fertilizers are associated with every aspect of this nutrient replacement process. The application of fertilizer is based on a knowledge of plant growth and development, soil chemistry, and plant-soil interactions.

Soil Nutrients

Plants require an adequate supply of both macronutrients (calcium, magnesium, sulfur, nitrogen, potassium, and phosphorus) and micronutrients (iron, copper, zinc, boron, manganese, chloride, and molybdenum) from the soil. If any one of these nutrients is not present in sufficient amounts, plant growth and, ultimately, yields will be reduced. Because micronutrients are required in small quantities, and deficiencies in these minerals occur infrequently, the majority of agricultural fertilizers contain only macronutrients. Although magnesium and calcium are utilized in large quantities, most agricultural soils contain an abundance of these two elements, either derived from parent material or added as lime. Most soils also contain sufficient amounts of sulfur from the weathering of sulfur-containing minerals, the presence of sulfur in other fertilizers, and atmospheric pollutants.

The remaining three macronutrients (nitrogen, potassium, and phosphorus) are readily depleted and are referred to as fertilizer elements. Hence, these elements must be added to most soils on a regular basis. Fertilizers containing two or more nutrients are called mixed fertilizers. A fertilizer labeled 10-10-10, for example, means that the product contains 10 percent nitrogen, 10 percent phosphorus, and 10 percent potassium. These elements can be supplied in a number of different forms, some of which may not be immediately useful to plants. In the United States, where fertilizer labeling is regulated for sales, most states require that the label reflect the percentage of nutrients available for plant use. Fertilizers are produced in a wide variety of single and mixed formulations, and the percentage of available nutrients generally ranges from a low of 5 percent to a high of 33 percent. Mixed fertilizers may also contain varying amounts of different micronutrients.

Sources and Production

Nitrogen fertilizers can be classified as either chemical or *natural organic*. Natural organic sources are derived from plant and animal residues and include such materials as animal manure, cottonseed meal, and soybean meal. Because natural organic fertilizers contain relatively small amounts of nitrogen, commercial operations rely on chemical fertilizers derived from sources other than plants and animals. Major chemical sources of nitrogen include ammonium compounds and nitrates. The chemical fixation of atmospheric nitrogen by the Claude-Haber ammonification process is the cornerstone of the modern nitrogen fertilizer manufacturing process. Once the ammonia is produced, it can be applied directly to the soil as anhydrous ammonia, or it can be mixed with water and supplied as a solution of aqueous ammonia and used in chemical reactions to produce other ammonium fertilizers, urea, or nitrates for nitrate fertilizers.

Some organic fertilizers contain small amounts of phosphorus, and organically derived phosphates from guano or acid-treated bone meal were used in the past. However, the supply of these materials is scarce. Almost all commercially produced agricultural phosphates are applied as either phosphoric acid or superphosphate derived from rock phosphate. The major phosphate component in commercially important deposits of rock phosphate is apatite. The apatite is mined, processed to separate the

Commercial fertilizers containing two or more nutrients are called mixed fertilizers. The numbers on the packaging refer to nitrogen, phosphorus, and potassium content. For example, the fertilizer above labeled 5-5-5 contains 5 percent nitrogen, 5 percent phosphorus, and 5 percent potassium, respectively.

phosphorus-containing fraction from inert materials, and then treated with sulfuric acid to break the apatite bond. The superphosphate precipitates out of the solution and sets up as a hard block, which can be mechanically granulated to produce a fertilizer containing calcium, sulfur, and phosphorus.

Potassium fertilizers, commonly called "potash," are also obtained from mineral deposits below the earth's surface. The major commercially available potassium fertilizers are potassium chloride extracted from sylvanite ore, potassium sulfate produced by various methods (including extraction from langbeinite or burkeite ores or chemical reactions with potassium chloride), and potassium nitrate, which can be manufactured by several different chemical processes. Although limited, there are sources of organic potassium fertilizers, such as tobacco stalks and dried kelp.

While the individual nitrogen, phosphorus, and potassium fertilizers can be applied directly to the soil, they are also commonly used to manufacture mixed fertilizers. Between two and ten different materials with widely different properties are mixed together in the manufacturing process. The three most common processes used in mixed fertilizer production are the ammonification of phosphorus materials and the subsequent addition of other materials, bulk blending of solid ingredients, and liquid mixing. Fillers and make-weight materials are often added to make up the difference between the weight of fertilizer materials required to furnish the stated amount of nutrient and the desired bulk of mixed products. Mixed fertilizers have the obvious advantage of supplying all the required nutrients in one application.

Application and Environment

The application of fertilizer to agricultural soil is by no means new. Farmers have been applying *manures* to improve plant growth for more than four thousand years. For the most part, this practice had little environmental impact. Since the development of chemical fertilizers in the late nineteenth century, however, fertilizer use has increased tremendously. During the second half of the twentieth century, the amount of fertilizer applied to the soil increased more than 450 percent. While this increase has more than doubled the worldwide crop production, it has also generated some environmental problems.

The production of fertilizer requires the use of a variety of natural resources, and some people have argued that the increased production of fertilizers has required the use of energy and mineral reserves that could have been used elsewhere. For every crop, there is a point at which the yield may continue to increase with the application of additional nutrients, but the increase will not offset the additional cost of the fertilizer. The economically feasible practice, therefore, is to apply the appropriate amount of fertilizer that produces maximum profit rather than maximum yield. Unfortunately, many farmers still tend to *overfertilize*, which wastes money and contributes to environmental degradation. Excessive fertilization can result in adverse soil reactions that damage plant roots or produce undesired growth patterns. Overfertilization can actually decrease yields. If supplied in excessive amounts, some micronutrients are toxic to plants and will dramatically reduce plant growth.

The most serious environmental problem associated with fertilizers, however, is their contribution to *water pollution*. Excess fertilizer elements, particularly nitrogen and phosphorus, are carried from farm fields and cattle feedlots by water runoff and are eventually deposited in rivers and lakes, where they contribute to the pollution of aquatic ecosystems. High levels of plant nutrients in streams and lakes can result in increased growth of phytoplankton, a condition known as *eutrophication*. During the summer months, eutrophication can deplete oxygen levels in lower layers of ponds and lakes. Excess nutrients can also be leached through the soil and contaminate underground water supplies. In areas where intense farming occurs, nitrate concentrations are often above recommended safe levels. Water that contains excessive amounts of plant nutrients poses health problems if consumed by humans and livestock, and it can be fatal if ingested by newborns.

Importance to Food Production

Without a doubt, the modern use of fertilizer has dramatically increased crop yields. If food and fiber production is to keep pace with the world's growing population, increased reliance on fertilizers will be required in the future. With increasing attention to the environment, future research will be aimed at finding fertilizer materials that will remain in the field to which they are applied and at improving application and cultivation techniques to contain materials within the designated application area. The use of technology developed from discoveries in the field of molecular biology to develop more efficient plants holds considerable promise for the future.

D. R. Gossett

Sources for Further Study

Altieri, Miguel A. *Agroecology: The Science of Sustainable Agriculture*. 2d ed. Boulder, Colo.: Westview Press, 1995. An excellent treatment of the ecological basis of agriculture. Altieri describes methods for restoring soil fertility as well as soil conservation, biological control of pests, and other aspects of organic farming.

Black, C. A. *Soil-Plant Relationships*. 2d ed. Malabar, Fla.: R. E. Krieger, 1984. Provides an excellent, in-depth discussion of soils and soil fertility. Although this book assumes that the reader has a background in basic chemistry, it contains much information that will be informative to the general reader.

Brady, Nyle C. *Elements of Nature and Properties of Soils*. Upper Saddle River, N.J.: Prentice Hall, 2000. An excellent book dealing with a wide range of soil characteristics, often used as a textbook in introductory soil classes.

Follett, Roy H., Larry S. Murphy, and Roy L. Donahue. *Fertilizers and Soil Amendments*. Englewood Cliffs, N.J.: Prentice Hall, 1981. An authoritative textbook dealing with the basic aspects of plant nutrition and the economic use of fertilizers in crop production.

Havlin, John L., ed. *Soil Fertility and Fertilizers: An Introduction to Nutrient Management*. 6th ed. Upper Saddle River, N.J.: Prentice Hall, 1999. Provides a basic introduction to the biological, chemical, and physical properties affecting soil fertility and plant nutrition. Covers all aspects of nutrient management for profitable crop production, with particular attention to minimizing the environmental impact of soil and fertilizer management.

Laegreid, M. *Agriculture, Fertilizers, and the Environment*. New York: CABI, 1999. A balanced scientific review of the environmental and sustainability issues relating to fertilizer use and how its environmental impact can be minimized.

See also: Agronomy; Biofertilizers; Composting; Eutrophication; Hormones; Nutrient cycling; Nutrients; Nutrition in agriculture; Soil.

FLAGELLA AND CILIA

Categories: Anatomy; bacteria; cellular biology; microorganisms

Flagella and cilia are hairlike structures, made primarily of protein, found on the surfaces of cells and used for movement by microorganisms and some specialized cells, such as the gametes of certain plants with motile sperm. Because flagella and cilia are so similar, many scientists use the term "undulipodia" for both in reference to eukaryotic organisms.

Although the term "flagellum" is used in reference to both prokaryotes (archaea and bacteria) and eukaryotes (fungi, protists, plants, and animals), the structure and mechanism of action of this structure in prokaryotes are quite different from the structure and mechanism of action in eukaryotes. Eukaryotic flagella and cilia, however, are structurally and functionally identical. The differences between them are in their number, length, and position. Flagella are less numerous, longer, and usually polar, while cilia are more numerous and shorter, covering much of the cell's surface. Because the dividing line between eukaryotic flagella and cilia is not precise, many scientists use the term *undulipodia* as a collective word for both eukaryotic flagella and cilia. In some algae, other protists, and the gametes of certain plants with motile sperm, flagella and cilia can occur.

Bacterial Flagella

Bacterial flagella are composed of a single protein called *flagellin*. Molecules of this globular protein are stacked to form a thin filament approximately 0.01-0.015 micrometer in diameter. The filament protrudes through the cell wall at the tip of the bacterium (polar flagella) or over the entire surface of the bacterium (peritrichous flagella). *Spirochetes* are unusual in this regard because the flagella do not pierce the cell wall but are located in the space between it and the plasma membrane. The base of each flagellum connects to a rotary motor anchored in the plasma membrane. The cell provides energy to the motor, which then rotates the flagellum to allow cell movement. Rotational movement may be counter-clockwise, which leads to generally straight-line motion, or clockwise, which leads to a more random tumbling motion.

Eukaryotic Flagella and Cilia

Eukaryotic flagella and cilia, or undulipodia, are more complex and larger (approximately 0.25 micrometer in diameter) than their prokaryotic counterparts. The main component of these eukaryotic structures is the *microtubule*; a long, cylindrical structure composed of *tubulin* proteins. In eukaryotic flagella and cilia, two central microtubules are surrounded by a circular arrangement of nine microtubule pairs. Eukaryotic flagella and cilia also contain more than five hundred other proteins, including dynein and kinesin, motor proteins that use cell energy to slide the microtubules past each other, causing an undulating motion (hence the name undulipodium). Unlike bacterial flagella, eukaryotic flagella and cilia are considered to be intracellular structures because they are covered by a continuation of the plasma membrane.

Although absent from fungi, undulipodia are found in many protists and in some plants. Unicellular algae (such as *Chlamydomonas* and *Euglena*) and colonial algae (*Volvox*) use undulipodia for locomotion. Multicellular algae (*Phaeophyta*, *Rhodophyta*) produce flagellated sperm. Among the true plants, bryophytes (*Hepatophyta*, *Anthocerotophyta*, and *Bryophyta*), ferns and their allies (*Psilotophyta*, *Lycophyta*, *Sphenophyta*, and *Pterophyta*), and some gymnosperms (*Cycadophyta* and *Ginkgophyta*) also produce flagellated sperm. Other gymnosperms (*Coniferophyta*) and angiosperms (*Anthophyta*) do not produce cells with flagella or cilia.

Richard W. Cheney, Jr.

See also: Algae; Bacteria; Chemotaxis; Cryptomonads; Cytoskeleton; Diatoms; Dinoflagellates; Euglenoids; Evolution of plants; Haptophytes; Heterokonts; Oomycetes; Phytoplankton; Prokaryotes; *Protista*; Reproduction in plants; *Ulvophyceae*.

Sources for Further Study

Amos, W. Bradshaw, and J. G. Duckett, eds. *Prokaryotic and Eukaryotic Flagella*. New York: Cambridge University Press, 1982. Serves to make the distinctions between the two forms clear.

Raven, Peter H., Ray F. Evert, and Susan E. Eichhorn. *Biology of Plants*. 6th ed. New York: W. H. Freeman/Worth, 1999. Describes briefly the flagellated sperm of several plant phyla for the beginning student, offering the context for botanical studies.

Satir, Peter. *Structure and Function in Cilia and Flagella*. New York: Springer, 1965. A brief monograph. Illustrations, bibliography.

Sleigh, Michael A., ed. *Cilia and Flagella*. New York: Academic Press, 1974. At five hundred pages and illustrated, a set of important studies. Illustrated; bibliographies.

Wilson, Leslie, William Dentler, and Paul T. Matsudaira, eds. *Cilia and Flagella*. Methods in Cell Biology 47. San Diego: Academic Press, 1995. At more than six hundred pages, this text is designed for biology students, researchers, and professionals working in laboratories, covering the roles of cilia and flagella in eukaryotic organisms, the cell cycle, cell-cell recognition and other sensory functions, methodologies, and applications to human diseases. Illustrated, with bibliography and index.

FLOWER STRUCTURE

Categories: Anatomy; angiosperms

Flowers are the modified shoots bearing modified leaves that serve as the sexual reproductive organs of angiosperms. This strategy for reproduction has been so successful that angiosperms now dominate the plant world, and accordingly there are many variations on the basic structure of a flower.

Flowers are organs of sexual reproduction produced by the angiosperms (phylum *Anthophyta*), the largest phylum of photosynthetic organisms, with roughly 250,000 species. This large number represents a great diversity of flower types, but all flowers have some common structural elements.

Flower Parts

Flowers are modified shoots bearing modified leaves. In the typical flower, the modified leaves can be grouped into four sets based on appearance and function: sepals, petals, stamens, and pistils. The *sepals* and *petals* are lowermost on the shoot toward the sides of the flower. The *stamens* and *pistils* are at the tip of the shoot at the inside. While sepals and petals are easy to see, stamens and pistils are often visible only when the flower is closely examined. Two other important parts are the *pedicel*, a stalk on which flowers are frequently borne, and the top of the pedicel, called the *receptacle*, to which the other flower parts are typically attached.

Of the four main parts, the sepals are generally the most leaflike and generally are attached to the bottom of the receptacle. Sepals protect the immature flower during the bud stage. Flowers typically have three to eight sepals, depending on the species. Collectively, the sepals in a flower are called the *calyx*. Above the sepals are the *petals*. Although flattened like the sepal, each petal is usually soft and colored—white, yellow, pink, blue, purple, orange, maroon, or even brown. Petals attract insects, hummingbirds, bats, or other animals, aiding the reproductive process. Usually, the number of petals in a flower will be the same as the number of sepals. Collectively, the petals in a flower are called the *corolla*.

The *stamens*, located inside the petals, are composed of a small *anther* (ball-shaped, egg-shaped, or tubular) and a threadlike *filament* connecting the anther to the rest of the flower. The anther, in turn, is composed of two or four tiny chambers, within which powdery *pollen* grains are produced and stored. Each grain of pollen contains the immature sperm of the plant. Thus, the stamens function as the male part of the flower in sexual plant reproduction, and they may number from one to dozens. The term *androecium* refers collectively to all the stamens within a flower.

The *pistils* form the final set of parts. Each pistil is often shaped like a vase, although the shape varies. The *ovary*, the base of the pistil, is swollen and hollow. The wall of the ovary, called the *pericarp*, is typically green, and the hollow space in the ovary is called the *locule*. Within the locule are one or more tiny globular *ovules*, each containing an egg nucleus and thus functioning as the female structure in sexual reproduction. In addition to the ovary, the pistil is typically composed of two or more parts: the *style*, a slender necklike structure above the ovary, and the *stigma*, a swollen area at the top of the style that traps pollen grains with minute hairs covered by a sticky, sugary film. While most flowers have only one pistil, many have several pistils, attached to the receptacle. The pistils within a flower are collectively called the *gynoecium*.

Functions of Flower Parts

Flowers and their parts function to achieve sexual reproduction, including pollination and seed formation. After pollination is finished, the flower begins the process of seed and finally fruit formation. During *pollination*, pollen grains are released from the anther and carried to the stigma, either by animals (such as insects, birds, and bats) or by wind. Animals, attracted by the flower's colors or aromas, visit flowers to obtain food—either the pollen itself or the *nectar*, a sugary liquid produced by small glands called nectaries at the base of the flower. The animal brushes up against the anthers, which deposit pollen on the animal's body. The animal transfers the pollen to the stigma of either the same flower (self-pollination) or a second flower (cross-pollination). During wind pollination, the anthers release their pollen, which is then borne by air currents. Some of the grains are deposited on a stigma of the same or another flower.

Each pollen grain germinates and produces a slender thread of protoplasm that grows downward through the style and into the ovary. This thread, the *pollen tube*, contains the sperm and grows toward an ovule, where it deposits its sperm. The sperm then fuses with the egg, achieving fertilization as the first cell of the new generation is produced. The ovule matures to form a seed. At the same time, the surrounding ovary enlarges greatly, becoming a fruit as other parts of the flower recede and die off.

Because the stamens and pistils are intimately involved in reproduction, botanists refer to these as *essential parts*. The sepals and petals are termed *nonessential parts*, though in fact they remain important. The sepals and petals are sometimes called the *perianth* because they are found on the periphery of the anthers. A *complete flower* is one that has all four sets of parts. A *perfect flower* is one that has both androecium and gynoecium and is thus bisexual.

Structural Variations

Many plant species produce flowers that deviate from the idealized format. Certain lilies, for example, do not have sepals and petals that are clearly distinguishable from each other. In magnolias and some water lilies, each flower produces perianth parts that intergrade from a sepal-like form toward the outside to a petal-like form toward the inside.

Many flowers have evolved to become simpler and are called *incomplete flowers*; they may lack one or more sets of parts. *Apetalous flowers* have only sepals, although some, such as those of the liverleaf and anemone, may be petal-like. Elms, mulberries, oaks, plantains, pigweeds, and goosefeet have sepals that remain green and are usually tiny. *Naked flowers*, including those of birches and willows, develop neither sepals nor petals. Grass flowers are associated with tiny green parts called *bracts*, which are neither sepals nor petals. The nature of the perianth is related to the way a plant undergoes

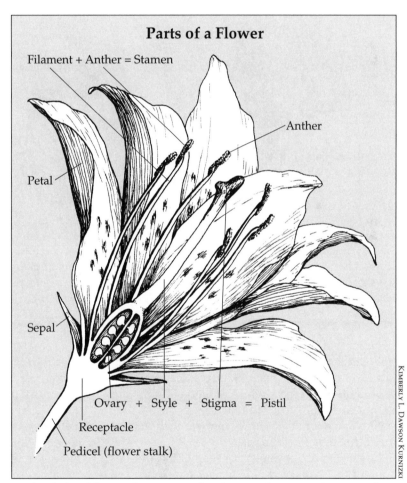

Parts of a Flower

Filament + Anther = Stamen

Anther

Petal

Sepal

Ovary + Style + Stigma = Pistil

Receptacle

Pedicel (flower stalk)

KIMBERLY L. DAWSON KURNIZKI

In the idealized flower, the parts are free down to the receptacle. Many flowers, however, exhibit connation, in which similar parts are fused above the receptacle: for example, the petals of the morning glory, fused to form a corolla, or the sepals of carnations, which form a calyx tube. Many plants have pistils composed of individual fused segments, called carpels, while others, such as the mallow, have connate stamens, forming a stamen tube.

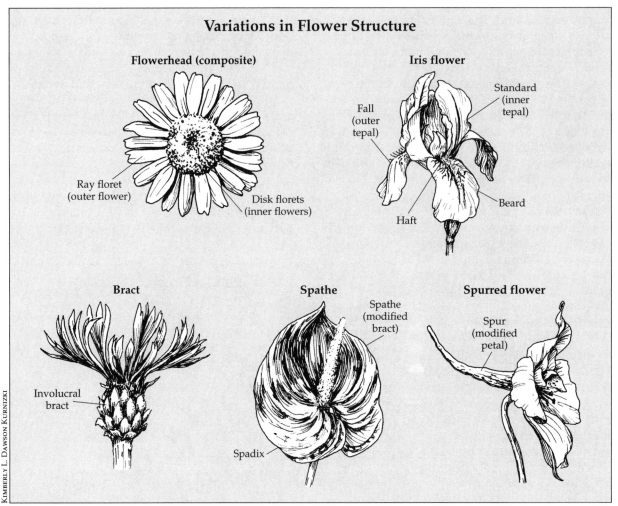

Variations in Flower Structure

Flowers take on many different forms, which have evolved to facilitate pollination by animals, wind, or water. Smaller, less showy flowers tend to be pollinated by wind (which can easily lift and carry their pollen), whereas animal-pollinated flowers have evolved colors, odors, and even structures that mimic insects or store nectar—all designed to attract the pollinators.

pollination. Flowers with a well-developed corolla or a calyx made up of petal-like sepals are attractive to animals and insects, which function to pollinate them. Apetalous and naked flowers are wind-pollinated; they do not need to waste their energy making showy flower parts.

Some incomplete flowers lack either the androecium or the gynoecium. These *imperfect flowers* are unisexual and fall into two categories: *staminate flowers* are male flowers, having only stamens and no pistils; *pistillate flowers* are female, having only pistils and no stamens. Forced into cross-pollination, imperfect flowers benefit the plant by preventing some of the harm inherent in self-pollination.

In the idealized flower, the parts are free down to the receptacle. Many flowers, however, exhibit *connation*, in which similar parts are fused above the receptacle. The petals of the morning glory are fused to form a funnel-shaped corolla. Carnations have connate sepals, forming a calyx tube. Many plants have pistils composed of individual fused segments, called *carpels*, while others, such as the mallow, have connate stamens, forming a stamen tube.

Other flowers show *adnation*, which involves the fusion of different parts. The stamens of phlox flowers are fused to the petals. The sepals, petals, and stamens of roses are all fused, forming a cup-shaped structure called a *hypanthium*. The presence of a hypanthium can be best observed in plum and cherry blossoms, whose individual sepals, petals, and stamens are attached to the rim of the hypanthium. Finally, many flowers have a hypanthium that is fused to the wall of the ovary. The result is that the sepals, petals, and stamens emerge from the top of the ovary, a good example being the apple blossom. Flowers of the latter category are said to have an inferior ovary, whereas the others have superior ovaries.

Flowers' corollas also vary: In flowers with a regular corolla, such as buttercups, lilies, and roses, all the petals are equal in size and shape, giving the flower a star shape. In flowers with irregular corollas, such as the snapdragon, pea, and orchid, one or more of the petals are unequal. Some irregular flowers, such as the violet, touch-me-not, and columbine, have a rounded, cone-shaped, or pointed extension of the corolla called a spur, which serves to store nectar.

Although not technically floral structures, color, shape, and inflorescences (the loose or dense clusters in which flowers appear on a plant) are other ways in which flowers differ, important because they allow certain pollinators to enter but exclude others. Bowl-shaped flowers are visited by a variety of insects, such as beetles and bees. Irregular flowers are typically pollinated by honeybees and bumblebees, and in some cases the insects fit the flower like a key fits a lock. Flowers with long spurs are pollinated by long-tongued insects such as moths. Color, determined by special molecules called pigments that occur within the cells of the plant, attracts different pollinators as well: red flowers are pollinated by birds, specifically hummingbirds and butterflies. White flowers are often open at night and are visited by moths. One group of plants have brown or maroon flowers and an odor of rotting flesh. These "carrion flowers" are pollinated by an array of insects, particularly beetles and flies. Interestingly, the way that humans perceive color is often different from the way that other animals perceive color. For example, xanthophylls reflect not only yellow but also a deep violet that bees can perceive but that humans cannot.

Kenneth M. Klemow

See also: Angiosperm cells and tissues; Angiosperm evolution; Angiosperm life cycle; Angiosperm plant formation; Angiosperms; Animal-plant interactions; Flower types; Flowering regulation; Fruit: structure and types; Garden plants: flowering; Hormones; Inflorescences; Pollination; Reproduction in plants; Seeds; Shoots; Stems.

Sources for Further Study

Bell, Adrian D. *Plant Form: An Illustrated Guide to Flowering Plant Morphology.* New York: Oxford University Press, 1991. Discusses the broad variety of morphological features of flowers. Illustrations, bibliographical references, index.

Lloyd, David G., and Spencer C. H. Barrett, eds. *Floral Biology: Studies on Floral Evolution in Animal-Pollinated Plants.* New York: Chapman & Hall, 1996. Focuses on the synthesis of pollination biology and plant mating systems which is rejuvenating floral biology. Illustrations, bibliographical references, indexes.

Raven, Peter H., Ray F. Evert, and Susan E. Eichhorn. *Biology of Plants.* 6th ed. New York: W. H. Freeman, 1999. This basic textbook presents floral structure with ample pictures and diagrams; covers flower evolution and the interplay between floral structure and pollinators.

Rudall, Paula. *Anatomy of Flowering Plants: An Introduction to Structure and Development.* 2d ed. New York: Cambridge University Press, 1993. This basic text focuses exclusively on angiosperm anatomy and physiology.

Russell, Sharman Apt. *Anatomy of a Rose: Exploring the Secret Life of Flowers.* New York: Perseus, 2001. Whimsical but scientifically sound discussion of how flowers—their colors, shapes, and scents—serve the sole purpose of sexual reproduction. What registers with human senses, however, may be completely different from what attracts bees and other pollinators.

FLOWER TYPES

Categories: Anatomy; angiosperms

The flower is the most distinctive feature of the phylum Anthophyta, *commonly referred to as angiosperms or flowering plants, and is responsible in making them the most dominant, diverse, and widespread of all groups of plants.*

There are already about 250,000 species of flowering plants that have been discovered and named. The basis for their diversity comes from their incredible reproductive success in a wide variety of habitats. The success of this group is also reflected by the diversity of their flowers that show astonishing displays of different forms, sizes, shapes, and colors—all of these to lure pollinators and effect sexual reproduction.

Flowers are considered as an organ system because they are made up of two or more sets, or whorls, of leaflike structures. A typical flower is composed of four whorls, which are the *sepals*, *petals*, *stamens*, and a *pistil* with one or more *carpels*. Much of the variation among flowers is based on variation of these basic parts.

Complete and Incomplete Flowers

A flower that has all four whorls of floral parts is said to be a *complete flower* (such as the hibiscus and the lily). An *incomplete flower* lacks any one or more of these parts (such as those of elms, willows, oaks, and plantains). With or without sepals and petals, a flower that has both stamen and pistil is called a *perfect flower*. Thus, all compete flowers are prefect, but not all perfect flowers are complete. In contrast, flowers that have only stamens or only pistils are called *imperfect flowers*.

Unisexual and Bisexual Flowers

Unisexual flowers are either *staminate* (bearing stamens only) or *pistillate* (bearing pistils only) and are said to be imperfect. *Bisexual flowers* are perfect because they have both stamens and pistil. When staminate and pistillate flowers occur on the same individual, the plant is called *monoecious* (examples include corn and the walnut tree). When staminate and pistillate flowers are borne on separate individual flowers, the plant is said to be *dioecious* (examples include asparagus and willow).

Superior or Inferior Ovaries

The position of the ovary also varies among different flower types. A flower has a *superior ovary* when the base of the ovary is located above where the sepals, petals, and stamens are attached. This point of attachment is referred to as the *receptacle* or *hypanthium*, the fused bases of the three floral parts (tulips and St. John's wort are examples). An *inferior flower* has an ovary below where the sepals, petals, and stamens are attached (as do daffodils and sabatia). Some flowers show an *intermediate type*, where the receptacle partly surrounds the ovary; the petals and stamens branch from the receptacle about halfway up the ovary (as in cherry, peach, and almond flowers).

Hypogynous, Epigynous, and Perigynous Flowers

The position of the ovary in relation to the attachment of floral parts also varies from superior to inferior ovaries. Flowers in which the sepals, petals, and stamens are attached below the ovary are called *hypogynous*, and the ovaries of such flowers are said to be superior (as in pelargonium and silene). Flowers in which the sepals, petals, and stamens appear to be attached to the upper part of the ovary due to the fusion of the hypanthium are called *epigynous*, and the ovaries of such flowers are said to be inferior (as in cornus and narcissus). Flowers in which the hypanthium forms a cuplike or tubular structure that partly surrounds the ovary are called *perigynous*. In such flowers, the sepals, petals, and stamens are attached to the rim of the hypanthium, and the ovaries of such flowers are superior.

Fused and Distinct Floral Parts

The parts of a flower may be *free* or *united*. Fusion of like parts (such as petals united to petals) is called *connation*. When like parts are not fused, they are said to be *distinct* (one petal is distinct from another petal). Fusion of unlike parts (stamens united to petals) is called *adnation*, and the contrasting condition is called *free* (stamens are free from petals). Fused structures may be united from the moment of origin onward, or they may initially be separate and grow together as one later in development.

Regular and Irregular Flowers

In many flowers, the petals of similar shape radiate from the center of the flower and are equidistant from one another. Such flowers are said to have *regular* or *radial symmetry*. In these cases, even though there may be an uneven number of sepals and petals, any line drawn through the center of the flower will divide it into two similar halves. The halves are either exact duplicates or mirror images of each other. Flowers with radial symmetry are also called *actinomorphic flowers* (examples: stonecrop, morning glory). Flowers with *irregular* or *bilateral symmetry* have parts arranged in such a way that only one line can divide the flower into equal halves that are more or less mirror images of each other. Flowers with bilateral symmetry are also called *zygomorphic flowers* (examples: mint, pea, snapdragon). A few flowers have no plane of symmetry and are referred to as *asymmetrical*.

Corolla Shapes

Corolla is the collective term for all the petals of a single flower. This is usually the showy part of the flower. In fused corollas, any extension of the petal beyond its fused part is called the *limb*. The tubelike structure where the petals are united at the bottom of the fused corollas is called the *tube*. The opening at the top of the tube in fused corollas is called the *throat*. In the following different types of corolla shapes, numbers 1 to 6 are actinomorphic, while numbers 7 to 11 are zygomorphic.

1. *Rotate*: wheel-shaped with a short tube and large limb (example: bluets).

2. *Campanulate*: bell-shaped with an extended, flaring tube (example: bellflower).

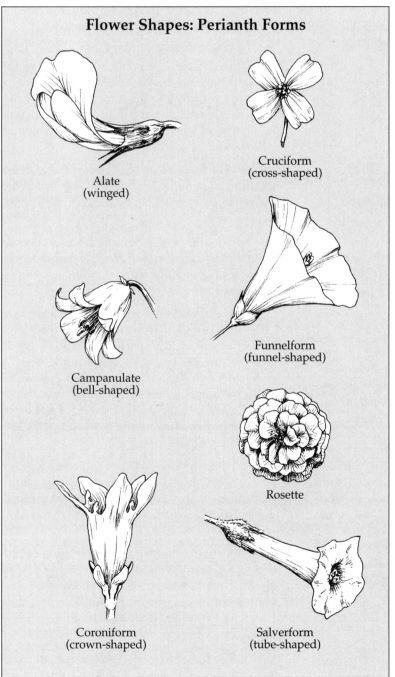

Flower Shapes: Perianth Forms

Alate
(winged)

Cruciform
(cross-shaped)

Campanulate
(bell-shaped)

Funnelform
(funnel-shaped)

Rosette

Coroniform
(crown-shaped)

Salverform
(tube-shaped)

3. *Funnelform*: funnel-shaped with a continuously expanding tube and little flaring (example: bindweeds).

4. *Tubular*: an elongated tube with minimal limb (example: trumpet vine).

5. *Salverform*: an elongated tube with a conspicuous limb, trumpet-shaped (examples: Russian olive, morning glory).

6. *Urceolate*: an inflated tube with a terminal constriction, urn-shaped (example: highbush blueberry).

7. *Bilabiate*: two-lipped, usually because of the presence of a landing platform formed by basal lobes (examples: snapdragon, salvia).

8. *Ligulate*: petals connate at the margins to form a strap-shaped corolla (example: asters).

9. *Galeate*: helmet-shaped (example: pedicularis).

10. *Spurred*: with an extension or spur that often produces nectaries (examples: impatiens, utricularia).

11. *Papilionaceous*: like a butterfly with a central standard petal and lateral wing petal (example: lupines).

Flowers of Monocots and Dicots

Floral variation provides part of the basis for dividing the flowering plants into two major groups: the *dicotyledons* and the *monocotyledons*. The informal name "dicot" is given to plants having two cotyledons (seed leaves) in each seed; "monocot" refers to plants that have one cotyledon in the seed. In monocots, the flower parts occur in threes or multiples of three; for example, three sepals, three petals, six stamens, and a pistil with three carpels. In dicots, flower parts usually occur in fours or fives or multiples of four or five. Although dicots and monocots may have other numbers of floral parts, many other features are unique to each group. Dicots include about 80 percent of all angiosperm species, including many herbaceous plants and all woody, flower-bearing trees and shrubs. Monocots are primarily herbaceous, but they also include some trees, such as palms and Joshua trees.

Types of Inflorescence

Flowers may be solitary, or they may be grouped together in an *inflorescence*, a cluster of flowers. An inflorescence has one main stalk, or *peduncle*. It may also bear numerous smaller stalks called pedicels, each with a flower at its tip. The arrangement of pedicels on a peduncle characterizes different kinds of inflorescences. Some of the common types of inflorescences are as follows:

Spike: The flowers, which are with a very short or with no pedicel, are attached along the elongate and unbranched peduncle of the inflorescence (examples: plantain, spearmint, tamarisk).

Raceme: The flowers are with pedicels of about the same length, which are attached along the elongate and unbranched peduncle of the inflorescence (examples: lily of the valley, snapdragon, mustard, currant). The oldest flowers are at the base of the inflorescence and the youngest at the apex.

Panicle: The flowers are with pedicels, which are attached along the branches arising from the peduncle of the inflorescence (examples: oats, rice, fescue).

Corymb: The flowers are with pedicels of unequal length, which are attached along an unbranched, elongate peduncle, forming a flat-topped inflorescence (examples: hawthorne, apple, dogwood).

Umbel: The flowers are with pedicles, which are all attached at about the same point at the end of the peduncle—this is specifically called a *simple umbel* (examples: onion, geranium, milkweed). A *compound umbel* is formed when the peduncle produces branches that end at approximately the same level, forming a flat top, and the ends of these branches arise from a common point (examples: carrot, dill, parsley).

Head: The flowers do not have pedicels, and they all cluster tightly on the expanded tip of the peduncle (examples: sunflower, daisy, marigold). This type of inflorescence is also referred to as *capitulum*.

Cyme: The flowers with pedicels are located at the ends of the peduncle and lateral branches as well as along the length of the lateral branches. The youngest flowers in any cluster occur farthest from the tip of the peduncle (example: chickweed).

Catkin: The flowers have no pedicels, are unisexual (either staminate or pistillate), and are attached along the length of the peduncle (examples: hazelnut, willow, birch, walnut). The flowers are usually very small and fall as a group. This type of inflorescence is also referred to as *ament*.

Spadix: The flowers have no pedicels and are attached along the length of the thickened or fleshy peduncle, which is enveloped by a conspicuously colored bract called a spathe (example: philodendron, anthurium).

Some types of inflorescences characterize different groups of plants. For example, nearly all members of the carrot family (*Apiaceae*) have compound umbels. All members of the sunflower family (*Asteraceae*) have heads, including chrysanthemums, zinnias, marigolds, and dandelions. All members of the arum family (*Araceae*) have a spadix inflorescence.

Danilo D. Fernando

See also: Angiosperms; Flower structure; Inflorescences; Plant tissues; Pollination; Shoots; Stems.

Sources for Further Study
Mauseth, James D. *Botany: An Introduction to Plant Biology*. 2d ed. Sudbury, Mass.: Jones and Bartlett, 1998. This textbook presents many flower types with pictures, diagrams, and examples.
Raven, Peter H., Ray F. Evert, and Susan E. Eichhorn. *Biology of Plants*. 6th ed. New York: W. H. Freeman/Worth, 1999. This textbook presents various floral types with pictures, diagrams and examples.
Rost, Thomas L., et al. *Plant Biology*. New York: Wadsworth, 1998. This textbook presents many different types of flowers with diagrams, pictures, and examples.

FLOWERING REGULATION

Categories: Angiosperms; physiology; reproduction and life cycles

All flowering is regulated by the integration of environmental cues into an internal sequence of processes. These processes regulate the ability of plant organs to produce and respond to an array of signals. The numerous regulatory switches permit precise control over the time of flowering.

Control over the time of flowering is essential for the survival of flowering plants (angiosperms). Insect pollinators may be present only at certain times; unless an insect-pollinated plant is flowering at that time, pollination and the production of the next generation cannot occur. Embryo and seed development may be successful only under certain climatic conditions. The ability to respond to environmental cues is an essential factor in the regulation of flowering.

While the basic biochemical sequence of events may be common to all angiosperms, the specific regulatory steps vary greatly among species. A floral *promoter* is produced by the leaves and is transported to the shoot apex, which results in the initiation and, ultimately, the production of flowers. To analyze control points in this sequence, it is helpful to focus separately on environmental signals such as temperature and photoperiod and the way organs perceive and respond to these signals.

Chemical Communication
The regulation of flowering requires interactions between the shoot apex and other organs and thus depends heavily on chemical signals. There is strong evidence for the existence of a floral promoter called *florigen*, which may be produced in the leaves. The existence of florigen was first proposed by M. Kh. Chailakhyan, a Soviet plant physiologist, in 1937. Florigen was believed to be produced in leaves, because if leaves were removed before the photoperiod was right for flowering (a process called photoinduction), no flowering occurred. Later work by Anton Lang showed that the

The role of day length in the regulation of flowering had been recognized by 1913. The impact of photoperiod on flowering in numerous species soon became apparent, but some plants, including sunflowers, are day-neutral; they flower independently of photoperiod.

internal signals has been most clearly established for photoperiod. The role of day length in the regulation of flowering had been recognized by 1913. The impact of photoperiod on flowering in numerous species soon became apparent.

In the 1930's W. W. Garner and H. A. Allard found an unusually large tobacco plant growing in a field. The plant stood out because it failed to flower; they named it the Maryland Mammoth. Maryland Mammoth cuttings flowered in a greenhouse that December, and subsequent experimentation demonstrated that flowering would occur only when days were short and nights long. The Maryland Mammoth is an example of a *short-day plant*. Short-day plants generally flower in the spring or fall, when day lengths are shorter. Other examples of short-day plants are poinsettias, cockleburs, Japanese morning glories, and chrysanthemums. Plants such as spinach, lettuce, and henbane will flower only if a critical day length is exceeded; they are categorized as *long-day plants* and generally flower during long summer days.

Photoperiodic control mechanisms may be more complex, as in the case of ivy, a *short-and-long-day plant*, which requires at least a twelve-hour photoperiod followed by a photoperiod of at least sixteen hours. Still other plants, including sunflowers and maize, are *day-neutral*: They flower independent of photoperiod. By the 1940's it was established that night length, not day length, is critical in the photoperiodic control of flowering. For example, flowering in the short-day Japanese morning glory can be prevented by a brief flash of light during the critical long night. In comparing short-day and long-day plants, the distinguishing factor is not the absolute length of night required; rather, the difference is whether that night length provides the minimum (short-day plants) or maximum (long-day plants) period of darkness required to permit flowering.

How a plant perceives night length and translates this into the appropriate response in terms

plant hormone *gibberellin* could induce flowering in certain plants, even without appropriate photoinduction. This prompted Chailakhyan to consider the possibility that florigen was actually composed of two different substances, gibberellin and a new substance he called *anthesin*.

In the late 1970's Lang, Chailakhyan, and I. A. Frolova, working with tobacco plants, discovered that there was also a floral inhibitor they called *antiflorigen*. Later, several genes controlling the production of an inhibitor in pea cotyledons and leaves were identified in other laboratories. In addition to leaf-derived inhibitors, root-derived inhibitors have been shown to regulate flowering in black currant and tobacco plants. Aside from the clear role of gibberellin in flowering, none of the other promoters and inhibitors has been identified. Nutrient levels and allocation throughout the plant may also control the time of flowering.

Photoperiod
One major role of environmental signals is to control the timing of the production of florigen and antiflorigen. This link between environmental and

of flowering is not fully understood. A pigment known as *phytochrome*, however, plays a critical role. Phytochrome exists in two forms (P_r and P_{fr}) that are interconvertible. P_r absorbs red light and is converted to P_{fr}, which absorbs far-red light and is subsequently converted back to the P_r form of the pigment. Sunlight contains both red and far-red light, and thus an equilibrium between the two forms is achieved. At noon, about 60 percent of the phytochrome is in the P_{fr} form. In the dark, some P_{fr} reverts to P_r, and some breaks down. Because of the absence of red light, no new P_{fr} is generated.

The relationship between phytochrome and photoperiodic control of flowering has been established using night-break experiments with red and far-red light. (In these experiments, darkness is interrupted by momentary exposure to light.) Flowering in the Japanese morning glory, a short-day plant, can be inhibited by a flash of red light (as well as light equivalent to sunlight) in the middle of a long night. Far-red light has no effect. A flash of far-red light following a flash of red negates the inhibitory effect of the red light. In long-day plants, flowering can be induced when the dark period exceeds the critical night length with a red-light night break. Far-red light flashes do not result in flowering. The effect of the red flash can be negated by a subsequent far-red flash. In these experiments, the light flashes alter the relative amounts of P_r and P_{fr}.

Circadian Rhythms

How the perception of light by phytochrome is linked to the production of gibberellin and anthesin in long-day and short-day plants is not clear. One idea is that plants measure the amount of P_{fr} present. Flowering in short-day plants would be inhibited by P_{fr}, and these plants would not flower until very little or no P_{fr} remained after a long night. To flower, long-day plants would require some minimum level of P_{fr}, which would not be available if the nights were too long—but this explanation is not viable because P_{fr} vanishes within a few hours after the dark period begins.

Alternately, levels of phytochrome may influence an internal biological clock that keeps track of time. The clock establishes a free-running *circadian rhythm* of about twenty-four hours; this clock needs to be constantly reset to parallel the natural changes in photoperiod as the seasons change. Phytochrome interacts with the clock to synchronize the rhythm with the environment, a prospect that is strength-ened by night-break experiments, where the time of the light flash during the night is critical. In the case of the Japanese morning glory, there are times during the night that a red light flash completely inhibits flowering and other times when it has no effect. In these experiments, the phase of the rhythm of the clock defines the nature of the interaction with phytochrome.

Studies on the relationship between flowering and day length have focused on the production of gibberellin and anthesin. There is evidence that the production of inhibitors by leaves is also under photoperiodic control. This has been demonstrated for photoperiodic tobacco plants and for some peas. In the case of the pea, the inhibitory effect is most obvious for short days, but lower levels of inhibitors continue to be produced as the days grow longer.

Temperature

Plants also use temperature as an environmental clue to ensure flowering. Assessing two environmental factors provides added protection. Some plants have a *vernalization* requirement—a chill that promotes or is essential for flowering. The control point regulated by vernalization may be different for different species. Vernalization has been shown to affect the sensitivity of leaves to respond to photoperiod. In some plants, only leaves initiated after the shoot apex has been chilled will induce flowering after the appropriate photoperiod.

Clearly, the competence of the leaves and shoot apex to respond to environmental and internal signals is crucial to the regulation of flowering. Pea mutants have been identified that have shoot apexes with differential sensitivity to floral induction or inhibition. Another pea mutant has an apex that is not competent to initiate flowers and remains perpetually vegetative. The competence of a day-neutral tobacco apex changes with age. In another species, the apexes of shoots cannot respond to vernalization early in development; this period of time is considered to be the juvenile phase.

A juvenile phase of development is most common in woody perennials. During this time, flowering cannot occur even under optimal environmental conditions. Maturation occurs gradually and may be accompanied by changes in leaf morphology and the ability of cuttings to root. The most significant occurrence is that the plant becomes competent to flower.

Genetic Control of Flowering

The Maryland Mammoth, discussed above, is an example of a short-day plant resulting from a mutation. Mutations that affect flowering time can lead to delayed flowering or to rapid flowering, regardless of photoperiod. Harmful effects of such mutations may include inadequate photosynthetic capability to sustain the crop (when flowering occurs too soon) or susceptibility to pests or cold temperatures at the end of the season (because of flowering too late). A focus of work on genetic modification of plants is the achievement of optimal flowering times.

Susan R. Singer, updated by Bryan Ness

See also: Angiosperm life cycle; Circadian rhythms; Dormancy; Genetics: mutations; Germination and seedling development; Hormones; Leaf abscission; Photoperiodism.

Sources for Further Study

Campbell, Neil A. *Biology*. 5th ed. Menlo Park, Calif.: Benjamin Cummings, 1999. The chapter on control systems in plants provides clear descriptions of circadian rhythms, photoperiodism, phytochrome, and florigen. Figures in this chapter are especially helpful in elucidating the effect of light on flowering.

Key, Joe L., and Lee McIntosh, eds. *Plant Gene Systems and Their Biology*. New York: Alan R. Liss, 1987. Two chapters on flowering may be especially helpful for college-level students. "Perspectives in Flowering Research" by Anton Lang provides insight into research on florigen and antiflorigen as well as molecular approaches to flowering regulation. Daphne Vince-Prue's chapter, "The Induction and Initiation of Flowers," provides more detailed information on the relation between phytochrome and the biological clock. Molecular approaches to studying photoperiodic induction in leaves are also considered.

Raven, Peter H., Ray F. Evert, and Susan E. Eichhorn. *Biology of Plants*. 6th ed. New York: W. H. Freeman/Worth, 1999. A well-known undergraduate botany textbook. The chapter on growth regulation and growth responses covers the topic in detail.

FLUORESCENT STAINING OF CYTOSKELETAL ELEMENTS

Categories: Cellular biology; methods and techniques

Fluorescent antibody staining is a precise technique for marking elements of the cytoskeleton so they become visible under a microscope. The technique has revealed much about the location and functions of the cytoskeleton within a cell. It also offers some of the most visually appealing of the microscopic images available to biologists.

Cells require an internal system of fibers in order to maintain and change their shape. Perhaps the most straightforward function of the cytoskeleton is to provide cell support, a scaffolding that gives each cell a distinctive three-dimensional shape. Without such support every cell would be shaped like a fried egg. The fibers found in the cytoskeleton serve other functions as well: for example, as rails along which substances shuttle from one part of the cell to another. Equally important is the role of the cytoskeleton in cell movement and change in cell shape. The various cytoskeletal fibers can be rearranged, dissolved, and reconstructed at new locations when need be. The cytoskeleton is also a key component of some highly specialized cell structures, such as *cilia* and *flagella*.

Fluorescent antibody staining techniques allow scientists to observe exactly how various cytoskeletal fiber types are oriented within the cell. This technique has greatly increased knowledge of how the cytoskeleton does its jobs. It is also among the most aesthetically beautiful of all the procedures used by cell biologists.

The Cytoskeleton

The cytoskeleton is composed of three different kinds of fibers, each with different, specialized functions. *Microfilaments* are the smallest (6 nanometers in diameter). They are made up of a protein similar to the actin protein in muscle. It is thus not surprising that microfilaments participate in cell movement and shape changes. They usually are found in bundles just inside the surface of the cell, where they are best situated to help the cell change shape.

Intermediate filaments are somewhat larger. Unlike microfilaments, they seem to serve as passive scaffolding elements within the cell. They typically form an interlaced meshwork in the cytoplasm. The proteins that make up these fibers are subtly different in different kinds of cells; the significance of such variation is not clear.

The largest fibers are called *microtubules* (25 nanometers in diameter). They are hollow spirals of protein building blocks. The microtubules have important, specialized roles in particular regions of cells and at certain times in a cell's life. For example, they help move the chromosomes around during division of one cell into two. They are important components of both cilia and flagella—whiplike structures on the surfaces of some cells that serve as oars to help them swim about or to move substances along their surfaces.

Technique

The properties of the cytoskeleton depend upon its precise three-dimensional architecture within the intact cell. Thus, it is important that the researcher use methods designed to analyze the intact cytoskeleton inside whole cells. The most popular method is to fasten a fluorescent dye onto an antibody. The most commonly used dyes are *rhodamine* (for red fluorescence) and *fluorescein* (for greenish-yellow fluorescence). When such a fluorescent antibody is then added to cells, it can be located using a special *fluorescence microscope*.

Although this procedure works reasonably well, it can be improved by amplifying the signal—that is, by devising a method to add more than one molecule of fluorescent dye to each anticytoskeleton antibody molecule, which will increase the amount

of fluorescent light emitted. The brighter light is much easier to see. This can be done using the so-called *secondary-antibody procedure*. A "secondary antibody" must be prepared using the first antibody (against the cytoskeletal protein) as an antigen. This time, the secondary antibody is made fluorescent. The cytoskeletal proteins are then tagged in two steps: First, the primary antibody is added to cells, then the fluorescent secondary antibody is added. Because of the nature of the secondary antibody, numerous molecules of it can adhere to each molecule of primary antibody. In this way, many fluorescent molecules can be attached to each molecule of primary antibody. The result is that the researcher can now see strikingly beautiful images in the fluorescence microscope: brilliantly colored glowing strands of the cytoskeleton, arranged in various patterns depending on what is happening inside the cell.

What Biologists Have Learned

Fluorescent antibody techniques have helped scientists learn that microfilaments and microtubules are dynamic fibers, made up of many kinds of protein (perhaps as many as several hundred), which grow and shrink as necessary. (Intermediate fibers are more stable.) In addition to proteins, whose primary role is to construct the filament itself, there are proteins associated with each fiber type whose role is to make the decisions about when, where, and how fast to assemble or disassemble the filaments. They can be assembled or disassembled like a set of building blocks. The building blocks can be moved from one part of a cell to another quite rapidly if required. Within a few seconds, the distribution of fibers can change dramatically within a living cell. This phenomenon occurs, for example, if a moving cell encounters an obstacle and changes direction. Under certain conditions, microtubule proteins can be added to one end of a microtubule at the same time that they are removed from the opposite end. The result is that the microtubule appears to "move" toward the growing end. This process has been likened to the movement of the tread on a Caterpillar-type tractor.

The first cell function to be definitely attributed to the cell's cytoskeleton was cell division. This process is an elaborate, highly choreographed minuet in which the microtubules move two sets of *chromosomes* (the structures that carry the genetic information) apart from one another and the microfilaments squeeze the cell in two between the chromosome sets. The result is two cells where there was only one, and each has a complete set of genes. How are the activities of the two kinds of fibers coordinated in both time and space, so that this elegant process occurs properly? It is now understood that changing concentrations of calcium atoms inside cells help to coordinate the actions of the several fiber types. Fluorescent antibody staining has revealed that the chromosome-moving microtubules change in length during movement and that they generate a force sufficient to drag the chromosomes through the cell.

Howard L. Hosick

See also: Cytoskeleton; Flagella and cilia.

Sources for Further Study

Becker, Wayne M. *The World of the Cell*. 4th ed. San Francisco: Benjamin/Cummings, 2000. This is a textbook for a beginning-level college course in cell biology. Because it is written in an unusually straightforward manner, however, an interested nonscientist should be able to learn from it about all aspects of cell biology. The use of fluorescent antibodies is described in chapter 10, "Membranes: Their Structure and Chemistry." The diagrams are particularly good. Includes references.

Bershadsky, Alexander D., and Juri M. Vasiliev. *Cytoskeleton*. New York: Plenum Press, 1988. A comprehensive summary of the cytoskeleton, providing the reader with some insight into the complexity of the cytoskeleton. Emphasis is on concepts rather than on details of experiments; thus, the main themes should be accessible to those with a solid background in high school or college chemistry and biology. Includes many excellent photographs and helpful drawings.

Carraway, K. L., and C. A. C. Carraway, eds. *The Cytoskeleton: A Practical Approach*. New York: IRL Press, 1992. Covers a wide range of methods for investigating the eukaryotic cytoskeletons, focusing on how the microfilaments, intermediate filaments, and microtubules interact with their associated proteins. Oriented toward practical applications of the methodologies.

Starr, Cecie, and Ralph Taggart. *Biology: The Unity and Diversity Of Life*. 9th ed. Pacific Grove, Calif.: Brooks/Cole, 2001. A biology text designed for college-level freshmen. Explains the structure of cells and how they work. Chapter 30, on immunity, is a straightforward, basic introduction to how the immune system is constructed and how antibodies are produced in response to antigens. The section on the cytoskeleton in chapter 5 has several good immunofluorescence photographs, drawings of the structure of each fiber type, photographs of cilia, and other useful information.

FOOD CHAIN

Categories: Ecology; food; nutrients and nutrition

The food chain concept allows ecologists to interconnect the organisms living in an ecosystem and to trace mathematically the flow of energy from plants through animals to decomposers.

The food chain concept provides the basic framework for production biology and has major implications for agriculture, wildlife biology, and calculating the maximum amount of life that can be supported on the earth. As early as 1789, naturalists such as Gilbert White described the many sequences of animals eating plants and themselves being eaten by other animals. However, the use of the term "food chain" dates from 1927, when Charles Elton described the implications of the food chain and food web concept in a clear manner. His solid exposition advanced the study of two important biological concepts: the complex organization and interrelatedness of nature, and energy flow through ecosystems.

Food Chains in Ecosystem Description

Stephen Alfred Forbes, founder of the Illinois Natural History Survey, contended in 1887 that a lake comprises a system in which no organism or process can be understood unless its relationship to all the parts is understood. Forty years later, Elton's food chains provided an accurate way to diagram these relationships. Because most organisms feed on several food items, food chains were cross-linked into complex webs with predictive power. For instance, algae in a lake might support an insect that in turn was food for bluegill. If unfavorable conditions eliminated this algae, the insect might also disappear. However, the bluegill, which fed on a wider range of insects, would survive because the loss of this algae merely increases the pressure on the other food sources. This detailed linkage of food chains advanced agriculture and wildlife management and gave scientists a solid overview of living systems. When Arthur George Tansley penned the term *ecosystem* in the 1930's, it was food-chain relationships that described much of the equilibrium of the ecosystem.

Today most people still think of food chains as the basis for the "balance of nature." This phrase dates from the controversial 1960 work of Nelson G. Hairston, Frederick E. Smith, and Lawrence B. Slobodkin. They proposed that if only grazers and plants are present, grazing limits the plants. With predators present, however, grazers are limited by predation, and the plants are free to grow to the limits of the nutrients available. Such explanations of the "balance of nature" were commonly taught in biology books throughout the 1960's and 1970's.

Food Chains in Production Biology

Elton's explanation of food chains came just one year after Nelson Transeau of Ohio State University presented his calculations on the efficiency with

Unlike calories, which are reduced at each step in a food chain, some toxic substances become more concentrated as the molecules are passed along. The pesticide DDT provides the most notorious example of biological magnification: DDT was found to be deposited in animal body fat in ever-increasing concentrations as it moved up the food chain to species such as pelicans. High levels of DDT in these birds broke down steroid hormones and interfered with eggshell formation.

which corn plants converted sunlight into plant tissue. Ecologists traced this flow of stored chemical energy up the food chain to herbivores that ate plants and on to *carnivores* that ate *herbivores*. Food chains therefore undergirded the new "production biology" that placed all organisms at various *trophic levels* and calculated the extent to which energy was lost or preserved as it passed up the food chain.

With data accumulating from many ecologists, Elton extended food chains into a pyramid of numbers. The *food pyramid*, in which much plant tissue supports some herbivores that are in turn eaten by fewer carnivores, is still referred to as an *Eltonian pyramid*. In 1939 August Thienemann added *decomposers* to reduce unconsumed tissues and return the nutrients of all levels back to the plants. Early pyramids were based on the amount of living tissues, or *biomass*.

Calculations based on the amount of chemical energy at each level, as measured by the heat released when food is burned (calories), provided even more accurate budgets. Because so much energy is lost at each stage in a food chain, it became obvious that this inefficiency was the reason food chains are rarely more than five or six links long and why large, fierce animals are uncommon. It also became evident that because the earth intercepts a limited amount of sunlight energy per year, there is a limit on the amount of plant life—and ultimately upon the amount of animal life and decomposers—that can be fed. Food chains are also important in the accounting of carbon, nitrogen, and water cycling.

Value of Food Chains in Environmental Science

Unlike calories, which are dramatically reduced at each step in a food chain, some toxic substances become more concentrated as the molecules are passed along. The concentration of molecules along the food chain was first noticed by the Atomic Energy Commission, which found that radioactive iodine and strontium released in the Columbia River were concentrated in tissue of birds and fish. However, the pesticide DDT provided the most notorious example of biological magnification: DDT was found to be deposited in animal body fat in ever-increasing concentrations as it moved up the food chain to ospreys, pelicans, and peregrine falcons. High levels of DDT in these birds broke down steroid hormones and interfered with eggshell formation.

Because humans are *omnivores*, able to feed at several levels on the food chain (that is, both plants and other animals), it has been suggested that a higher world population could be supported by humans moving down the food chain and becoming vegetarians. A problem with this argument is that much grazing land worldwide is unfit for cultivation, and therefore the complete cessation of pig or cattle farming does not necessarily free up substantial land to grow crops.

While the food chain and food web concepts are convenient theoretical ways to summarize feeding interactions among organisms, real field situations have proved far more complex and difficult to measure. Animals often switch diet between larval and adult stages, and they are often able to shift food sources widely. It is often difficult to draw the boundaries of food chains and food webs.

John Richard Schrock

See also: Animal-plant interactions; Biomass related to energy; Community-ecosystem interactions; Community structure and stability; Ecology: concept; Ecosystems: overview; Ecosystems: studies; Trophic levels and ecological niches.

Sources for Further Study

Colinvaux, Paul A. *Why Big Fierce Animals Are Rare*. New York: Penguin Books, 1990. Colinvaux summarizes the work of Norman Transeau in the chapter "The Efficiency of Life."

Elton, Charles. *Animal Ecology*. New York: October House, 1966. Elton first used the term "food chain" in this work. Food chains were a key concept in the development of ecology.

Golley, Frank B. *A History of the Ecosystem Concept in Ecology*. New Haven, Conn.: Yale University Press, 1993. The basic early work of Gilbert White, Stephen Alfred Forbes, and Charles Elton is clearly described.

"Something's Fishy." *Science News* 146, no. 1 (July 2, 1994): 8. Researchers are building evidence in a case linking pollution to viral epidemics that are devastating animals at the top of the marine food chain. Toxic organochlorines may affect vulnerability to disease in animals.

Svitil, Kathy A. "Collapse of a Food Chain." *Discover* 16, no. 7 (July, 1995): 36. Describes how warmer water off California's coast is producing fewer plankton, and the ecosystem's entire food chain is suffering as a result. Global warming could be responsible, but so could natural factors like the El Niño cycle.

Transeau, Norman. "Presidential Address to the Ohio Academy of Sciences." *The Ohio Journal of Science* 26 (1926). Transeau calculates the efficiency of converting sunlight into corn.

FOREST AND RANGE POLICY

Categories: Economic botany and plant uses; ecosystems; forests and forestry

Forest and range policies are laws protecting forests and rangelands. Such policies usually seek to sustain and protect biodiversity while setting guidelines for the sustainable use of natural resources.

Many national governments have established forest and range policies. *Rangeland*, land that supplies forage for grazing and browsing animals, covers almost one-half of the ice-free land on earth. More than three billion cattle, sheep, goats, camels, buffalo, and other domestic animals graze on rangelands. These animals are important in converting forages into milk and meat, which provide nourishment for people around the world. Forests cover almost 30 percent of the earth and provide humans with lumber, fuel woods, food products, latex rubber, and valuable chemicals that constitute prescription and nonprescription drugs. Rangelands and forests also function as important ecosystems that help provide food and shelter for wildlife, control erosion, and purify the atmosphere. Forests and rangelands have been undergoing destruction and degradation at alarming rates at the hands of humans.

Protecting Forests and Rangelands

The nearly 15 billion acres of forest that originally existed on the earth have been reduced to approximately 11 billion acres by human conversion of land to cropland, pastureland, cities, and nonproductive land. Forests, if properly maintained or left alone, are the most productive and self-sustaining ecosystems that land can support. Tropical rain forests are the natural habitat for at least 50 percent, and possibly up to 90 percent, of the species on earth. In the late 1990's, Harvard biologist Edward O. Wilson stated that 25 percent of the earth's species could become extinct by the year 2050 if the current rate of tropical forest destruction were not stopped.

Many national governments have established policies for protecting forest habitats and the biological diversity found within them. National parks and reserves provide protection for both forests and rangelands. Some countries have laws that prohibit clearing, burning, or logging of particular forests.

China, which suffered from erosion and floods as a result of centuries of deforestation, began an impressive *reforestation* campaign during the 1990's, planting almost 11 million acres of new trees. Korea attained 70 percent reforestation after losing almost all its forested land in a civil war during the 1950's. Japan has enacted strict environmental laws, which have allowed it to reforest 68 percent of its land area. Japan has relied upon imported timbers in order to allow its new forest projects to flourish. Even with such worldwide success in reforestation, it is estimated that protection and sustainable management of forests and rangelands still need to be increased by a factor of three if forests are to be saved.

Multiple Use

Protecting forestland involves an interdisciplinary approach. In the United States, 191 million acres of forestland are managed by the U.S. Forest Service. The Forest and Rangeland Renewable Resource Planning Act (RPA) of 1974 and the National Forest Management Act (NFMA) of 1976 mandated management plans for forests and rangelands to ensure that resources would be available on a sustained basis. Management policies must sustain and protect biodiversity; old-growth forests; riparian areas; threatened, endangered, and sensitive species; rangeland; water and air quality; access to forests; and wildlife habitat.

The Forest Service provides inexpensive grazing lands for more than three million cattle and sheep every year, supports multimillion-dollar mining operations, maintains a network of roads eight times longer than the U.S. interstate highway system, and allows access to almost one-half of all national forest land for commercial logging. The Forest Service is responsible for producing plans for the multiple use of national lands.

Sustainability policies require that the net productive capacity of the forest or rangeland does not decrease with multiple use. This involves making sure that soil productivity is maintained by keeping erosion, compaction, or displacement by mining or logging equipment or other motorized vehicles within tolerable limits. It further requires that a large percentage of the forest remains undeveloped so that soils and habitats, as well as tree cover, will remain undisturbed and in their natural state.

The RPA and NFMA, along with the Endangered Species Act (ESA) of 1973, mandate policies that encourage the proliferation of species native to and currently living in the forest. Natural ecosystem processes are followed to ensure their survival. Even though forests and rangelands are required to be multiple-use areas, policy maintains that there can be no adverse impact to threatened, endangered, or sensitive species. Species habitats within the forest are to remain well distributed and free of barriers that can cause fragmentation of animal populations and ultimately species loss. If a forest contains fragmented areas created by human activity, corridors that connect the forest patches are constructed. In this way species are not isolated from one another, and viable populations can exist.

The Forest Service creates artificial habitats to encourage the survival of species in cases of natural disaster. When Hurricane Hugo devastated the Francis Marion National Forest in South Carolina in 1989, winds snapped 90 percent of the trees with active woodpecker cavities in some areas of the forest. The habitat destruction caused 70 percent of the red-cockaded woodpecker population to disappear. The Forest Service and university researchers created nesting and roosting cavities to save the woodpeckers. Within a four-year period, the population had dramatically recovered.

AP/WIDE WORLD PHOTOS

Cattle are herded into Cooper's Meadow in the mountains above Pinecrest, California. The issue of grazing rights on public lands in the western United States is a controversial one.

Timber, Oil, and Mineral Leasing

Logging activities in forests are covered by the Resource Planning Act of 1990. Forested land must be evaluated for its ability to produce commercially usable timber without negative environmental impact. There must be reasonable assurance that stands managed for timber production can be adequately restocked within five years of the final harvest. Further, no irreversible resource damage is allowed to occur. Policy further requires use of the *silviculture* practices that are best suited to the land management objectives of the area. Cutting practices are then monitored. The 1990's were characterized by a trend toward restricting logging methods in order to protect habitats and preserve older stands of trees. In the 1993 Renewable Resource Assessment update, the Forest Service found that timber mortality, at 24.3 percent, was still interfering with biological diversity. Some forested areas were withdrawn from timber production because of their fragility.

Multiple use under the NFMA allowed forests to be available for oil and gas leasing. Certain lands were exempted from mineral exploration by acts of Congress or executive authority. However, the search for and production of mineral and energy sources remained under the jurisdiction of the Forest Service, which was charged to provide access to national forests for mineral resources activities. The Federal On-Shore Oil and Gas Leasing Reform Act of 1987 gave the Forest Service more authority in making lease decisions.

Pest and Weed Control

Pesticides are sometimes used during attempts to ensure the health of forestland. Policy in the United States requires the use of safe pesticides and encourages the development of an *integrated pest management* (IPM) plan. Any decision to use a particular pesticide must be based on an analysis of its effectiveness, specificity, environmental impact, economic efficiency, and effects on humans. The application and use of pesticides must be coordinated with federal and state fish and wildlife management agencies. Pesticides can be applied only to areas that are designated as wilderness when their use is necessary to protect or restore resources in the area. Other methods of controlling disease include removing diseased trees and vegetation from the forest, cutting infected areas from plants and removing the debris, treating trees with antibiotics,

and developing disease-resistant plant varieties.

Forest Service policy on integrated pest management was revised in 1995 to emphasize the importance of integrating noxious weed management into the forest plan for ecosystem analysis and assessment. Noxious weed management must be coordinated in cooperation with state and local government agencies as well as private landowners. Noxious weeds include invasive, aggressive, or harmful nonindigenous or exotic plant species. They are generally opportunistic, poisonous, toxic, parasitic, or carriers of insects or disease. The Forest Service is responsible for the prevention, control, and eradication of noxious weeds in national forests and grasslands.

In North Dakota, one strategy for promoting weed-free forests uses goats to help control leafy spurge. The goats graze on designated spurge patches during the day and return to portable corrals during the night. A five-year study found that the goats effectively reduced stem densities of spurge patches to the extent that native livestock forage plants were able to reestablish themselves.

A strategy that has been implemented in Wyoming, Colorado, Idaho, Utah, and Montana requires pack animals on national forest land to eat state-certified weed-free forage. Another strategy involves the use of certified weed-free straw and gravel in construction and rehabilitation efforts within national forests. *Biocontrols*, *herbicides*, and *controlled burning* are also commonly used during IPM operations in forests.

Other Protection Issues

Natural watercourses and their banks are referred to as *riparian areas*. The plant communities that grow in these areas often serve as habitats for a large variety of animals and birds and also provide shade, bank stability, and filtration of pollution sources. It is therefore important that these areas remain in good ecological condition. Riparian areas and streams are managed according to legal policies for *wetlands*, floodplains, water quality, endangered species, and wild and scenic rivers.

Dirt roads in national forests are often closed when road sediment pollutes riparian areas and harms fish populations. Forest and rangeland roads are also closed to prevent disruption of breeding or nesting colonies. Seemingly harmless human endeavors—such as seeking mushrooms, picking berries, or hiking in the forest—can cause problems for calving elk and nesting eagles. Therefore, the amount of open roads in the forest is being reduced in order to preserve habitat and return land to a more natural state.

Fire management is important to healthy forests. In many cases fires are prevented or suppressed, but prescribed fires are used to protect and maintain ecosystem characteristics. Some conifers, such as the giant sequoia and the jack pine, will release their seeds for germination only after being exposed to intense heat. Lodgepole pines will not release their seeds until they have been scorched by fire. Ecosystems that depend on the recurrence of fire for regeneration and balance are called *fire climax ecosystems*. Prescribed fires are used as a management tool in these areas, which include some *grasslands* and pine habitats.

In 1964 the U.S. Congress passed the Wilderness Act, which mandates that certain federal lands be designated as wilderness areas. These lands must remain in their natural condition, provide solitude or primitive types of recreation, and be at least 5,000 acres in area. They usually contain ecological or geological systems of scenic, scientific, or historical value. No roads, motorized vehicles, or structures are allowed in these areas. Furthermore, no commercial activities are allowed in wilderness areas except livestock grazing and limited mining endeavors that began before the area received wilderness designation.

Grazing Practices and Problems

Approximately 42 percent of the world's rangeland is used for grazing livestock; much of the rest is too dry, cold, or remote to serve such purposes. It is common for these rangelands to be converted into croplands or urban developments. The rate of loss for grazing lands worldwide is three times that of tropical forests, and the area lost is six times that of tropical forests. There are more threatened plant species in North American rangelands than any other major biome.

Rangeland grasses are known for their deep, complex root systems, which makes the grasses hard to uproot. When the tip of the leaf is eaten, the plant quickly regrows. Each leaf of grass on the rangeland grows from its base, and the lower half of the plant must remain for the plant to thrive and survive. As long as only the top half of the grass is eaten, grasses serve as renewable resources that can provide many years of grazing. Each type of grassland is evaluated based on grass species, soil type, growing season, range condition, past use, and climatic conditions. These conditions determine the herbivore *carrying capacity*, or the maximum number of grazing animals a rangeland can sustain and remain renewable.

Overgrazing occurs when herbivore numbers exceed the land's carrying capacity. Grazing animals tend to eat their favorite grasses first and leave the tougher, less palatable plants. If animals are allowed to do this, the vegetation begins to grow in patches, allowing cacti and woody bushes to move into vacant areas. As native plants disappear from the range, weeds begin to grow. As the nutritional level of the forage declines, hungry animals pull the grasses out by their roots, leaving the ground bare and susceptible to damage from hooves. This process initiates the desertification cycle. With no vegetation present, rain quickly drains off the land and does not replenish the groundwater. This makes the soil vulnerable to erosion. Almost one-third of rangeland in the world is degraded by overgrazing. Among the countries suffering severe range degradation are Pakistan, Sudan, Zambia, Somalia, Iraq, and Bolivia.

The United States has approximately 788 million acres of rangeland. This represents almost 34 percent of the land area in the nation. More than one-half of the rangeland is privately owned, while approximately 43 percent is publicly owned and managed by the Forest Service and the Bureau of Land Management (BLM). State and local governments manage the remaining 5 percent. Efforts to preserve rangelands include close monitoring of carrying capacity and removal of substandard ranges from the grazing cycle until they recover. New grazing practices, such as cattle and sheep rotation, help to preserve the renewable quality of rangelands. Grazing is managed with consideration to season, moisture, and plant growth conditions. Noxious weed encroachment is controlled, and native forages and grasses are allowed to grow.

Most rangelands in the United States are short-grass prairies located in the western part of the nation. These lands are further characterized by thin soils and low annual precipitation. They undergo numerous environmental stresses. Woody shrubs, such as mesquite and prickly cactus, often invade and take over these rangelands as overgrazing or other degradation occurs. Such areas are especially susceptible to desertification. Recreational vehicles, such as motorcycles, dune buggies, and four-wheel-drive trucks, can damage the vegetation on ranges. According to the 1993 Renewable Resource Assessment update, many of the rangelands in the United States were in unsatisfactory condition.

Steps to restore healthy rangelands include restoring and maintaining riparian areas and priority watersheds. These areas are monitored on a regular basis, and adjustments are made if their health is jeopardized by sediment from road use or degradation of important habitats caused by human activity. The Natural Resources Conservation program is teaching private landowners how to burn un-wanted woody plants on rangelands, reseed with perennial grasses that help hold water in the soil, and rotate grazing of cattle and sheep on rangelands so that the land is able to recover and thrive. Such methods have proven to be successful.

Toby R. Stewart and Dion Stewart

See also: Biological invasions; Deforestation; Desertification; Erosion and erosion control; Forest fires; Forest management; Grasslands; Grazing and overgrazing; Integrated pest management; Logging and clear-cutting; Sustainable agriculture; Sustainable forestry.

Sources for Further Study

Davis, Lawrence S. *Forest Management: To Sustain Ecological, Economic, and Social Values*. 4th ed. Boston: McGraw-Hill, 2001. Thorough coverage of forest management, including issues, theories, and techniques in forest management and harvest scheduling.

McNeely, Jeffrey A. *Conserving the World's Biological Diversity*. Washington, D.C.: International Union for Conservation of Nature and Natural Resources, World Resources Institute, Conservation International, World Wildlife Fund-U.S., and the World Bank, 1990. Examines deforestation around the world.

Robbins, William G. *American Forestry: A History of National, State, and Private Cooperation*. Lincoln: University of Nebraska Press, 1985. A good history of the National Forest Service and its cooperation with state and private landowners.

Stalcup, Brenda, ed. *Endangered Species: Opposing Viewpoints*. San Diego: Greenhaven Press, 1996. Contains interesting articles on forest management priorities as well as other biodiversity issues.

FOREST FIRES

Categories: Ecology; ecosystems; environmental issues; forests and forestry

Whether natural or caused by humans, fires destroy life and property in forestlands but are also vital to the health of forests.

Evidence of forest fires is routinely found in soil samples and tree borings. The first major North American fires in the historical record were the Miramichi and Piscataquis fires of 1825. Together, they burned 3 million acres in Maine and New Brunswick. Other U.S. fires of significance were the Peshtigo fire in 1871, which raged over 1.28 million acres and took fourteen hundred human lives in Wisconsin; the fire that devastated northern Idaho and northwestern Montana in 1910 and killed at least seventy-nine firefighters; and a series of fires that joined forces to sweep across one-third of Yellowstone National Park in 1988.

Fire Behavior

Fires need heat, fuel, and oxygen. They spread horizontally by igniting particles at their edge. At first, flames burn at one point, then move outward, accumulating enough heat to keep burning on their own. Topography and weather affect fire behavior. Fires go uphill faster than downhill because warm air rises and preheats the uphill fuels. Vegetation on south- and west-facing slopes receives maximal sunlight and so is drier and burns more easily. Heat is pulled up steep, narrow canyons, as it is up a chimney, increasing heat intensity. For several reasons, only one-third of the vegetation within a large fire usually burns. This mosaic effect may be caused by varied tree species that burn differently, old burns that stop fire, strong winds that blow the fire to the leeward side of trees, and varied fuel moisture.

Forest Management

One of the early criteria of forest management was fire protection. In the second quarter of the twentieth century, lookout towers, firebreaks, and trails were built to locate fires as quickly as possible. Low fires that otherwise would have burned through the forest at ground level and cleared out brush every five to twenty-five years were suppressed. As a result, the natural cycle of frequent fires moving through an area was broken. Fallen trees, needles, cones, and other debris collected as kindling on the forest floor, rather than being incinerated every few years.

It took foresters and ecologists fifty years to realize that too much fire suppression was as bad as too little. Infrequent fires cause accumulated kindling to burn hot and fast and explode into treetops. The result is a devastating *crown fire*, a large fire that advances as a single front. Burning embers of seed cones and sparks borne by hot, strong winds created within the fire are tossed into unburned areas to start more fires.

In the 1970's *prescribed burning* was added to forest management techniques used to keep forests healthy. Fires set by lightning are allowed to burn when the weather is cool, the area isolated, and the risk of the fire exploding into a major fire low. More than 70 percent of prescribed burning takes place in the southeastern states, where natural fires burn through an area more frequently than in the West.

Causes of Fires

Forest fires may be caused by natural events or human activity. Most natural fires are started by lightning strikes. Dozens of strikes can be recorded from one lightning storm. When a strike seems likely, fire spotters watch for columns of smoke, and small spotter planes will fly over the area, looking for smoke. Many of the small fires simply smoulder and go out, but if the forest is dry, multiple fires can erupt from a single lightning storm.

The majority of forest fires are human-caused, and most are the result of carelessness rather than arson. Careless campers may leave a campsite without squelching their campfire completely, and winds may then whip the glowing embers into flames. A smoker may toss a cigarette butt from a car window. Sparks from a flat tire riding on the rim may set fire to vegetation alongside the highway. The sun shining through a piece of broken glass left by litterers may ignite dry leaves.

In some areas, prescribed fires are set in an attempt to re-create the natural sequence of fire. In Florida, prescribed burns provide wildlife habitat by opening up groves to encourage healthy growth. Other fires start accidentally but are allowed to burn until they reach a predetermined size.

Benefits of Fire

Some plant species require very high temperatures for their seed casings to split for germination. After fire periodically sweeps through the forest, seeds will germinate. Other species, such as the fire-resistant ponderosa pine, require a shallow layer of decaying vegetable matter in which to root. Fires burn excess debris and small trees of competing species and leave an open environment suitable for germination. Dead material on the forest floor is processed into nutrients more quickly by fire than by decay, and in a layer of rich soil, plants will sprout within days to replace those destroyed in the fire.

Fire's Disadvantages

Erosion is one of the devastating effects of a fire. If the fuels burn hot, tree oils and resins can be baked into the soil, creating a hard shell that will not absorb water. When it rains, the water runoff gathers mud and debris, creating flash floods and extreme stream *sedimentation*. Culverts and storm drains fill with silt, and streams flood and change course. Fish habitat is destroyed, vegetation sheltering stream banks is ripped away, and property many miles downstream from the forest is affected.

When a fire passes through timber it generally leaves pockets of green, although weakened, stands. Forest pests, such as the bark beetle, are attracted to the burned trees and soon move to the surviving trees, weakening them further. Healthy trees outside the burn area may also fall to pest infestation unless the burned trees are salvaged before pests can take hold. The ash and smoke from hot, fast-burning forest fires can be transported for

A firefighter works to contain a forest fire near Tehuacán in the state of Puebla, Mexico.

miles, affecting air quality many miles from the actual fire.

Relationship to Timber Resources

Although a prescribed fire is an attempt to duplicate natural fire, it is not as efficient, because private and commercial property within the fire path must be protected. Once a fire has occurred, burned timber deteriorates quickly, either through insect infestation or blueing—a mold that stains the wood. Private landowners can move quickly to salvage fire-damaged trees and plant new seedlings to harness erosion. On federal land, regulations governing the salvage of trees can delay logging of the burned snags until deterioration makes it uneconomical to harvest them.

Fire Fighting

In fire fighting, bulldozers are used to cut fire lines ahead of the approaching fire, and fuels between fire lines and the fire are backburned. Heli-copters and tanker planes drop water with a fire-retardant additive, or bentonite, a clay, at the head of the fire to smother fuels. Firefighters are equipped with fire shelters in the form of aluminized pup tents, which they can pull over themselves if a fire outruns them. Despite technological advances, one of the best tools for fighting fires—along with the shovel—remains the pulaski, a combination ax and hoe, first produced commercially in 1920. This tool, in the hands of on-the-ground firefighters, is used to cut fire breaks and to throw dirt on smoldering debris.

Public Policy and Public Awareness

Since the early twentieth century, forest fires have engendered public policy in the United States. In the aftermath of major fires in 1903 and 1908 in Maine and New York, state fire organizations and private timber protective associations were formed to provide fire protection. These, in turn, contributed to the Weeks Act of 1911, which permitted cooperative fire protection between federal and state governments.

People who make their homes in woodland settings in or near forests face the danger of forest fire, and government agencies provide information to help people safeguard themselves and their property. Homes near forests should be designed and landscaped with fire safety in mind, using fire-resistant or noncombustible materials on the roof and exterior. Landscaping should include a clear safety zone around the house. Hardwood trees, less flammable than conifers, and other fire-resistant vegetation should be planted.

J. A. Cooper

See also: Forests; Forest management; Sustainable forestry; Timber industry; Wood and timber.

Sources for Further Study

Fuller, Margaret. *Forest Fires: An Introduction to Wildland Fire Behavior, Management, Firefighting, and Prevention.* New York: Wiley, 1991. An extensive bibliography is included.

Johnson, Edward A. *Fire and Vegetation Dynamics: Studies from the North American Boreal Forest.* New York: Cambridge University Press, 1992. It is almost dogma that the boreal forest in North America is a fire-dependent forest, yet ecologists often do not consider in any technical detail how forest fires affect individual plants and plant populations. This text sets out to correct this deficiency by assembling the relevant studies.

Kasischke, Eric S., and Brian J. Stocks, eds. *Fire, Climate Change, and Carbon Cycling in the Boreal Forest.* New York: Springer, 2000. Discusses the mechanisms by which fire and climate interact to influence carbon cycling in North American boreal forests. Summarizes the information needed to understand and manage fire's effects on the ecology of boreal forests and its influence on global climate change issues. Following chapters detail the role of fire in the ecology of boreal forests.

Pringle, Laurence P. *Fire in the Forest: A Cycle of Growth and Renewal.* New York: Simon and Schuster, 1995. Reveals the role of fire in the growth and maintenance of a forest. An introduction to this type of organic recycling explains how fire provides new food sources for wildlife while clearing the way for new generations of trees.

Pyne, Stephen J. *Fire in America: A Cultural History of Wildland and Rural Fire.* 2d ed. Seattle: University of Washington Press, 1997. Pyne, who spent fifteen seasons on a fire crew, gives an environmental history.

Whelan, Robert J. *The Ecology of Fire.* New York: Cambridge University Press, 1995. Examines the changes wrought by fires with reference to general ecological theory. Aimed at senior undergraduate students, researchers, foresters, and other land managers.

FOREST MANAGEMENT

Categories: Disciplines; economic botany and plant uses; forests and forestry

Forest management includes reforestation programs as well as techniques to manage logging practices, provide grazing lands, support mining operations, maintain infrastructure networks, or slow the destruction of rain forests.

Forests provide lumber for buildings, wood fuel for cooking and heating, and raw materials for making paper, latex rubber, resin, dyes, and essential oils. Forests are also home to millions of plants and animal species and are vital in regulating climate, purifying the air, and controlling water runoff. A 1993 global assessment by the United Nations Food and Agriculture Organization (FAO) found that three-fourths of the forests in the world still have some tree cover, but less than one-half of these have intact forest ecosystems. *Deforestation* is occurring at an alarming rate, and management practices are being sought to try to halt this destruction.

Thousands of years ago, forests and woodlands covered almost 15 billion acres of the earth. Approximately 16 percent of the forests have been cleared and converted to pasture, agricultural land, cities, and nonproductive land. The remaining 11.4 billion acres of forests cover about 30 percent of the earth's land surface. Clearing forests has severe environmental consequences. It reduces the overall productivity of the land, and nutrients and biomass stored in trees and leaf litter are lost. Soil once covered with plants, leaves, and snags becomes prone to erosion and drying. When forests are cleared, habitats are destroyed and biodiversity is greatly diminished. Destruction of forests causes water to drain off the land instead of being released into the atmosphere by transpiration or percolation into groundwater. This can cause major changes in the hydrologic cycle and ultimately in the earth's climate. Because forests remove a large amount of carbon dioxide from the air, the clearing of forests causes more carbon dioxide to remain, thus upsetting the delicate balance of atmospheric gases.

Rain Forests

Rain forests provide habitats for at least 50 percent (some estimates are as high as 90 percent) of the total stock of plant, insect, and other animal species on earth. They supply one-half of the world's annual harvest of hardwood and hundreds of food products, such as chocolate, spices, nuts, coffee, and tropical fruits. Tropical rain forests also provide the main ingredients in 25 percent of prescription and nonprescription drugs, as well as 75 percent of the three thousand plants identified as containing chemicals that fight cancer. Industrial materials, such as natural latex rubber, resins, dyes, and essential oils, are also harvested from tropical forests.

Tropical forests in Asia, Africa, and Latin America are rapidly being cleared to produce pastureland for large cattle ranches, establish logging operations, construct large plantations, grow narcotic plants, develop mining operations, or build dams to provide power for mining and smelting operations. In 1985 the FAO's Committee on Forest Development in the Tropics developed the Tropical Forestry Action Plan to combat these practices, develop sustainable forest methods, and protect precious ecosystems. Fifty nations in Asia, Africa,

and Latin America have adopted the plan.

Several management techniques have been successfully applied to slow the destruction of tropical forests. *Sustainable logging* practices and *reforestation* programs have been established on lands that allow timber cutting, with complete bans of logging on virgin lands. Certain regions have set up extractive reserves to protect land for the native people who live in the forest and gather latex rubber and nuts from mature trees. Sections of some tropical forests have been preserved as *national reserves*, which attract tourists while preserving trees and biodiversity.

Developing countries have been encouraged to protect their tropical forests by using a combination of *debt-for-nature swaps* and *conservation easements*. In debt-for-nature swaps, tropical countries act as custodians of the tropical forest in exchange for foreign aid or relief from debt. Conservation easement involves having another country, private organization, or consortium of countries compensate a tropical country for protecting a specific habitat.

Another management technique involves putting large areas of the forest under the control of indigenous people who use *slash-and-burn agriculture* (also known as swidden or milpa agriculture). This traditional, productive form of agriculture follows a multiple-year cycle. Each year farmers clear a forest plot of several acres in size to allow the sun to penetrate to the ground. Leaf litter, branches, and fallen trunks are burned, leaving a rich layer of ashes. Fast-growing crops, such as bananas and papayas, are planted and provide shade for root crops, which are planted to anchor the soil. Finally, crops such as corn and rice are planted. Crops mature in a staggered sequence, thus providing a continuous supply of food. Use of mixed perennial *polyculture* helps prevent insect infestations, which can destroy *monoculture* crops. After one or two years, the forest begins to take over the agricultural plot. The farmers continue to pick the perennial crops but essentially allow the forest to reclaim the plot for the next ten to fifteen years before clearing and planting the area again. Slash-and-burn agriculture can, however, post hazards: A drought in Southeast Asia in 1997 caused fires to burn for months when monsoon rains did not materialize, polluting the air and threatening the health of millions of Indonesians. In 1998, previous abuse of the technique resulted in flooding and mudslides in Honduras after the onset of Hurricane Mitch.

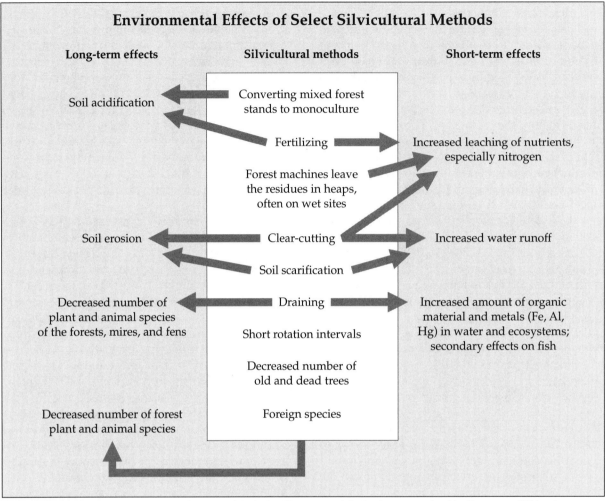

Environmental Effects of Select Silvicultural Methods

Source: Adapted from I. Stjernquist, "Modern Wood Fuels," in *Bioenergy and the Environment*, edited by Pasztor and Kristoferson, 1990.

U.S. Forest Management

Forests cover approximately one-third of the land area of the continental United States and comprise 10 percent of the forests in the world. Only about 22 percent of the commercial forest area in the United States lies within national forests. The rest is primarily managed by private companies that grow trees for commercial logging. Land managed by the U.S. Forest Service provides inexpensive grazing lands for more than three million cattle and sheep every year, supports multimillion-dollar mining operations, and consists of a network of roads eight times longer than the U.S. interstate highway system. Almost 50 percent of national forest land is open for commercial logging. Nearly 14 percent of the timber harvested in the United States each year comes from national forest lands. Total wood production in the United States has caused the loss of more than 95 percent of the old-growth forests in the lower forty-eight states. This loss includes not only high-quality wood but also a rich diversity of species not found in early-growth forests.

National forests in the United States are required by law to be managed in accordance with principles of *sustainable yield*. Congress has mandated that forests be managed for a combination of uses, including grazing, logging, mining, recreation, and protection of watersheds and wildlife. Healthy forests also require protection from pathogens and insects. *Sustainable forestry*, which emphasizes biological diversity, provides the best management. Other management techniques include removing only infected trees and vegetation, cutting infected areas and removing debris, treating trees with antibiotics, developing disease-resistant species of trees, using insecticides and fungicides, and developing integrated pest management plans.

Two basic systems are used to manage trees: even-aged and uneven-aged. *Even-aged management* involves maintaining trees in a given stand that are about the same age and size. Trees are harvested, then seeds are replanted to provide for a new even-aged stand. This method, which tends toward the cultivation of a single species or monoculture of trees, emphasizes the mass production of fast-growing, low-quality wood (such as pine) to give a faster economic return on investment. Even-aged management requires close supervision and the application of both fertilizer and pesticides to protect the monoculture species from disease and insects.

Uneven-aged management maintains trees at many ages and sizes to permit a natural regeneration process. This method helps sustain biological diversity, provides for long-term production of high-quality timber, allows for an adequate economic return, and promotes a multiple-use approach to forest management. Uneven-aged management also relies on selective cutting of mature trees and reserves *clear-cutting* for small patches of tree species that respond favorably to such logging methods.

Harvesting Methods

The use of a particular tree-harvesting method depends on the tree species involved, the site, and whether even-aged or uneven-aged management is being applied. *Selective cutting* is used on intermediate-aged or mature trees in uneven-aged forests. Carefully selected trees are cut in a prescribed stand to provide for a continuous and attractive forest cover that preserves the forest ecosystem.

Shelterwood cutting involves removing all the mature trees in an area over a period of ten years. The first harvest removes dying, defective, or diseased trees. This allows more sunlight to reach the healthiest trees in the forest, which will then cast seeds and shelter new seedlings. When the seedlings have turned into young trees, a second cutting removes many of the mature trees. Enough mature trees are left to provide protection for the younger trees. When the young trees become well established, a third cutting harvests the remaining mature trees, leaving an even-aged stand of young trees from the best seed trees to mature. When done

correctly, this method leaves a natural-looking forest and helps reduce soil erosion and preserve wildlife habitat.

Seed-tree cutting harvests almost every tree at one site, with the exception of a few high-quality seed-producing and wind-resistant trees, which will function as a seed source to generate new crops. This method allows a variety of species to grow at one time and aids in erosion control and wildlife conservation.

Clear-cutting removes all the trees in a single cutting. The clear-cut may involve a strip, an entire stand, or patches of trees. The area is then replanted with seeds to grow even-aged or tree-farm varieties. More than two-thirds of the timber produced in the United States, and almost one-third of the timber in national forests, is harvested by clear-cutting. A clear-cut reduces biological diversity by destroying habitat. It can make trees in bordering areas more vulnerable to winds and may take decades to regenerate.

Forest Fires

Forest fires can be divided into three types: surface, crown, and ground fires. *Surface fires* tend to burn only the undergrowth and leaf litter on the forest floor. Most mature trees easily survive, as does wildlife. These fires occur every five years or so in forests with an abundance of ground litter and help prevent more destructive crown and ground fires. Such fires can release and recycle valuable mineral nutrients, stimulate certain plant seeds, and help eliminate insects and pathogens.

Crown fires are very hot fires that burn both ground cover and tree tops. They normally occur in forests that have not experienced fires for several decades. Strong winds allow these fires to spread from deadwood and ground litter to treetops. They are capable of killing all vegetation and wildlife, leaving the land prone to erosion.

Ground fires are more common in northern bogs. They can begin as surface fires but burn peat or partially decayed leaves below the ground surface. They can smolder for days or weeks before anyone notices them, and they are difficult to douse.

Natural forest fires can be beneficial to some plant species, including the giant sequoia and the jack pine trees, which release seeds for germination only after being exposed to intense heat. Grassland and coniferous forest ecosystems that depend on fires to regenerate are called fire climax ecosystems.

They are managed for optimum productivity with prescribed fires.

The Society of American Foresters has begun advocating a concept called *new forestry*, in which ecological health and biodiversity, rather than timber production, are the main objectives of forestry. Advocates of new forestry propose that any given site should be logged only every 350 years, wider buffer zones should be left beside streams to reduce erosion and protect habitat, and logs and snags should be left in forests to help replenish soil fertility. Proponents also wish to involve private landowners in the cooperative management of lands.

Toby R. Stewart and Dion Stewart

See also: Agriculture: traditional; Deforestation; Forest fires; Forests; Old-growth forests; Rainforest biomes; Rain forests and the atmosphere; Reforestation; Savannas and deciduous tropical forests; Sustainable forestry.

Sources for Further Study

Davis, Kenneth P. *Forest Management*. New York: McGraw-Hill, 1996. Deals with forest management techniques.

Davis, Lawrence S., and K. Norman Johnson. *Forest Management*. 3d ed. New York: McGraw-Hill, 1987. A good general discussion of the biological and economic principles of forest management.

Hunter, Malcolm L., Jr. *Wildlife, Forests, and Forestry*. Englewood Cliffs, N.J.: Prentice-Hall, 1990. Discusses forest management principles with an emphasis on wildlife and biological diversity.

McNeely, Jeffrey A. *Conserving the World's Biological Diversity*. Washington, D.C.: WRI, 1990. An excellent book on forest management problems throughout the world.

Nyland, Ralph D. *Silviculture: Concepts and Applications*. 2d ed. Boston: McGraw-Hill, 2002. Defines and describes the different methods and treatments that comprise modern silviculture and characterizes the kinds of conditions and responses that silviculture can create.

Robbins, William G. *American Forestry*. Lincoln: University of Nebraska Press, 1985. Provides a good history of cooperation among U.S. national, state, and privately owned forests.

Smith, David M. *The Practice of Silviculture*. 9th ed. New York: Wiley, 1997. A source of ecologically and economically sound information on the long-term treatment of forests. Provides information necessary to understand, analyze, and apply various techniques available for growing stands under any circumstances.

Spurr, Stephen H., and Burton V. Barnes. *Forest Ecology*. 4th ed. New York: Wiley, 1998. Field teachers and researchers of forest ecology and professionals in forest ecology in both public and private arenas share their expertise. This revised edition emphasizes an earth science perspective as well as that of forest biology.

FORESTS

Categories: Ecosystems; forests and forestry

Forests are complex ecosystems in which trees are the dominant type of plant. There are three main forest biomes: tropical, temperate, and boreal.

Both humans and animals depend on forests for food, shelter, and other resources. Forests once covered much of the world and are still found from the equator to the Arctic regions. A forest may vary in size from only a few acres to thousands of square miles, but generally any natural area in which trees are the dominant type of plant can be considered a forest. For a plant to be called a tree, the standard

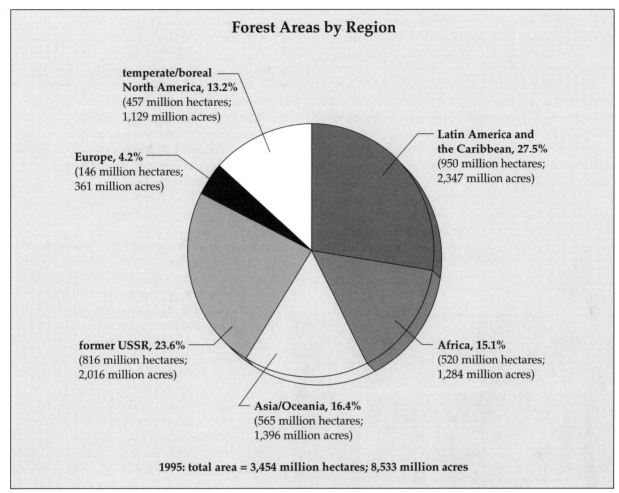

Forest Areas by Region

temperate/boreal
North America, 13.2%
(457 million hectares;
1,129 million acres)

Europe, 4.2%
(146 million hectares;
361 million acres)

Latin America and
the Caribbean, 27.5%
(950 million hectares;
2,347 million acres)

former USSR, 23.6%
(816 million hectares;
2,016 million acres)

Africa, 15.1%
(520 million hectares;
1,284 million acres)

Asia/Oceania, 16.4%
(565 million hectares;
1,396 million acres)

1995: total area = 3,454 million hectares; 8,533 million acres

Source: Data are from United Nations Food and Agriculture Organization (FAOSTAT Database, 2000).

definition requires that the plant must attain a mature height of at least 8 feet (about 3 meters), have a woody stem, and possess a distinct crown. Thus, size makes roses shrubs and apples trees, even though apples and roses are otherwise close botanical relatives. Foresters generally divide the forests of the world into three general categories: *tropical*, *temperate*, and *boreal*.

Tropical Rain Forest

The tropical rain forest is discussed in depth elsewhere (see "Rain-forest biomes"). In brief, it is a forest consisting of a dizzying variety of trees, shrubs, and other plants that remain green year-round. The growth is lush and usually includes both a dense *canopy* formed by the crowns of the largest trees and a thick *understory* of smaller trees and shrubs. Growth is often continuous, rather

than broken into periods of dormancy and active growth, so that fruiting trees are occasionally seen bearing blossoms and mature fruit simultaneously.

Temperate Forest

The temperate forest lies between the tropical forest and the boreal, or northern, forest. The forests of the Mediterranean region of Europe as well as the forests of the southern United States are temperate forests. Trees in temperate forests can be either *deciduous* or *coniferous*. Although coniferous trees are generally thought of as evergreen, the distinction between types is actually based on seed production and leaf shape. Coniferous trees, such as spruces, pines, and hemlocks, produce seeds in cones and have needle-like leaves. Deciduous trees, such as maples, poplars, and oaks, have broad leaves and bear seeds in other ways. Some conifers, such as tamarack, do change color and drop their needles in the autumn, while some deciduous trees, particularly in the southerly regions of the temperate forest, are evergreen.

Deciduous trees are also referred to as *hardwoods*, while conifers are *softwoods*, a classification that refers more to the typical density of the wood than to how difficult it is to nail into it. Softwoods are lower in density and will generally float in water while still green. Hardwoods are higher in density on average and will sink.

Like tropical forests, temperate forests can be quite lush. While the dominant species vary from area to area, depending on factors such as soil types and available rainfall, a dense understory of shade-tolerant species often thrives beneath the canopy. Thus, a mature temperate forest may have thick stands of rhododendrons 20 to 30 feet (6 to 9 meters) high thriving in the shade of 80-foot (24-meter) oaks and tulip poplars. As the temperate forest approaches the edges of its range and the forest makes the transition to boreal, the understory thins out, disappearing almost completely or consisting only of low shrubs. Even in temperate forests, the dominant species may prevent an understory from forming. Stands of southern loblolly pine, for example, often have a parklike feel, as the thick mulch created by fallen needles chokes out growth of other species.

Boreal Forest

The boreal forest, which lies in a band across the northern United States, Canada, northern Europe, and northern Asia, is primarily a coniferous forest. The dominant species are trees such as white spruce, hemlock, and white pine. Mixed stands of northern hardwoods, such as birch, sugar maple, and red oak, may be found along the southern reaches of the boreal forest. As the forest approaches the Arctic, trees are fewer in type, becoming primarily spruce, birch, and willows, and smaller in size. The understory is generally thin or nonexistent, consisting of seedlings of shade-tolerant species, such as maple, and low shrubs. Patches of boreal-type forest can be found quite far south in higher elevations in the United States, such as the mountains of West Virginia. The edge of the temperate forest has crept steadily northward following the retreat of the glaciers at the end of the Ice Age twenty thousand years ago.

Forest Ecology and Resources

In all three types of forest a complex system of interrelationships governs the ecological well-being of the forest and its inhabitants. Trees and animals have evolved to fit into particular environmental niches. Some wildlife may need one resource provided by one species of tree in the forest during one season and a resource provided by another during a different time of year, while other animals become totally dependent on one specific tree. Whitetail deer, for example, browse on maple leaves in the summer, build reserves of fat by eating acorns in the fall, and survive the winter by eating ever-

DIGITAL STOCK

A rain forest in Washington State.

greens. Deer are highly adaptable in contrast with other species, such as the Australian koala, which depends entirely on eucalyptus leaves for its nutritional needs. Just as the animals depend on the forest, the forest depends on the animals to disperse seeds and thin new growth. Certain plant seeds, in fact, will not sprout until being abraded as they pass through the digestive tracts of birds.

Humans also rely on the forest for food, fuel, shelter, and other products. Forests provide wood for fuel and construction, fibers for paper, and chemicals for thousands of products often not immediately recognized as deriving from the forest, such as plastics and textiles. In addition, through the process of transpiration, forests regulate the climate by releasing water vapor into the atmosphere while removing harmful carbon compounds. Forests play an important role in the hydrology of watersheds. Rain that falls on a forest will be slowed in its passage downhill and is often absorbed into the soil rather than running off into rivers and lakes. Thus, forests can moderate the effects of severe storms, reducing the dangers of flooding and preventing soil erosion along stream and river banks.

Threats to the Forest

The primary threat to the health of forests around the world comes from humans. As human populations grow, three types of pressure are placed on forests. First, forests are cleared to provide land for agriculture or for the construction of new homes. This process has occurred almost continuously in the temperate regions for thousands of years, but it did not become common in tropical regions until the twentieth century. Often settlers level the forest and burn the fallen trees to clear land for farming (*slash-and-burn agriculture*) without the wood itself being used in any way. Tragically, the land thus exposed can become infertile for farming within a few years. After a few years of steadily diminishing crops, the land is abandoned. With the protective forest cover removed, it may quickly become a barren, eroded wasteland.

Second, rising or marginalized populations in developing nations often depend on wood or charcoal as their primary fuel for cooking and for home heat. Forests are destroyed as mature trees are removed for fuel wood faster than natural growth can replace them. As the mature trees disappear, younger and younger growth is also removed, and eventually the forest is gone completely.

Finally, growing populations naturally demand more products derived from wood, which can include everything from lumber for construction to chemicals used in cancer research. Market forces can drive forest products companies to harvest more trees than is ecologically sound as stockholders focus on short-term individual profits rather than long-term environmental costs. The challenge to foresters, ecologists, and other scientists is to devise methods that allow humanity to continue to utilize the forest resources needed to survive without destroying the forests as complete and healthy ecosystems.

Nancy Farm Männikkö

See also: Forest and range policy; Forest fires; Forest management; Old-growth forests; Rain-forest biomes; Rain forests and the atmosphere; Slash-and-burn agriculture; Sustainable forestry; Timber industry; Wood; Wood and charcoal as fuel resources.

Sources for Further Study

Holland, Israel I. *Forests and Forestry.* 5th ed. Danville, Ill.: Interstate, 1997. A standard text in the field that is clear, concise, and easily accessible to the general reader. Includes sections on computer applications, ecosystem management, ethical pesticide use, fire management, global positioning systems, and genetic techniques.

Kimmins, J. P. *Balancing Act: Environmental Issues in Forestry.* 2d ed. Vancouver: UBC Press, 1997. Discusses sustainable forestry, respect for nature, ecosystem management, and health and integrity of forests. A good discussion of the pressures on the world's forests.

Page, Jake. *Forest.* Rev. ed. Alexandria, Va.: Time-Life Books, 1987. Provides lavishly illustrated views of the forest with a global perspective.

Walker, Laurence C. *The Southern Forest.* Austin: University of Texas Press, 1991. Fascinating memoir by a retired forester. Gives the reader a personal history of changes in the forest industry in the twentieth century.

FOSSIL PLANTS

Categories: Evolution; paleobotany

Fossil plants are remnants, impressions, or traces of plants from past geologic ages preserved in the earth's crust.

The rise of land-dwelling animals paralleled the rise of plants, which have always been the basis for animal life. Fossil plants are a valuable source of information regarding such phenomena as changes in climate, ancient geography, and the evolution of life itself.

Thallophytes

The earliest fossil plants are represented by a phylum called the *thallophytes*. The geological record of the thallophytes is incomplete. Of seven large groups, only a few are represented by fossils. Although some records from the Paleozoic era have been found, the earliest identifiable specimens found are from the Jurassic period. The dearth of fossils from this group of plants can be attributed to their minute size and the fragile nature of their remains.

The thallophytes are in the most primitive plants, lacking roots, stems, leaves, and conducting cells. The simplest thallophytes are in the subphylum of *autophytic thallophytes*, which include blue-green bacteria (or cyanobacteria, formerly known as blue-green algae), diatoms, and algae. All these plants produce chlorophyll. Cyanobacteria are unicellular organisms occurring in colonies held together by a jellylike material. Diatoms are one-celled plants enclosed in a wall consisting of two overlapping valves. The next class, called simply algae, consists of several different types of seaweed, such as chara, or stonewort, which secretes lime with which it encrusts its leaves and is responsible for many freshwater limestones of the past. Many fossils that have been described as algae were actually molds of burrows or tracks of animals.

The second subphylum of the thallophytes is called the *heterophytic thallophytes*. These plants are distinguished by the absence of chlorophyll; as in animals, their principal source of energy is organic. The heterophytic thallophytes are subdivided into three classes. *Bacteria*, one-celled plants without definite nuclei, are the chief agents of the decomposition of organic matter; without bacteria, more prehistoric plant and animal remains would have been preserved. The next class, *slime fungi*, are sticky masses enclosing many nuclei but without cell walls. Slime fungi have never been found as fossils. The final class is the *fungi*, which are composed essentially of a branching mass of threads called the mycelium, which penetrate the cell walls of their "host"—plant or animal—and live upon its substance. *Lichens* are made up of a fungus and an alga living together in symbiosis. Fossil lichens have been recognized only from very recent formations.

Bryophytes

The next phylum to emerge, the *bryophytes*, exhibits a distinct advance over the thallophytes: Bryophytes adapted more successfully to the terrestrial environment. They were able to take water and other necessary substances from the soil by means of rootlike hairs called rhizoids. The most distinct advance of the bryophytes over the thallophytes is in their method of reproduction. The spores produced by these plants germinate by sending out a mass of green threads, the protonema.

The simplest bryophytes are the *liverworts*. The *mosses*, which are more abundant today than the liverworts, possess leaves consisting of many small chlorophyll-bearing cells. Because the ancient members of the bryophyte group were more delicate than the modern forms, they have been preserved only under exceptional conditions, such as those provided by the silicified peat beds at Rhynie, Scotland, which contain fossils from the Devonian period.

Pteridophytes

The *pteridophytes* were much more advanced than the bryophytes. While the structure of the bryophytes was primarily cellular, that of the fern plant is vascular. Unlike the bryophytes, the pteridophytes originate from a fertilized egg and produce spores. Pteridophytes are well represented by the ferns, which have existed from the Devonian period. Another class of pteridophytes is the horsetails (*Equisetales*), which also have existed from the Devonian to the present.

The third class of pteridophytes is the club mosses, which are largely creeping, many-branched plants with numerous tiny, mosslike leaves spirally arranged on the stem. The final class of pteridophytes, *Sphenophyllales*, consisted of slender plants with jointed stems and leaves in whorls. These climbing plants are known from the Devonian to the Permian periods.

Spermatophytes: Gymnosperms

The fourth phylum, the *spermatophytes*, are distinguished by the production of seeds, although the lower groups have the same alternation of the vegetative (asexual) and reproductive (sexual) generations as is seen in the pteridophytes. The chief distinguishing characteristics of the spermatophytes are the formation of a pollen tube and the production of seeds.

The first class, the *gymnosperms*, are typified by the pines, mostly evergreens. Members of one order of gymnosperms, *Cycadofilicales*, were fernlike in habit but were not actually ferns. Because the leaf and stem remained practically unchanged, it is very easy to mistake the early seed plants for ferns. One of the most familiar of the fernlike fronds of the Pennsylvanian coal deposits is *Neuropteris*, which had large, compound leaflets.

The stem in most forms was thick and short and covered with an armor of leaf bases. It represents an advance over previous plants in that it had a true flower because both male and female organs were borne on the same axis and were arranged in the manner of later flowering plants. Thus *Cycadales* is an intermediary in the line of development of the angiosperms (flowering plants) from their fern ancestors. This order formed the dominant vegetation of the Mesozoic, ranging from the Triassic into the Lower Cretaceous.

The next order of gymnosperms, *Cordaitales*, is an extinct group of tall, slender trees that thrived throughout the world from the Devonian to the Permian period. The leaves of these trees were swordlike and distinguished by their parallel veins and great size, reaching up to 1 meter. The *Cordaitales* were the dominant members of the gymnosperm forests during the Devonian period. The fourth order of gymnosperms, *Ginkgoales*, resembles the conifers in general appearance. The leaves, however, are fanlike and are shed each year. Like the cycads and ferns, the male cells are motile in fertilization.

The order *Coniferales* includes mostly evergreen trees and shrubs, with needles or scalelike leaves and with male and female cones. Derived from *Cordaitales* of the Paleozoic, *Coniferales* possesses fewer primitive characters than *Ginkgoales*. The yews, which are comparatively modern, have fruit with a single seed surrounded by a scarlet, fleshy envelope. Another family, *Pinaceae*, having cones with woody or membranous scales, are represented by *Araucaria*, which is very common in the Petrified Forest in eastern Arizona.

The *Abietae*, one of the more common families of evergreens, includes pines, cedars, and hemlocks dating back to the Lower Cretaceous. One of the most extraordinary members of the conifers was the family *Taxodiaceae*, which includes the genus *Sequoia*, represented today only by the redwood and the *Sequoia gigantea*, which grow in California. These species' twigs, cones, and seeds were abundant in the Lower Cretaceous of North America. Finally, the family *Cupressaceae* includes the junipers and is known from the Jurassic.

Spermatophytes: Angiosperms

The second class of spermatophytes is the *angiosperms*, commonly known as flowering plants. The angiosperms contain the plants of the highest rank. This group comprises well over one-half of all known living species of plants. The typical flower is composed of an outer bud-covering portion, the stamens, and the pistil. When the wind or an insect brings the pollen into contact with the pistil, the pollen is held in place by a sugary solution. After the pollen penetrates an ovule, the nucleus divides several times. This fusion is called fertilization. The embryo, consisting of a stem with seedling leaves, is called a seed.

Both subclasses of the angiosperms first appeared in the upper part of the Lower Cretaceous. *Dicotyledoneae* (the dicotyledones, or dicots) comprises a primitive subclass that begins with two seedling leaves that are usually netted-veined. The stem is usually thicker below than above, with the vascular bundles arranged to form a cylinder enclosing a pith center. As growth proceeds, new cylinders are formed. The last of the dicots to appear was the sassafras tree, flourishing throughout North America and Europe since the Lower Cretaceous.

The second subclass, *Monocotyledoneae* (monocotyledones, or monocots), descended from the dicots. These plants are distinguished by the fact that they begin with a single leaflet, or *cotyledon*. The veins of the leaves are parallel, the stem is cylindrical, and the roots are fibrous. This subclass is represented by the grasses and grains. Fossils from this subclass date back to the upper part of the Lower Cretaceous of eastern North America. The fossil record of the palm goes back to the mid-Cretaceous.

Evolution of Plants

The evolution of plants is the story of their struggle to adapt themselves to land. One of the changes necessary in the development of land flora was the change from a cellular structure to a vascular one, which opened up possibilities for increase in size and laid the foundation for the trees. In order to adapt to land, plants also had to develop a resistance to the dehydrating quality of the air.

The earliest plants, the thallophytes, were closely tied to water. One of the first examples of flora adapting to land were the freshwater algae. The change from a cellular to a vascular structure led to the development of roots; the pteridophytes were the first plants to take this step. The mosses and ferns adapted to land but still required rain or dew for the union of the gametes. It is only the spermatophytes that developed a device that freed them from the necessity of external water for fertilization to occur. This ability permitted the spermatophytes to proliferate throughout the earth.

Alan Brown

See also: Angiosperm evolution; Cycads and palms; Evolution of plants; Paleobotany; Paleoecology; Petrified wood.

Sources for Further Study

Agashe, Shripad N. *Paleobotany: Plants of the Past, Their Evolution, Paleoenvironment, and Application in Exploration of Fossil Fuels.* Enfield, N.H.: Science Publishers, 1997. Describes the geologic history of the earth and the origin and evolution of different groups of plants on the earth's surface. Covers principles and applications of paleobotany. With illustrations.

Cleal, Christopher J., and Barry A. Thomas. *Plant Fossils: The History of Land Vegetation.* New York: Boydell Press, 1999. Relates the history of land vegetation, illustrated with a series of plant photographs.

Gensel, Patricia G., and Dianne Edwards, eds. *Plants Invade the Land: Evolutionary and Environmental Perspectives.* New York: Columbia University Press, 2001. Collection of essays on what is known about the origins of plants on land, integrating paleobotanical information with physical, chemical, and geological data.

Poinar, George O., and Roberta Poinar. *The Amber Forest: A Reconstruction of a Vanished World.* Princeton, N.J.: Princeton University Press, 1999. A fascinating, detailed picture of tropical forests as they were fifteen million to forty-five million years ago. Includes drawings and photographs.

Ross, Andrew. *Amber.* Cambridge, Mass.: Harvard University Press, 1998. Describes how amber, the fossilized resin of ancient trees, preserves organic material, mostly insects and other invertebrates, and with it the shape and surface detail that most fossils do not show.

Stewart, Wilson N., and Gar W. Rothwell. *Paleobotany and the Evolution of Plants.* 2d ed. New York: Cambridge University Press, 1993. Describes the origin and evolution of plants as revealed by the fossil record. Summarizes relationships among major plant groups.

FRUIT CROPS

Categories: Agriculture; economic botany and plant uses; food

Fruits are mature, or ripened, ovaries of angiosperms, their contents, and any accompanying accessory structures. From a botanical point of view, common foods such as grains, nuts, dried beans, squash, eggplant, and tomatoes are fruits, but in common usage the term is usually restricted to fleshy fruits that are commonly grown commercially as crops and are eaten, primarily raw, for their fleshy and juicy pulp.

Fertilization of ovules and the initiation of seed development lead to hormone production that triggers fruit development. Consequently, fruits usually contain seeds, but seeds can form without fertilization (*parthenogenesis*), and fruits can develop without seeds (*parthenocarpy*). Many fruits that are cooked or eaten as part of a main course are usually classed as vegetables. This dichotomy is reflected in the origins of the two words: fruit, from the Latin *fruor*, "to enjoy," and vegetable, from the Latin *vegetare*, "to enliven."

Ecology of Fruits

In nature, fleshy fruits serve as a reward for seed-dispersing animals. In keeping with the evolutionary principle that selection tends to minimize the cost of structures while maximizing their function, the flesh of these fruits contains comparatively few calories and basically consists of colored, flavored sugar water. Using animals as dispersal agents carries a risk, however, of seed destruction. Consequently, fleshy fruits that are dispersed by animals exhibit a number of mechanisms that protect seeds. One of these is the production of a large, hard seed that an animal cannot eat, such as a peach pit or a mango seed. Another protective characteristic is small seeds that go through an animal's digestive system without being crushed or digested, such as strawberry seeds. Over the last ten thousand years that humans have been practicing agriculture, many fleshy fruited species have been domesticated and bred for improved fruit production and quality. Several of the most marketed fruits worldwide are discussed below.

Temperate Fruits

The rose family (*Rosaceae*) contains a wide array of fruits grown in the cool regions of the world: ap-

ples, pears, plums, peaches, cherries, strawberries, and raspberries. Apples are the most important fruit tree crop of temperate regions. Apple and pear fruits are known as *pomes* because the edible fleshy part of the fruit is a combination of the outer ovary wall and the basal part of the flower. Cultivated apples are believed to have originated in western Asia and were enjoyed in prehistoric times. Apples were brought to North America about 1620 and are now the most widely grown fruit in the United States. Most of the apples grown today are diploids, but many are triploids. Orchards are usually planted with grafted trees, to ensure uniformity of the crop. Literally thousands of varieties of apples have been developed over the centuries since the species was domesticated.

Plums, peaches, and cherries come from different species of the genus *Prunus*. They share a fruit type known as a *drupe*, consisting of a fleshy mesocarp and a single seed inside of a hard endocarp. While there are native species of *Prunus* in the New World, the domesticated species are native to Eurasia.

The modern cultivated strawberry is a hybrid that apparently formed spontaneously in a European garden between a species of *Fragaria* from Chile and one from Virginia. Europeans had eaten native strawberries for centuries before the discovery of the New World, but the hybrid (*Fragaria ananassa*) was larger in size, as flavorful, and produced more fruit. A strawberry is actually an aggregation of fruits, or *aggregate fruit*. Each tiny seed is itself a fruit. The large, succulent, mass is the swollen top of the stem on which the flower was borne. Raspberries are also aggregated fruits, but each globular segment of the raspberry is itself a fruit, called a *drupelet*. The caps of drupelets pull free of the stem tips when the berry is picked.

Grapes are the second most widely cultivated fleshy fruit (on a tonnage-produced basis). However, the majority of grapes are not eaten as fruit but are turned into other foods, such as vinegar, liqueurs, raisins, and wine. The most widely cultivated species of grape is *Vitis vinifera* (Vitaceae), a woody perennial vine native to middle Asia. There are hundreds of varieties of grapes that vary in the color of the skin, flesh, flavor, and sweetness of the berries.

Nuts are dry fruits, each of which contains a single seed that is free inside the ovary wall, except for an attachment at one end called the funiculus. The pericarp (the walls of the ovary) is hard and fibrous. Commercially grown nuts include filberts, pecans, walnuts, and macadamia nuts, sold both for eating and for cooking.

Tropical Fruits

Many species of the genus *Citrus* (including sweet orange, tangerine, grapefruit, lemon, and lime) are grown for their edible fruits. Sweet oranges (*Citrus sinensis*) are the most widely grown fruit in the world, but in the nineteenth century they were considered luxuries and prescribed as cold remedies by physicians. Like other citrus fruits, the fruit of an orange is technically a *hesperidium*, a berry with a leathery rind and a juicy pulp that is formed of juice sacks borne on the inner layer of the fruit wall. The juice sacks fill the sections of the fruit and surround the seeds. The watery solution in the sacks is high in vitamin C. There are three main classes of oranges: Valencias, navels, and blood oranges. Valencias, with their deep orange color and rich flavor, are the source of most or-

PhotoDisc

Over the last ten thousand years that humans have been practicing agriculture, many fleshy fruited species have been domesticated and bred for improved fruit production and quality. Strawberries are one of the wide array of fruits grown in the cool regions of the world.

ange juice. Navel oranges, favored for eating, are the result of a mutation that produces a second ovarylike structure (the navel) instead of seeds. Because they are seedless, navel oranges are all propagated by grafting. Blood oranges are seeded oranges named for the patches of deep red-purple color in the fruit.

Bananas became a major fruit crop in the twentieth century; prior to the use of refrigerated ships, bananas spoiled before they could reach markets outside the tropics. Wild bananas, native to eastern Asia, have seeds, but the common domesticated banana (*Musa paradisiaca*) is seedless and is the product of several cycles of hybridization followed by increases in chromosome number. Whether the fruit will be a tender and sweet yellow or red or a tough, starchy green plantain depends on the particular combination of chromosomes in the hybrid. Banana plants are giant herbs, not trees, and are propagated vegetatively by planting a piece of stem. Over the course of a year, the stem grows and produces a long terminal inflorescence of many clusters of female flowers along the flowering stalk and male flowers at the tip. The female flowers spontaneously mature, with each cluster forming a hand, or bunch, of bananas. An entire inflorescence can produce more than three hundred bananas, weighing 110 pounds. After fruiting, the shoot dies and is cut, allowing a sprout from the base to produce the next flowering stem.

Mangoes (*Mangifera indica*) are extremely common and important fruits in tropical areas, particularly in their native region of Southeast Asia, where the fruit pulp and even the seeds have been used for food. Mangoes are borne on trees that can grow only within tropical regions where there is adequate water in the summer. The fruit is a berry with musky yellow flesh surrounding a single large seed. Mangoes belong to the poison ivy family; some people are allergic to the latex produced in the skins. Mangoes were introduced into Brazil by the Portuguese in the early 1700's and subsequently spread to other areas of the New World tropics.

Melons, both common melons (*Cucumis melo*) and watermelons (*Citrullus lanatus*) belong to the same family (*Cucurbitaceae*) as squashes, but the latter are all native to the New World and the former to the Old World. All these species share the same kind of fruit, a *pepo*, which consists of a hard rind derived in part from the basal parts of the flower, and an edible fleshy layer of inner ovary tissue. Melons are monoecious vines with showy male and female flowers that require pollination to set fruit. Melons are native to Africa, where they were undoubtedly prized for their high water content and fresh flavor. Selection has led to numerous varieties, including cantaloupe, Crenshaw, honeydew, Persian, musk, and a variety of other melons that differ in the color and surface of the rind, color of the flesh, taste, and degree of sweetness.

Watermelons are native to sub-Saharan Africa, but they can be grown in temperate regions because they are annuals. The flesh is 87-92 percent water and is acidic enough to curdle milk. Recently, seedless types have been produced by artificially making triploid plants. The pollen of these triploids is sterile, and the seeds abort early. Farmers plant the triploids with fertile diploids. Pollen from the diploids fertilizes the ovules of the triploids and triggers fruit production. The seeds quickly die, but the fruit continues to mature into a seedless watermelon.

Beryl B. Simpson

See also: Angiosperm life cycle; Flower structure; Fruit: structure and types; Pollination; Reproduction in plants; Vegetable crops.

Sources for Further Study

Morton, Julia F. *Fruits of Warm Climates*. Winterville, N.C.: Creative Resource Systems, 1987. Covers almost all tropical and subtropical fruits, even the obscure ones, including origins, varieties, cultivation, and diseases of each.

Simpson, B. B., and M. C. Ogorzaly. *Economic Botany: Plants of Our World*. 3d ed. Boston: McGraw-Hill, 2000. This text covers all major crops of the world, with a chapter devoted to temperate fruits and one to tropical fruits.

Smartt, J., and N. W. Simmonds, eds. *Evolution of Crop Plants*. 2d ed. London: Longman Scientific and Technical, 1995. This edited volume provides an account of almost every crop species, with each article written by an authority on that species.

FRUIT: STRUCTURE AND TYPES

Categories: Anatomy; angiosperms; economic botany and plant uses; reproduction and life cycles

Fruits are the seed-containing reproductive organs, including nuts and grains, produced by angiosperms (flowering plants).

When most people think of a fruit, what typically comes to mind is a juicy, edible object, such as an apple, orange, or banana. To botanists, however, fruit includes many plant-derived structures, such as grains, nuts, and many vegetables. In essence, a fruit is an enlarged *ovary*, often with some accessory tissue, that develops after a flower has been pollinated. After pollination, seed development begins, and soon the peripheral parts of the flower fall away, leaving the immature fruit. The fruit subsequently enlarges and then ripens to maturity. It is then often edible.

Fruit Structure

Almost all fruits have a general structure that consists of an outer layer called the *pericarp*. The pericarp, in turn, encloses the *seed* or seeds. Usually there is a space between the seed and the pericarp, called a *locule*. The pumpkin is a good illustration of this structure, with orange rind as the pericarp, the hollow space within the locule, and the seeds inside the locule.

There are many different kinds of fruits. Some, such as cherries, tomatoes, and apples, have fleshy, juicy pericarps. Others, such as peanuts, milkweed pods, and acorns, have dry pericarps. The variability in fruits represents different seed dispersal strategies. In some plants, the seeds are dispersed while still enclosed within the fruit. Seeds in fleshy fruits are often dispersed by animals that eat the fruit and then either discard the seeds or later defecate them. Other fruits have barbs or hooks that catch on to fur or feathers and then travel with an animal until they are removed or drop off. Still other seeds, such as those of maple and ash, have "wings" for wind dispersal. A few, such as those of the coconut palm and many sedges, have fruits that float and are dispersed by water. In most instances, the fruit merely opens, and the seeds drop onto the ground. Some plants, such as witch hazel and the touch-me-not, produce fruits that open explosively and can disperse their seeds great distances.

Fruit Types

Fruits can be classified, based on the nature of the pericarp, into two groups: fleshy and dry. Fleshy fruits, in turn, are classified into several types, including *drupes, berries, pomes, hesperidia,* and *pepos*. Dry fruits are also subdivided into several categories, including *follicles, legumes, capsules, achenes, nuts, samaras, schizocarps,* and *caryopses*.

The three most familiar types of *fleshy fruits* are drupes, berries, and pomes. A drupe is a fleshy fruit that contains a single seed surrounded by a hard, bony inner wall of the pericarp (called the *endocarp*). The middle and outer walls of the pericarp (called the *mesocarp* and *exocarp*, respectively) are juicy and often sweet. Drupes include all the pitted fruits, such as cherries, plums, peaches, and olives. A berry typically has several seeds, and the pericarp is fleshy throughout. Familiar examples include tomatoes, eggplants, and grapes. A pome is a fleshy fruit, often with many seeds, that has a thick layer of *accessory tissue* immediately surrounding the pericarp. The accessory tissue is generally juicy, sweet, and often edible. Representative pomes include apples and pears.

Two other fleshy fruits, the hesperidium and the pepo, are characterized by a leathery rind. Hesperidia, also known as a citrus fruits, have rinds rich in aromatic oils surrounding a juicy interior composed of wedge-shaped segments that have a sugary, acidic sap. Hesperidia include oranges, grapefruits, lemons, and limes. The pepo has a tough exocarp which is either smooth or variously sculptured and normally contains many seeds. Examples include cucumbers, cantaloupes, and squash.

Some common *dry fruits* are follicles, legumes,

Fruit Types

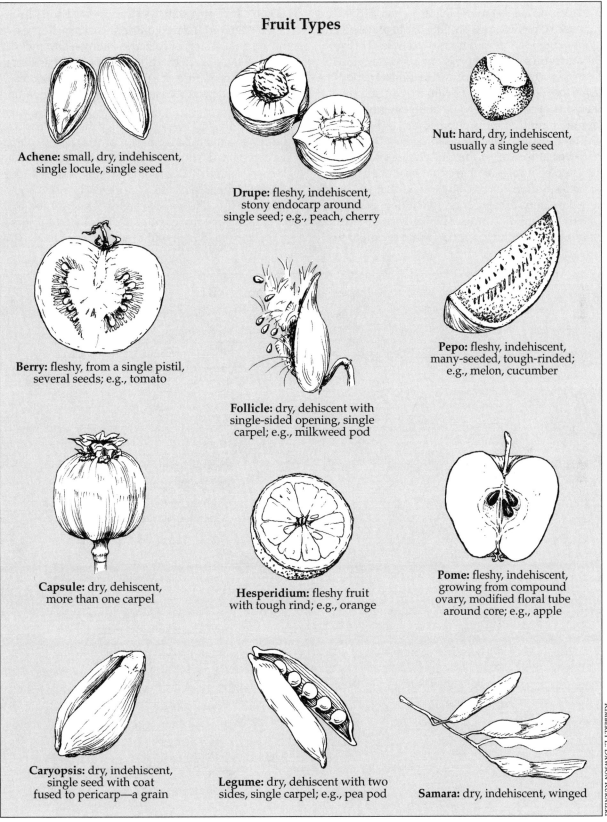

Achene: small, dry, indehiscent, single locule, single seed

Drupe: fleshy, indehiscent, stony endocarp around single seed; e.g., peach, cherry

Nut: hard, dry, indehiscent, usually a single seed

Berry: fleshy, from a single pistil, several seeds; e.g., tomato

Follicle: dry, dehiscent with single-sided opening, single carpel; e.g., milkweed pod

Pepo: fleshy, indehiscent, many-seeded, tough-rinded; e.g., melon, cucumber

Capsule: dry, dehiscent, more than one carpel

Hesperidium: fleshy fruit with tough rind; e.g., orange

Pome: fleshy, indehiscent, growing from compound ovary, modified floral tube around core; e.g., apple

Caryopsis: dry, indehiscent, single seed with coat fused to pericarp—a grain

Legume: dry, dehiscent with two sides, single carpel; e.g., pea pod

Samara: dry, indehiscent, winged

KIMBERLY L. DAWSON KURNIZKI

capsules, achenes, nuts, samaras, schizocarps, and caryopses. Follicles are podlike fruits that open up along one side, revealing numerous seeds. Examples include milkweed pods and the aggregate follicles of magnolias. In contrast, legumes are podlike fruits that open up along two lines, releasing several seeds. They are produced by many members of the legume family, such as peas, beans, and peanuts.

The capsule opens along three or more lines or by pores at the top of the fruit. Lilies and poppies are good examples of plants that produce capsules. Achenes each contain a single seed that is free inside the cavity, except for an attachment at one end called the *funiculus*, and are typically small. A good example is a sunflower "seed." A nut is similar to an achene, except that the pericarp is hard and fibrous and is derived from a compound ovary. Representative nuts include acorns, hazelnuts, and hickory nuts. A samara is a modified achene that has part of the pericarp flattened to form a wing. Examples of plants that produce samaras include ashes and elms.

Maples have a winged fruit, called a *schizocarp*, which is often mistaken for a samara. Close observation reveals that the schizocarps come in attached pairs that later split into single-seeded portions. Schizocarps, which are generally not winged, also occur in the parsley family, where they may split into more than two parts. Finally, a *caryopsis* is a single-seeded fruit whose seed coat is fused to the

Fruit Types and Characteristics

Classification	Examples	Type[1]	Fleshy/ Dry[2]	Dehiscent/ Indehiscent[3]	Comments
Accessory	Strawberry	Accessory	F	—	Develops from receptacle; ovaries are achenes on surface.
Achene	Dandelion, sunflower	Simple	D	I	Single locule with single seed; not especially hard.
Aggregate	Raspberry	Aggregate	F	—	Drupelets from several pistils on a single flower.
Berry	Grape, tomato	Simple	F	—	Not enclosed by a receptacle, many seeds, soft skin.
Capsule	Lily, poppy	Simple	D	D	Formed by two or more carpels not separated by a septum.
Caryopsis (grain)	Wheat, corn, other grains	Simple	D	I	Single seed fused to pericarp.
Drupe	Peach, plum, cherry	Simple	F	—	Larger fruit with single seed (stone).
Drupelet	Raspberry drupelet	Simple on aggregate	F	I	Small, single seeded; occurs as a segment of an aggregate fruit.
Follicle	Milkweed pod	Simple	D	D	Single carpel opening on one side.
Hesperidium	Orange, other citrus	Simple	F	—	Berry-shaped, with several seeds and a tough rind.
Hip	Rose hips	Simple	F/D	—	Enlarged hypanthium surrounding many achenes.
Legume	Bean, pea pod	Simple	D	D	Composed of one carpel with seeds.

pericarp. Caryopses are produced only by plants in the grass family and include the familiar grains wheat, corn, rice, and oats.

The fruits listed above commonly fall under the category of *simple fruits*. In other words, they are identifiable as individual structures. Other plants produce fruits in dense clusters, and these are termed either *aggregate fruits* or *multiple fruits*. Aggregate fruits are produced by a single flower that has numerous pistils. One example is a raspberry, which is an aggregate of drupes which are often referred to as drupelets because of their small size. Another is the strawberry, not a berry in the botanical sense, which is an aggregate of tiny achenes attached to the surface of a swollen, juicy, *receptacle*

(originally the base of the flower where all the flower parts were attached). In contrast, multiple fruits are produced by clusters of small flowers, each of which produces a single fruit. Representative multiple fruits include mulberries, figs, and pineapples.

Development and Maturation

In nature, fruits develop only after the flower is effectively pollinated. If pollination does not occur, the entire flower shrivels up, and no fruit is formed. Fruit development is apparently stimulated when the developing seeds produce hormones that diffuse into the ovary wall, causing it to enlarge. Two hormones are particularly implicated in fruit for-

Classification	Examples	Type[1]	Fleshy/ Dry[2]	Dehiscent/ Indehiscent[3]	Comments
Loment	Tick trefoil, sweet vetch	Simple	D	I	Legume but with individual seeds constricted by the pod.
Multiple	Pineapple, mulberry	Multiple	F	—	Formed by many flowers and ovaries on one receptacle.
Nut, nutlet	Acorn, hazelnut	Simple	D	I	Usually single seed inside a hard, woody pericarp.
Pepo	Melon, cucumber	Pepo	F	—	Not enclosed by receptacle; many seeds; tough rind.
Pome	Apple	Simple	F	—	Floral tube (core) containing multiple seeds, surrounded by fruit.
Samara	Maple, ash, elm fruits	Simple	D	I	Not splitting; winged.
Schizocarp	Carrot, geranium fruits	Simple	D	I	Fruit splitting but not releasing seeds.
Silicle (silique)	Mustard (*Brassicaceae*) fruits	Simple	D	D	Two carpels separated by septum, less than twice longer than wide.
Synconium	Fig	Multiple	F	—	From entire inflorescence (many flowers) with inverted receptacle bearing flowers internally; fruit formed of ripened ovaries (receptacle tissue).

— Not applicable.
1. Four main types: Simple (formed from a single mature ovary from a single flower); aggregate (several mature ovaries from a single flower); multiple (several mature ovaries from several flowers united on one receptacle, forming a mass); accessory (from tissues surrounding the ovary or ovaries, generally from flowers with inferior ovaries).
2. All fruits are either dry or fleshy, characterizing the pericarp at maturity.
3. Dehiscent fruits are dry fruits that split open when mature to release their contents, along either one or two sides. Indehiscent fruits are dry fruits that do not split open at maturity; often they are single-seeded.

mation: *auxin* and *gibberellin*. Many fruit growers routinely spray their plants with auxin to induce the formation of seedless, or *parthenocarpic*, fruits.

Fruit ripening is an important process that must occur properly in order for the seeds to be effectively dispersed. In fleshy fruits, such as tomatoes, cherries, apples, oranges, and bananas, fruit ripening involves several important changes in the pericarp that make the fruit more visible and palatable to a potential animal disperser. Perhaps the most visible change is in the color of the fruit. Immature fruits are green because of the presence of the pigment *chlorophyll* in the cells of the outer layer. Potential dispersers fail to notice the immature fruit because it blends in with the surrounding leaves. As fruits ripen, the chlorophyll breaks down, and other colors, such as orange, yellow, red, or blue, become evident. Those colors are the result of pigments that either are present in the unripe fruit and masked by the chlorophyll or develop as the fruit ripens. The texture and chemical composition of the pericarp change as well. Most fruits soften as they ripen, a result of the degradation of the cell walls in the pericarp. At the same time, starches or oils in the pericarp are chemically transformed into simple sugars such as *fructose*. That change causes the fruit to become better-tasting, more digestible, and thus more attractive to a hungry animal.

The physiology of fruit ripening has been well studied. Fruits such as grapes, citrus fruits, and strawberries ripen gradually. Others, such as tomatoes, apples, and pears, exhibit a transitional event called *climacteric*, which is marked by a dramatic increase in the rate at which oxygen is absorbed by the fruit, followed by a rapid change in the color and physical nature of the pericarp. Studies of climacteric fruit have shown that the onset of ripening can be delayed by storing the fruit at low temperatures or in an atmosphere devoid of oxygen. On the other hand, climacteric can be induced by exposing the fruit to *ethylene*, a plant hormone. Interestingly, ethylene is produced by ripening fruits, and thus a ripe fruit promotes the development of any unripe fruits nearby.

Finally, many plants drop their fruits at some point after they become ripened. Botanists use the term *abscission* to refer to the dropping process. Fruit abscission, like leaf abscission, occurs when a layer of cells at the base of the *pedicel* (the stem that attaches to the plant) become weakened. Studies have shown that abscission is influenced by two hormones: ethylene and auxin. Ethylene promotes abscission of fruits in many plant species, such as cherries, blueberries, and blackberries. Auxin, on the other hand, has effects that vary depending on the dose: Low concentrations promote fruit retention, while high doses cause fruits to drop.

Kenneth M. Klemow

See also: Hormones; Pigments in plants.

Sources for Further Study

Attenborough, David. *The Private Life of Plants: A Natural History of Plant Behavior.* Princeton, N.J.: Princeton University Press, 1995. A companion volume to the 1995 PBS series hosted by Attenborough, abundantly illustrated and aimed at a general audience.

Bowes, Bryan G. *A Color Atlas of Plant Structure.* Ames: Iowa State University Press, 2000. A fundamental guide to plant structure from the cellular level to complete plants, featuring nearly four hundred color photographs and concise scientific text. Chapter 8 covers the structure of fruit.

Mauseth, James D. *Botany: An Introduction to Plant Biology.* Enhanced 2d ed. Sudbury, Mass.: Jones & Bartlett, 1998. A very thorough introductory botany textbook. Enhanced second edition contains a CD-ROM tutor, and the text is linked to a Web site maintained by the publisher and author with continually updated information on subjects covered in the text.

Raven, Peter, Ray F. Evert, and Susan E. Eichhorn. *Biology of Plants.* 6th ed. New York: W. H. Freeman/Worth, 1999. An excellent introduction to all phases of modern botanical science, including sections on cell biology, energetics, genetics, diversity, structure, evolution, and ecology. Chapter 24 gives an authoritative review of the role that various hormones play in fruit set and ripening. The diversity of fruits is reviewed in chapter 29.

Simpson, Beryl Brintall, and Molly Conner-Ogorzaly. *Economic Botany: Plants in Our World.* 3d ed. Boston: McGraw-Hill, 2001. A highly readable and informative account concerning the numerous uses of plants by various societies since the advent of recorded history. The book is copiously illustrated throughout, with photographs and line drawings. Chapters 4 and 5 provide an excellent account of fruits and nuts from temperate and tropical regions.

FUNGI

Categories: Fungi; taxonomic groups

The kingdom of nonphotosynthetic eukaryotic organisms, fungi are heterotrophic organisms, feeding on other materials rather than making their own food. They live on dead organisms by secreting digestive enzymes and absorbing the breakdown products. Although these organisms are stationary like plants and thus traditionally studied in botany courses, the heterotropic fungi are fundamentally different from plants, which are autotrophic organisms.

Although some unicellular forms of fungi exist, most fungi are characterized by a mycelial growth form; that is, they generally are made up of a mass of *hyphae* (tubular filaments). All fungi live in their food and have an absorptive mode of nutrition by which they secrete digestive enzymes and absorb the breakdown products. They are therefore heterotrophs. Fungi are also characterized by possession of cell walls made of chitin, synthesis of the amino acid lysine via the amino adipic acid (AAA) pathway, and possession of a ribosomal DNA sequence that classifies them more closely with other fungi than with any other group of organisms. Fungi produce spores by either asexual or sexual means. The way they produce their spores constitutes one of the main taxonomic criteria for classifying them within the kingdom *Fungi*.

Ecology and Habitats

Because of their absorptive, heterotrophic mode of acquiring nutrients, fungi are important members of the *decomposer* community in ecosystems. Because fungi break down dead organic matter (that is, because they live as *saprophytes*), they simultaneously release nutrients that are taken up by other members of their ecosystem. Fungi exist in terrestrial and aquatic habitats. Because the saprophytic fungi grow in a radial fashion from a point of origin, the mushroom types of fungi sometimes form so-called fairy rings: At the advancing edge of the mycelia, mushroom fruiting bodies appear in the shape of an irregular ring. One mass of mycelia, for example, became so large that it occupied 37 acres (15 hectares) in the Upper Peninsula of Michigan and became known as the Humongous Fungus.

Fungi are major *parasites*, living on live plants, animals, and other fungi. Fungal plant pathogens have evolved a variety of mechanisms to enable the fungus to penetrate the host plant and overcome the host's defenses. Next, the pathogen absorbs food from the host by establishing *haustoria*, which form a highly specialized absorbing system. These structures do not actually penetrate the host plasma membrane, but they reside within pockets of the host cell, where they secrete extrahyphal enzymes and absorb the soluble result. Consequently, the host plant develops a series of symptoms characteristic of the infection. Farmers can suffer severe loss of crops if an infestation is left unchecked. Plant breeders attempt to breed disease resistance into crop plants as a means of combating fungal diseases.

In contrast to plant fungal pathogens, some fungi grow within plants and do not cause disease symptoms. These fungi are called *endophytes*. They appear to protect host plants from herbivores and from certain pathogenic microbes.

Interactive Associations

Mycorrhizae are mutualistic associations between fungi and vascular plants whereby both members of the association benefit. A fungus that grows on plant roots facilitates nutrient and water uptake for the plant, and the plant provides organic nutrients to the fungus. About 90 percent of higher

Mushrooms are some of the best-known representatives of the kingdom Fungi.

cyanobacterium (blue-green bacteria, formerly thought to be blue-green algae). Of all fungal partners, 98 percent are members of the phylum *Ascomycota*. In this controlled parasitism, the fungus obtains minerals and water and develops a physical structure to house the algal partner. The algal partner provides photosynthetic products to the fungal partner. These organisms can live in extreme environments such as deserts or Arctic regions. Because they have no way to eliminate toxic materials, they are extremely sensitive to air pollution. Indeed, the condition of lichens is sometimes used as an indicator of air pollution.

Some fungi have trapping mechanisms that allow them to prey upon invertebrate organisms such as nematodes, rotifers, and copepods. These trapping mechanisms involve networks of adhesive hyphae, adhesive branches, adhesive nets, adhesive knobs, nonconstricting rings, and constricting rings.

Chytrids

The type of sexually produced spore determines the taxonomy of fungal phyla. *Chytridiomycota*, or chytrids, reproduce sexually by a variety of means, including isogamy, anisogamy, oogamy, contact between gametangia (structures that contain the gametes), and contact between vegetative structures. However, the chytrids are better defined by their asexually produced, motile, uniflagellated zoospores; in fact, the chytrids are the only phylum of fungi that produce such spores. There are about 790 species in this phylum.

Zygomycota

Zygomycota reproduce sexually by producing zygospores within a zygosporangium. These fungi have a mycelial growth form that forms crosswalls only at reproductive structures. The zygomycetes produce spores asexually within a sporangium. Sexual reproduction leads to the production of a zygospore within a zygosporangium.

Mycelial cultures of a single strain may mate (in homothallic species) or may require a different mycelial strain (in heterothallic species) in order to

plants have mycorrhizal associates. The fungus will form one of two types of associations with the plant. *Endotrophic mycorrhizae* penetrate cortical root cells with specialized hyphae that are finely branched. Other fungi grow around the root and between the cortical cells but never penetrate the cells. More than 80 percent of fungal genera that form mychorrizae are in the phylum *Basidiomycota*.

Lichens are an association that appears to be a controlled parasitism of an alga partner by a fungal partner. Ninety percent of lichens have one of three genera of algae as the photosynthetic partner: *Trebouxia, Trentepohlia*, and *Nostoc*. The first two are *Chlorophyta* (green algae), and the latter is a mate. In both cases, copulation of two multinucleated gametangia occurs. Two lateral branches grow toward each other, sometimes attracted by the hormone trisporic acid. When these branches touch, a wall is laid down near the tip of each branch, creating the multinucleated gametangia. The wall separating the two gametangia at the point of contact breaks down, the protoplasmic contents mix, and the nuclei pair up and fuse. As the nuclei pair, a wall surrounding the fused gametangia develops into the zygosporangium. This thick-walled structure is resistant to environmental abuse. At or just prior to germination, the nuclei within the zygospore undergo meiosis. A stalk grows out of the zygospore bearing a sporangium. Included in this phylum of about 1,060 species are *Rhizopus*, the black bread mold; *Pilobolus*, the "cap" throwing fungus; and *Entomophthora*, the fly fungus.

Ascomycota

The ascomycetes reproduce sexually by producing ascospores within an ascus. Ascomycetes have a mycelial growth form with incomplete crosswalls. The septal hole permits the migration of cellular contents from cell to cell. *Yeasts* are single-celled ascomycetes that normally do not produce mycelial forms. Asexually, the *Ascomycota* reproduce by budding, fission, or production of chlamydospores or conidiospores. The unicellular yeasts reproduce using fission and budding methods. The majority of ascomycetes reproduce asexually, using conidiospores. A cell that is going to generate a conidium grows directly from hyphae or at the end of a stalk. The conidiogenous cell produces a swelling at the tip, into which protoplasm and a nucleus migrate. The nucleus is produced by mitotic means and is haploid. After swelling is complete and migration finishes, a wall is formed, separating the conidiospore from the conidiogenous cell.

Sexual reproduction occurs by gametangial contact between the female structure (the ascogonium) and the male structure (the antheridium). Usually, opposite mating types are required for sexual reproduction to occur. When contact between the gametangia occurs, nuclei from the antheridium migrate into the ascogonium and pair up with nuclei in the ascogonium. Alternatively, conidia or microconidia may land on the ascogonium and fuse with it. In both cases, fusion of paired nuclei is delayed until development of the ascus. Instead, special hyphae grow out from the ascogonium and

will eventually generate the ascus. These ascogenous hyphae possess two nuclei, one of each mating type, in each cell.

Frequently the development of the ascogenous hyphae and ascus is associated with the development of a fruiting body. When the fruiting body is at the appropriate developmental stage, a hooklike structure develops at the end of the ascogenous hyphae. At this time the nuclei fuse, producing the only diploid cell in the life cycle of these organisms. As the saclike ascus elongates, the diploid nucleus undergoes meiosis, producing four haploid nuclei. Usually one mitotic division then occurs, resulting in eight nuclei in the elongated sac. Eventually walls are formed around each nucleus, creating the ascospore. As noted above, many *Ascomycota*—but not the yeasts—produce a fruiting body, called the ascoma. The ascoma may take a variety of shapes, from a closed, ball-like structure (cleistothecium) to a pear-shaped structure with an opening at the top (perithecium) to a cup-shaped open structure (apothecium) or a stromatic structure containing cavities (ascostroma). In all cases, the asci with the ascospores are contained within these structures. The type of ascoma, the number of walls in the ascus, and the presence or absence of a fertile layer from which the asci arise are criteria used to distinguish among the about 32,300 species of *Ascomycota*.

Basidiomycota

Basidiomycota reproduce sexually by producing basidiospores on a basidium. Like the *Ascomycota*, the *Basidiomycota* fungi also have a mycelial growth form with incomplete crosswalls. However, the structure of these crosswalls is different in that there is a swelling surrounding the pore where the wall is incomplete. There is also a curved membrane on each side of the hole, together looking much like parentheses. This septal structure is called the dolipore septum and is a secondary characteristic of the *Basidiomycota*.

The mycelia of most *Basidiomycota* pass through three stages. The primary mycelium has cells with a single nucleus, all derived from the germinated basidiospore; it is said to be *homokaryotic*. Later in development, fusion of hyphae of opposite mating types occurs, establishing the secondary mycelium, in which each cell has two nuclei; the secondary mycelium is therefore *dikaryotic*. Clamp connections occur on the secondary mycelium to facilitate division of the two nuclei in limited space. *Clamp connections* are another secondary characteristic of the *Basidiomycota*. Tertiary mycelium develops in the specialized organized tissues of the fruiting bodies, the *basidioma*.

Asexual reproduction takes place by means of budding, fragmentation of mycelium, production of chlamydospores, or conidia. Chlamydospores are fragmented sections of mycelia that have rounded up and formed thick walls. Sexual reproduction in the *Basidiomycota* occurs primarily by fusion of genetically compatible hyphae, thus establishing the secondary mycelium. Fusion of these nuclei is delayed until the production of the ba-

sidium. Nuclei of the dikaryotic cell that is to become the basidium fuse to produce one diploid nucleus that immediately undergoes meiosis, resulting in four haploid nuclei. Depending on the type of basidiomycete fungus, the haploid nucleus and associated cytoplasm migrate into a swelling that develops at the tip of the basidium. When full-sized, a wall separates the basidiospore from the basidium. Eventually the basidiospore falls off or is shot off the basidium. Classification of the approximately 22,250 species of *Basidiomycota* depends on the presence or absence of a fruiting body, septation, and the number of cells in the basidium.

Fungi: Phyla and Characteristics

Group	Species	Characteristics
Ascomycota	32,000+	Members of this phylum produce sexual spores in a specialized cell called an ascus. Includes *Neospora* (powder mildew), *Morchella* (morels), and *Tuber* (truffles). Causes mildews, fruit rots, chestnut blight, Dutch elm disease.
Basidiomycota	22,000+	Members of this phylum produce sexual spores on a specialized cell called a basidium. Includes three classes: *Basidiomycetes* (mushrooms, stinkhorns, puffballs, bird's nest fungi), *Teliomycetes* (rusts), and *Ustomycetes* (smuts). Causes rusts and smuts; includes poisonous varieties of mushrooms such as *Amanita*.
Chytridiomycota	800+	Members of this phylum are mainly aquatic organisms with motile spores with a whiplike flagellum. Includes gut chytrids, which live in the rumina of herbivores. Includes *Allomyces*, *Batrachochytrium* (parasite of frogs), and *Coelomomyces* (parasite of various insect hosts). Causes corn spots, crown wart in alfalfa, black wart in potatoes.
Zygomycota	750+	Members of this phylum are recognized by their rapidly growing hyphae. Includes bread molds and fruit rots (*Rhizopus stolonifer*); *Glomus versiforme* is often involved in mycorrhizae ("fungus roots").
Deuteromycetes	15,000+	Also called "fungi imperfecti," an artificial group based on the members' purely asexual form of reproduction. Includes fungi in the genus *Penicillium*, involved in cheese making and in the antibiotic penicillin. Also cause rots, molds, diseases of grains, vegetables, and fruits. *Aspergillus flavus* produces the carcinogenic toxin found on peanuts, aflatoxin.
Lichens	17,000+	Symbiotic relationships of fungal (mycobiont) and algal (photobiont) partners, familiar on substrates such as rocks, wood, tree trunks. Most of the mycobionts are members of *Ascomycota* or *Basidiomycota*. Some species live more than a century. Important as environmental indicators, food for tundra animals, sources of dyes, potential medicinal properties.
Yeasts	600+	Not a formal taxon but an artificial group of unicellular fungi that grow by budding. Yeasts in genus *Saccharomyces* are used in food manufacturing (bread dough, alcohol fermentation); also cause disease, such as fungal infections (*Candida albicans*) and pneumonia (*Pneumocystis carini*). Most are members of *Ascomycota*; a quarter are from *Basidiomycota*.

Source: Data on species are from Peter H. Raven et al., *Biology of Plants*, 6th ed. (New York: W. H. Freeman/Worth, 1999).

Deuteromycota

Deuteromycota are fungi that reproduce only asexually, by production of conidiospores. The majority of these fungi are thought to be derived from the *Ascomycota* by evolutionary loss of sexual stages; either that, or the sexual stages have yet to be discovered. Because the sexual stages, also called the "perfect" stages, have been lost, these fungi are also referred to as the "imperfect fungi," or "fungi imperfecti." About fifteen thousand form-species are grouped into larger form-genera and form-classes, based on the morphological characteristics of their asexual reproductive structures. Because this is an artificial classification (not based on evolutionary relationships), no basis for conclusions about relatedness within groups can be implied or inferred.

Evolutionary History

Because of a lack of sufficient fossil evidence, phylogenetic relationships have been inferred based on morphological features associated with cell structure and sexually produced structures. With the advent of sequencing analysis of proteins and nucleic acids, observations of some relationships have been confirmed, and new relationships are being discovered.

Small subunit rDNA sequence analysis shows that the fungi are derived from a flagellated animal ancestor. These data show fungi to be a *monophyletic* group, with the *Chytridiomycota* and *Zygomycota* as the earliest branches within the group. The facts that all fungi utilize the AAA synthetic pathway for lysine and possess cell walls of chitin support this monophyletic view of all fungi. Sequence analysis supports a relationship between the *Ascomycota* and *Basidiomycota*, perhaps both being derived from yeasts. The evidence for the monophyletic evolution of the *Basidiomycota* is strong, based on sequence analysis and morphological features such as ballistospores, basidia, and clamp connections. Sequence analysis, however, appears to contradict morphologically based phylogeny groupings, which use structure and number of cells in the basidium and presence or absence of a fruiting body as key features. Within the *Ascomycota*, phylogeny appears to be monophyletic, but the evidence is not strong.

John C. Clausz

See also: Ascomycetes; *Basidiomycetes*; Basidiosporic fungi; Chytrids; Deuteromycetes; Lichens; Rusts; *Ustomycetes*; Yeasts; Zygomycetes.

Sources for Further Study

Alexopoulos, Constantine J., C. W. Mims, and M. Blackwell. *Introductory Mycology.* 4th ed. New York: John Wiley, 1996. The primary thrust of the book is morphology and taxonomy, but it also includes topics about activities of fungi.

Hudler, George W. *Magical Mushrooms, Mischievous Molds.* Princeton, N.J.: Princeton University Press, 1998. A readable book about fungi in human affairs. Intended for a nontechnical audience.

Kendrick, Bryce. *The Fifth Kingdom.* 3d ed. Newburyport, Mass.: Focus Information Group, 2001. Deals with classification, ecology, and genetics of fungi, fungi as plant pathogens, predatory fungi, biological control, mutualistic relationships with animals and plants, fungi as food, food spoilage and mycotoxins, poisonous and hallucinogenic fungi, medical mycology, antibiotics, and organ transplants.

GARDEN PLANTS: FLOWERING

Categories: Angiosperms; economic botany and plant uses; gardening

The most popular flowering garden plants worldwide include roses, lilies, tulips, irises, and daffodils. These plants range in form from herbs to bushes to trees. Their leaves, flowers, and optimum growth needs differ greatly.

Flowering plants (*angiosperms*) are grown in gardens for their beauty and their fragrant aromas. They and their relatives can also be grown for food. Myriad flowering plant types are grown in the world's gardens. Most popular flowering garden plants are in the rose family (*Rosaceae*, about three thousand species) and the lily family (*Liliaceae*, about forty-five hundred species), which together make up a wide variety of herbs, bushes, and trees. Also quite popular are members of the iris family, *Iridaceae*, which includes crocuses. Lilies and irises belong to the order *Liliales* (about eight thousand species) of herbaceous flowering plants. Other well-known plants of this order are tulips, daffodils, and hyacinths.

Roses and Lilies

The most popular garden plants are roses and lilies. Rose is the common name for members of the family *Rosaceae*, a family of one hundred genera and three thousand species. Included are important fruit and ornamental species, including the familiar genus *Rosa* (true roses). *Rosaceae* and more than twenty other families belong to the order *Rosales*. *Rosaceae* grow as trees, shrubs, or perennial herbs. Within this family, food is produced by apple, pear, peach, plum, cherry, apricot, almond, and nectarine trees. Many berries, including raspberries, blackberries, and strawberries, are *Rosaceae*.

Since antiquity, the true rose has been among the most popular garden flowers in the world. Roses evolved from sweetbriers (wild roses). This genus of perennials, with about one hundred species, is mostly native to the north temperate zone. Experts recognize two main classes of the approximately thirteen thousand cultivated rose varieties (*cultivars*) which have arisen from hybridization of a few original species, mostly from Asia. Members of the original class, such as brier, damask, and moss roses, bloom once a year, in early summer. The others, called *perpetual roses*, bloom more than once a season. They include the tea roses, polyanthas, and rugosas. Tea roses smell like tea or fruit. Other roses have the distinctive rose smell or no smell at all. True rose flowers are white or various shades of yellow, orange, pink, or red. A perpetual rose bush can grow up to 6 feet (2 meters) tall. Polyantha bushes are low and bear flower clusters; shrub roses grow up to 15 feet (4.5 meters) tall. The leaves of rose plants have stipules (leaflike appendages) at stalk bases and are most often compound.

The name "lily" indicates any of forty-five hundred species of the family *Liliaceae*. This is one of the largest, most important plant families in the order *Liliales*. The herbaceous flowering plants have beautiful, showy flowers. True lilies, *Liliaceae* of the genus *Lilium* (one hundred species), are native to temperate areas of the Northern Hemisphere and are among the oldest of cultivated plants. Examples are Colchis, tiger, Madonna, and Easter lilies. Within the same family are onions, garlic, and asparagus. Also included in *Liliales* are tulips, daffodils, hyacinths, and amaryllis. Among the eight thousand *Liliales* species are herbs, climbing shrubs, succulents, and trees. Their thick, fleshy stems grow from underground storage organs, and all have narrow, upright leaves with parallel veins. Most grow worldwide but flourish only in temperate and subtropical areas. These perennials bloom once yearly and store food and water in scaly bulbs, corms, or rhizomes. Stems and leaves may be storage organs, too, and have thick bark to prevent water loss. Many plants in the group can carry out asexual reproduction via bulblets on parent bulbs or flower clusters (for example, garlic).

Tulips, Daffodils, and Irises

Tulips and irises share some characteristics. The *Liliaceae* genus *Tulipa* contains about one hundred tulip species. They are native to Asia and the eastern Mediterranean region, and thousands of tulip cultivars are popular garden flowers. A tulip plant produces two to three thick, blue-green leaves, clustered at its base.

Irises, also in the *Liliales* order, belong to the family *Iridaceae*, which includes some of the world's most popular garden flowers. Irises are indigenous to the north temperate zone, Asia, and the Mediterranean region. "Iris" is the common name for spring-flowering, bulbous herbs. The leaves of iris plants rise directly from the bulb or rootstock; they are very narrow, erect, and swordlike. Iris plants arise from rhizomes (stems underground) or bulbs and have large flowers. *Iris*, the Greek word for "rainbow," refers to the flowers' rainbowlike color combinations.

Garden Plants and Places of Original Cultivation

Plant	Region	Plant	Region
Acacia	Australia	Hydrangea	China
African violet	Africa	Impatiens	Africa
Azalea	Japan	Japanese iris	Japan
Bellflower	Europe	Lily of the valley	Europe
Bird of paradise	Africa	Lobelia	Africa
Black-eyed Susan	North America	Lupine	North America
Bleeding heart	Japan	Marigold	Mexico
Bottle brush	Australia	Michaelmas daisy	North America
California poppy	North America	Morning glory	South America
Calla lily	Africa	Nasturtium	South America
Camellia	China	Nemesia	Africa
Candytuft	Mediterranean	Oleander	Mediterranean
Carnation	Mediterranean	Pansy	Europe
Checkered lily	Europe	Pelargonium	Africa
China aster	China	Penstemon	North America
Chrysanthemum	China	Peony	China
Clarkia	North America	Petunia	South America
Clematis	China	Phlox	North America
Columbine	North America	Plumbago	Africa
Coreopsis	North America	Poinsettia	Mexico
Cosmos	Mexico	Polyanthus	Europe
Crocus	Europe	Portulaca	South America
Dahlia	Mexico	Rose	Europe
Day lily	China	Salpiglossis	South America
Forget-me-not	Europe	Scabious	Europe
Forsythia	China	Snapdragon	Mediterranean
Foxglove	Europe	Snowdrop	Europe
Frangipani	Mexico	Stock	Europe
Freesia	Africa	Strawflower	Australia
Fuchsia	South America	Sunflower	North America
Gardenia	China	Sweet alyssum	Mediterranean
Gladiolus	Africa	Sweet pea	Mediterranean
Gloxinia	South America	Verbena	South America
Grape hyacinth	Mediterranean	Wallflower	Europe
Hollyhock	China	Wisteria	Japan
Hyacinth	Mediterranean	Zinnia	Mexico

Source: Data are from Brian Capon, *Botany for Gardeners: An Introduction and Guide* (Portland, Oreg.: Timber Press, 1990), pp. 81-82.

DIGITAL STOCK

"Iris" is the common name for spring-flowering, bulbous herbs. Irises are indigenous to the north temperate zone, the Mediterranean region, and Asia.

Another interesting and popular group of flowering garden plants is daffodils, of the genus *Narcissus*. They are sold as bulbs. Daffodils are so popular that they have been widely hybridized. A daffodil plant usually produces five to six lance-shaped leaves grouped about the base of its stem. Each plant has one large flower.

Blooming Habits: Roses and Lilies

Wild rose plants have regular, single flowers, with five petals. In most cultivars *double flowers*, having petals numbered in multiples of five, are produced. The flower also has a calyx with five lobes, many stamens, and one or more carpels. Rose sprouts have two seed leaves, so the plants are eudicots. Flowers of most cultivars bear few seeds, and the majority of them are sterile. The number of seeds is small because in double roses, flower parts that would otherwise produce seeds become extra petals. Therefore, most roses are grown from cuttings. All new rose varieties begin as seedlings, raised from fertile seeds.

Lily flowers grow one per stalk or in clusters. In contrast to roses, they have six petal-like segments,

causing the flowers to resemble trumpets or cups. The flowers range from white to shades of almost all other colors except blue. Lily flowers all have three-chambered ovaries with nectaries between the chambers. They produce large, well-developed seeds which hold plenty of food-storage tissue and embryos. Plants of most species are 1 to 4 feet (0.3 to 1.3 meters) tall, though a few grow up to 8 feet (nearly 3 meters) tall. Lilies are usually raised from bulbs but can be grown from seed. Most species of these perennials bloom once, in July or August. However, flowering periods of some species begin in May or late autumn.

Blooming Habits: Tulips, Irises, and Daffodils

Tulip flowers can be single or double, and most, called "self-colored," grow in a huge range of solid colors, from pure white to many shades of yellow, red, brown, and purple. Some, called "broken tulips," have varicolored flowers caused by a harmless virus carried by aphids. The garden tulip, introduced into Western Europe from Turkey in the 1600's, is commercially cultivated most in the Netherlands and the United States, especially Michigan and Washington. Bell-shaped lilies are usually solitary. They have three petals and three sepals, six stamens, and a triple ovary that ends in a three-lobed stigma. A tulip fruit holds numerous seeds, but many of the four thousand tulip cultivars can be propagated only via their scaly bulbs.

Iris flowers are more asymmetric than those of roses, lilies, or tulips. They possess six petallike floral segments: three inner, erect *standards* forming an arch atop the flower and three outermost, drooping, often multicolored *falls*. A set of three petallike stigmas also cover the stamens. Iris colors include white, yellow, bronze, mauve, purple, and red. Best-known are rhizomatious bearded (German) irises, which have multicolored falls. These cultivars arose from European species. They have stems up to 3 feet (1 meter) tall and yearly bear at least three flowers per plant. The twentieth century saw the development dwarf bearded irises and fragrant cultivars. The best-known beardless, rhizomatous irises are Japanese and Siberian irises, which have clusters of blooms. English and Spanish irises grow from bulbs.

Daffodils, the best-known members of the genus *Narcissus*, are indigenous to northern Europe and have been widely cultivated there and in North America. Usually daffodil plants grow to heights of about 1.5 feet (about 0.5 meter). Each plant produces one large blossom on its centrally located stem. The blossom has a corolla split into six lobes and a centrally located trumpet, the *corona*. The corona is frilled at its edges, contains flower stamens, and leads to its pistil. The flowers were originally yellow; however, they have been hybridized into cultivars in which the trumpet and petals are often of contrasting yellows, whites, pinks, or oranges.

Rose and Lily Cultivation

Although found all over the world, roses grow best in mild climates, such as southern France and the U.S. Pacific coast. Roses' excellent growth in many different kinds of soil and climate is a result of the availability of myriad cultivars. A rose garden should be protected from cold wind and be exposed to sunlight for several hours a day. Deep, rich loam is best for roses, but most cultivars grow in sandy and gravelly soil. The soil must be well drained. Roses are planted in spring, about 2 feet (0.6 meter) apart. In the United States, about sixty-five million commercial rose plants are cultivated yearly. About 35 percent are grown for cut flowers, and the rest are used in gardens or landscaping. For the best growth, rose plants require severe pruning, which is adapted to the intended use of the flowers. Most varieties are grown by budding on understocks. Roses are susceptible to diseases such as rust and black-spot disease, so pests should be watched for and discouraged.

Lilies grow best in well-drained, deep, sandy loam, sheltered from winds and hot sunlight. Their bulbs are planted 6 or more inches (15 or more centimeters) underground, in late fall. This deep placement is used because they send out their roots well above the bulbs. As soon as blooms wilt, their seed pods should be removed. Lilies can be made to bloom early by putting the bulbs in pots and covering them with peat moss or soil. When kept at 50 degrees Fahrenheit for two weeks and then stored at 60 degrees Fahrenheit, they bloom in three months. Lilies are susceptible to a number of diseases. The most serious is mosaic, carried from plant to plant by aphids. Infected plants should be uprooted at once and burned to prevent an epidemic. A second severe disease of lilies is botrytis blight.

Tulip, Daffodil, and Iris Growth

Tulips are early bloomers, flowering in early spring; mid-season-bloomers; or late bloomers. Late bloomers are the largest group, with the widest range of growth habits and colors. Tulips flourish in any good soil but, like roses, do best in well-drained loam. The bulbs are planted in autumn at depths of 6 or more inches. These perennial plants flower annually for a few years, but eventually flowering diminishes. For best flower yields, after four to five years it is necessary to dig up bulbs after the flowers are gone and the foliage yellows. The bulbs should then be stored in a cool, dry place and replanted in autumn. Tulips are rarely attacked by garden pests.

Daffodil bulbs are planted in the fall, in loose soil, about 4 inches (10 centimeters) deep. The plants appear in mid-February. Their blossoms open in early April, announcing spring. Like tulips, the plants flower well, annually, for a few years and then diminish. For best flower yields, it is useful to store the bulbs every four to five years in the manner described above for tulips.

Irises bloom from March to July and may be planted in the spring before blooming or in the autumn. These perennials give the best flowers if plants are replanted every four to five years to eliminate the problem of overcrowding. Bearded irises do best in sunny areas where the soil is not too rich, but beardless ones prefer damp, rich soil. Iris diseases include crown rot, soft rot, and leaf spot. If the disease is serious, roots and soil may have to be treated. The worst iris insect enemy is the iris borer. Its larvae eat through leaves and roots, bringing on soft rot.

Sanford S. Singer

See also: Angiosperms; Bulbs and rhizomes; Flower structure; Flower types; Fruit crops; Hybridization.

Sources for Further Study

Heath, Brent, and Becky Heath. *Daffodils for American Gardens.* Washington, D.C.: Elliott and Clark, 1995. Discusses and illustrates daffodils thoroughly and clearly. Includes illustrations, bibliography, and index.

Heywood, Vernon H., ed. *Flowering Plants of the World.* New York: Oxford University Press, 1993. Discusses many flowering plants, including morphology, growth, and disease. Illustrated; includes bibliography and index.

McRae, Edward A. *Lilies: A Guide for Growers and Collectors*. Portland, Oreg.: Timber Press, 1998. Discusses lily issues, including cultivation, species, diseases, and hybridizing. Illustrated; includes glossary, bibliography, and index.

Mattock, John, ed. *The Complete Book of Roses*. London: Ward Kock, 1994. Describes many roses, their morphology, cultivation, diseases, hybrids. Illustrated; includes bibliography, index.

Pavord, Anna. *The Tulip*. New York: Bloomsbury, 1999. Covers many aspects of tulip history, morphology, cultivation, and hybridization. Illustrated; includes bibliography, index.

Stebbings, Geoff. *The Gardener's Guide to Growing Irises*. Portland, Oreg.: Timber Press, 1997. Discusses iris types, cultivation, species, and hybridization. Illustrated; includes glossary, bibliography, and index.

GARDEN PLANTS: SHRUBS

Categories: Angiosperms; economic botany and plant uses; gardening

Shrubs are perennial, woody plants that have multiple ecological uses and are economically important for a variety of landscape purposes. Like trees, the shrubs consist of both primary and secondary cambium (cells that ultimately form wood or bark), but unlike trees, shrubs typically have multiple stems and are somewhat shorter in height.

Native and nonnative shrubs have many uses, both in private gardens and as elements in commercial and civic landscaping. Shrubby willows and alders function as soil stabilizers, preventing erosion naturally along waterways. Other shrubs are natural windbreaks that reduce wind flow over open soils. Farmers value shrubby windbreaks that preserve barren farm soils during fall and winter months, when the land is barren and topsoil is especially liable to be blown away during winter storms. For much the same reason, homeowners may plant certain types of shrubs at strategic locations along the boundaries of the yard to reduce air flow.

Many shrubs are economically important as landscape ornamentals. They are prized for the beauty, variety, pattern, and color of their leaves, buds, flowers, or fruits. The color, shape, and texture of their leaves are all valued qualities for which shrubs are appreciated and used. Their shape, texture, pattern, and even the intricate contrast and pattern of their leafless twigs in winter can enhance the beauty and interest of a garden landscape.

Food Sources

In addition to their uses in gardens, many shrubs are important sources of foods for wildlife and for humans. Fruits of blackberries, blueberries, huckleberries, currents, gooseberries, plums, and hazelnuts are but a few of the foods obtained from shrubs. Many, such as blueberries and raspberries, are made into pies. Some currents and gooseberries are key components of jams and spreads. Still others, such as plums, cherries, and grapes, are eaten as fruits or used in the preparation of jams, jellies, pies, as cooking ingredients, or used for other baked goods. Some fruits are gathered only in the wild, but many others are cultivated. Cranberries, blueberries, and blackberries are cultivated varieties derived from hundreds of years of selecting and cultivating wild native shrubs. Fruits of plums, cherries, and especially the serviceberries, or Juneberries, are used to make fine and natural wines, and it may be that serviceberries were named because they provided the earliest fruits that could be made into wines for the Eucharist during church services.

Ecologically, shrubs are also important for wildlife. They provide food, shelter, and resting and nesting areas for a wide variety of birds. For example, at least forty-six species of birds feed on elderberry fruits.

Medicinal Uses

By 1812 Johann Jacob Paul Moldenhawer had shown that plant tissue was composed of independent cells, and in 1826 Henri Milne-Edwards determined that all animal tissues were formed from globules. From time immemorial, humans have enjoyed the medicinal advantages of many species of shrubs. An extract derived from the bark of willows, called salicin, is the active ingredient in one of the most important and useful of all household drugs: aspirin. The distilled liquid from the bark of witch hazel is sold as a skin ointment. A soothing ointment is made of the oil of teaberry. Tannic acid, used for tanning leather and for medicinal purposes, was for many years obtained from the fruits of native sumacs.

Barberries

The barberries (*Berberis*) are a worldwide group of deciduous shrubs best known for their thorns and bright red berries. They have yellow wood and simple alternate or whorled leaves clustered on short spurs. Barberries were very popular landscape ornamentals, but their reputation and use drastically decreased when they were found to be the alternate host for the black stem rust, a fungus that infects wheat. The use of hybrid varieties eliminates this ecological problem, and the barberries have once again become common as interesting and colorful ornamentals that provide good hedges, ground covers, and living barriers. The bright red leaves make them an attractive fall shrub, while the persistent, bright red berries attractively color a yardscape through the winter months. The berries are also an important wildlife food used by a variety of birds and small mammals through the winter months.

Burning Bushes and Dogwoods

The various species of burning bushes, or *Euonymous*, are attractive and bushy shrubs that offer good landscape contours and contrast, but they are probably best known for their gorgeous scarlet, fire-red, or fire-pink colors during the fall season. Dogwoods (*Cornus*) are shrubs or small trees that have opposite deciduous leaves and branching. A number of dogwood shrubs are useful and colorful as garden ornamentals. Chief among these are the silky dogwood (*Cornus amomum*) and red-osier dogwood (*Cornus stolonifera*), which grow in moist soil conditions. Red-osier dogwood is the more colorful of the two shrubs, with bright red branches that provide attractive year-round landscape color, especially in winter. Almost all of the dogwoods are valuable garden shrubs for wildlife, both as nesting cover and for their fruits.

Dwarf Trees as Evergreen Shrubs

Evergreen shrubs add beauty, texture, and year-round cover as ornamentals. Many species are noted for their broad leaves and showy flowers, others for the protective cover that they offer wildlife. Many dwarf varieties of trees have also become landscape shrubs that offer year-round color, variety, and cover. Another of their properties is their remarkably slow growth, usually only a few inches a year. Dwarf trees are used as ground covers, creeping along the ground and over stone walls and fences. Some of the many types include the cypress (*Chamaecyparis*), which provides dense and colorful foliage; junipers (*Juniperus*), which range from low growth forms to spreading forms and are able to tolerate urban and suburban conditions of low moisture and poor soils; dwarf pines (*Pinus*), whose long needles offer landscape and rock garden diversity; and yews (*Taxus*), which are commonly used for hedges and topiaries.

Of these, the prostrate or ground junipers are especially useful. These low-growing evergreen shrubs rarely reach a height greater than 3 feet but may spread in a circular mat several feet in diameter. The horizontal stems of ground juniper burrow into the soil surface to send up numerous upright-arching stems that form a dense ground cover. The sharp and prickly points of the leaves also make this shrub useful as a protective mat or hedge.

Fruiting Shrubs

In addition to cover and color, many species and varieties of shrubs offer fruits for humans and for wildlife. These include the blueberries (*Vaccinium*), which occur in ornamental forms and also in low, spreading varieties, and the currants and gooseberries (*Ribes*), which offer black, purple, red, and white berries, depending on variety or species, that can be harvested and made into jellies and jams or simply left on the shrub to attract wildlife. The blackberries and raspberries (*Rubus*) are a diverse group of fruiting shrubs famous for their fruits but less desirable as garden shrubs because of their tenacious ability to spread if left untended. However, the fruits are harvested for preserves, pies, wines, and as ingredients in breads and other baked goods. The tangled thickets or briar patches that they form can be useful both as living hedges and as wildlife cover for quail, songbirds, and a variety of mammals.

Hollies and Hydrangeas

The hollies (*Ilex*) and inkberries are reasonably hardy shrubs that offer year-round beauty, color, and texture as landscape ornamentals. They are best known for their lustrous, bright green leaves, intricate spiny leaf shapes, and bright red berries, and they are especially desirable as plants during the winter holidays, when holly sprigs and berries brighten both outdoor yards and indoor holiday festivities. Many species offer cover and seasonal nesting habitat for a variety of bird species. Another favorite shrub is the hydrangea (*Hydrangea*), with its large clusters of flowers; it is available in both a compact growth form and larger varieties. Hydrangeas are most often used as decorative shrubs for shady or semishady and moist corners of lots and buildings.

Rhododendrons and Azaleas

Taking their name from two Greek words, *rhodo* for "rose" and *dendron* for "tree," the rhododendrons and azaleas (*Rhododendron*) are best known for their bright and showy flowers, but these shrubs also offer glossy green leaves. The large and showy flowers of rhododendrons come in purple, pink, red, and white, while azalea flowers offer additional yellow, orange, and apricot shades. Many rhododendrons are huge shrubs that are especially valuable as wildlife cover and as nesting habitat for desired species, such as cardinals.

Roses

Roses (*Rosa*) have been cultivated for centuries and celebrated in poetry, songs, and artwork for an equally long time. They occur in an enormous number of species and varieties and in every shade of flower color except blue. Roses have been bred into many varieties, from miniatures to ground-covering polyanthas to bush roses and floribundas to tall hybrid teas, grandifloras, and climbers. "Old roses" and wild roses have returned to commercial markets in recent years. Prolific and relatively easy to grow in most climates, roses fill many landscape needs, from garden color to cut flowers and even living fences to protect property.

Spireas and Sumacs

Spireas (*Spirea*) are deciduous shrubs with small, alternate, toothed, or lobed leaves. They are popular deciduous shrubs because they are generally easy to grow and make useful hedges, mass plantings, and borders. They are also popular for their dainty clusters of spring flowers. Many varieties are desired for their long blooms from spring through most of summer, while others are sought for their fall colors or yellowish-green foliage. Some of the spirea varieties offer low, spreading ground covers, while others are compact decorative shrubs. The sumacs (*Rhus*) are shrubs or small trees that often grow in thickets. Although infrequently used in gardens, several species offer both rapid growth cover in poor soils and rich fall colors.

Dwight G. Smith

See also: Fruit: structure and types; Garden plants: flowering; Leaf arrangements; Leaf shapes.

Sources for Further Study

Kellog, John, ed. *The Hillier Gardner's Guide to Trees and Shrubs*. Pleasantville, N.Y.: Reader's Digest Assocation, 1995. Colorful and well-illustrated volume includes descriptions of more than four thousand trees and shrubs. Also includes chapters on basic biology, propagation and care, pests, and diseases.

Krussman, Gerd. *Manual of Cultivated Broad-Leaved Trees and Shrubs*. 3 vols. Portland, Oreg.: Timber Press, 1986. Technical but comprehensive descriptions include thousands of species of shrubs. Keys, photos, and general information also included.

Symonds, George W. D. *The Shrub Identification Book*. New York: William Morrow, 1963. This shrub identification book has been reprinted many times and is still a mainstay for students who need a pictorial guide to shrubs. Good visuals organized in many sections, including thorns, leaves, flowers, fruit, twigs, and bark. All are profusely illustrated.

Whitman, Ann. *Trees and Shrubs for Dummies*. Foster City, Calif.: IDG Books WorldWide, 2000. Written with the collaboration of the editors of the National Gardening Association, this interesting and entertaining book is aimed at gardeners and landscapers but still has much to offer the scientist.

GAS EXCHANGE IN PLANTS

Categories: Photosynthesis and respiration; physiology

Gas exchange is the process whereby water vapor and oxygen leave and carbon dioxide enters plant leaves. The gaseous balance in plants is quite complex because plant cells carry on both respiration and photosynthesis.

All living organisms continually produce gases via metabolic and cellular activities, and the vast majority of living things are in one way or another in intimate contact with a gaseous medium. In most instances, therefore, there is ample opportunity for all organisms to exchange gases with the environment. The gaseous balance in plants is quite complex because plant cells carry on both *respiration* and *photosynthesis*. Plants respire in much the same way as animals; oxygen is used to oxidize carbohydrates, and carbon dioxide and water are produced as waste products. The photosynthetic process requires an input of carbon dioxide and water. These two *reactants* are used to produce carbohydrates, and oxygen is released as a waste product. Under normal conditions, photosynthetic rates are higher than respiration rates; thus, there is a net increase in oxygen production, accompanied by a net increase in the usage of carbon dioxide. On balance, therefore, plants use carbon dioxide and produce oxygen.

Stomata and Guard Cells

The gases move into and out of the plants through specialized openings located along the lower surface of the leaf. These openings, called *stomata*, are of optimum size, shape, and distribution for the efficient diffusion of gases. Each stoma (or stomate) is surrounded by two specialized structures called *guard cells*. These two cells are attached together at each end of both cells. The lateral edges of the two cells are not attached to each other, but, when flaccid, the sides of the guard cells do touch each other and effectively close the stomate. Specialized structural components prevent the guard cells from increasing in diameter as expansion occurs. Hence, when guard cells take up water, expansion takes place only along the longi-tudinal axis. Because the ends of the cells are connected to each other, the expanding of the cells forces the sides apart and results in the opening of the stomate.

Role of Water

The opening of stomata is dependent on how well hydrated the plant is. The water initially comes from the soil. The water enters the root by *osmotic* processes, then moves across the root and into the *xylem* tissues, which transport it up the stem to the leaves. From the xylem in the leaves, the water moves into the *palisade* and spongy *parenchyma cells*, which make up the bulk of the leaf tissue. The water then moves into the subsidiary cells that immediately surround the guard cells.

When the leaf is exposed to light, the process of photosynthesis begins. As the photosynthetic reactions proceed in the guard cells, the residual carbon dioxide is converted to carbohydrates. The disappearance of carbon dioxide from the cytosol of the guard cell results in an increase in the cellular pH. As the pH rises, the activity of the enzymes that convert starch and sugars to organic acids increases. The higher concentration of organic acids results in a higher concentration of hydrogen ions. The hydrogen ions of the guard cells are then exchanged for potassium ions in the subsidiary cells. This increased concentration of potassium, combined with the higher levels of organic acids, lowers the osmotic potential of the guard cells, and, since water moves from regions of high osmotic potential to regions of lower osmotic potential, water will move from the subsidiary cells into the guard cells. This movement of water increases the *turgor pressure* (inner pressure) of the guard cells and causes them to swell. Thus, the stomata open.

Oxygen Out, Carbon Dioxide In

Once the stomata open, the intercellular free space around both the palisade and spongy parenchymas is put into continuous contact with the outside atmosphere. As the water within the parenchyma cells moves across the cellular membranes, it evaporates into the free space and diffuses through the stomata into the atmosphere. Oxygen produced during photosynthesis exits the plant in much the same manner as the water vapor.

Carbon dioxide, however, follows the reverse path. It enters through and across cell membranes into the parenchyma tissues. In each case, the gas involved moves down a concentration or pressure gradient. The pressure of water vapor and oxygen is higher inside the leaf's free space than in the atmosphere, whereas the partial pressure of carbon dioxide is greater in the atmosphere than within the free space. Thus, the impetus is for the former two gases to move out of the plant and for the latter to enter it.

This exchange will take place as long as the stomata remain open and the pressure gradient is in the right direction. As a general rule, stomata close in the dark. Without an input of solar energy, the light-mediated reactions of photosynthesis stop. In the absence of these reactions, the carbon dioxide level increases, thus decreasing the pH. The lower pH activates the enzymatic conversion of organic acids to sugars and starch. This causes the potassium ions to move from the guard cells into the subsidiary cells. As a result, the osmotic potential of the guard cells is raised, the water moves out, the cells become flaccid, and the stomata close.

External Influences

Environmental conditions can affect stomatal openings. Drought conditions, which induce water stress, can affect gas exchange because the lack of water moving through the plant causes the guard cells to lose turgor and close the stomata. When the temperature becomes too warm, stomata also tend to close. In some instances, the higher temperature causes water to leave the leaf more rapidly, which leads to water stress.

In other cases, the increase in temperature causes an increase in cellular respiration that, in turn, increases carbon dioxide levels. Internal high carbon dioxide concentrations both reverse the carbon dioxide pressure gradient and cause the stomata to close.

The percentage of relative humidity can drastically affect the rate of water evaporating from the leaf surface. As the humidity increases, the higher water content of the air decreases the rate of water loss from the leaf because the water pressure gradient no longer favors evaporation from the leaf surface. The amount of solar radiation can also influence gas exchange. As the amount of light increases, the stomata open faster and wider, resulting in a more rapid rate of gas exchange. Wind currents will also increase gas exchange rates: As the wind blows across the leaf, it carries water vapor away and, in a sense, reduces the humidity at the leaf surface. Because of this lower humidity, the water evaporates from the leaf surface more rapidly.

Ecological Impact

The exchange of gases between living plants and the atmosphere is critical to the survival of all living organisms. Without the release of the oxygen produced during photosynthesis, the atmosphere would contain very little of this necessary gas. Furthermore, the vast majority of organisms on earth depend on the organic materials supplied by plants. The carbon dioxide taken from the atmosphere is photosynthetically fixed into the more complex carbon molecules that eventually serve as food not only for the plants but also for all those organisms that consume plants. The amount of carbon dioxide fixed in this fashion is tremendous: It is estimated that an average of approximately 191 million metric tons of carbon dioxide is fixed daily.

The flux of oxygen out of and carbon dioxide into the plant is possible only because of the opening of the stomata, which in turn is dependent on the flow of water through the plant. Studies have shown that for every kilogram of grain (such as corn) produced, as much as 600 kilograms of water will transpire through the stomata in a process called *transpiration*. This represents tremendous water loss and raises the question: What is the selective advantage of transpiration that outweighs its wastefulness?

The most logical explanation is that water loss by transpiration is the price plants pay to absorb carbon dioxide, essential to the life of the plant. There is the additional possibility that transpiration may serve some purpose beyond opening the stomata, such as mineral transport. Plant cell growth appears to be partially dependent on the existence of turgor pressure within the cell. Hence, the transpirational flow of water through the plant could supply the turgidity necessary for plant cell growth. Transpiration may also serve the same purpose in plants that perspiration does in humans—that is, to cool the leaf surface through water evaporation.

Gas exchange and transpiration in plants are very dynamic and interrelated processes. A thorough knowledge of both processes and of the inter-action between them could one day lead to increasing maximum crop production while decreasing the amount of water required for the process.

D. R. Gossett

See also: Active transport; C_4 and CAM photosynthesis; Calvin cycle; Liquid transport systems; Osmosis, simple diffusion, and facilitated diffusion; Photosynthesis; Respiration; Vesicle-mediated transport; Water and solute movement in plants.

Sources for Further Study

Campbell, Neil A., and Jane B. Reece. *Biology*. 6th ed. San Francisco: Benjamin Cummings, 2002. An introductory college-level textbook for science majors. The chapter "Transport in Plants" provides a clear, concise, and somewhat detailed description of water movement through plants. The well-written text, combined with superb graphics, furnishes the reader with a clear understanding of transpiration. List of suggested readings at the end of the chapter. Includes glossary.

Hopkins, William G. *Introduction to Plant Physiology*. 2d ed. New York: John Wiley, 1999. Well-illustrated textbook with a practical approach to the study of plant physiology. Emphasizes the roles of plants within their environment and ecosystem. The second chapter covers plant respiration and gas exchange.

Raven, Peter H., Ray F. Evert, and Susan E. Eichhorn. *Biology of Plants*. 6th ed. New York: W. H. Freeman/Worth, 1999. An introductory college-level textbook for science students. Chapter 23, "The Shoot: Primary Structure and Development," provides an excellent discussion of leaf structure. The profusely illustrated text furnishes an excellent pictorial study of stomatal arrangement within the leaf. Includes glossary.

Stern, K. R. *Introductory Plant Biology*. 8th ed. Boston: McGraw-Hill, 2000. An introductory college-level textbook. Chapter 9, "Water in Plants: Soils," provides a very good general discussion of the control of transpiration. Very readable text accessible to the nonscience student. Suggested readings at the end of each chapter. Includes glossary.

Willmer, Colin M., and David Fricker. *Stomata*. 2d ed. New York: Chapman & Hall, 1995. Upper-level science textbook on the function of stomata in gas exchange. Discusses structure and function of the stomata along with theoretical aspects of gas exchange and environmental factors affecting the process.

GENE FLOW

Categories: Ecology; genetics

Gene flow represents a recurrent exchange of genes between populations. This exchange results when immigrants from one population interbreed with members of another.

Charles Darwin published *On the Origin of Species by Means of Natural Selection* in 1859. Since then, scientists have modified and added new concepts to the theory of evolution by natural selection. One of those concepts, which was only dimly understood in Darwin's lifetime, is the importance of genetics in evolution, especially the concepts of migration and gene flow.

Genes

Genes are elements within the cells of a living organism that control the transmission of hereditary characteristics by specifying the structure of a particular protein or by controlling the function of other genetic material. Within any species, the exchange of genes via reproduction is constant among its members, ensuring genetic similarity. If a new gene or combination of genes appears in the population, it is rapidly dispersed among all members of the population through inbreeding. New *alleles* (forms of a gene) may be introduced into the *gene pool* of a breeding population (thus contributing to the evolution of that species) in two ways: *mutation* and *migration*. Gene flow is integral to both processes.

A mutation occurs when the DNA code of a gene becomes modified so that the product of the gene will also be changed. Mutations occur constantly in every generation of every species. Most of them, however, are either minor or detrimental to the survival of the individual and thus are of little consequence. A very few mutations may prove valuable to the survival of a species and are spread to all of its members by migration and gene flow.

Separation and Migration

In nature, gene flow occurs on a more or less regular basis between *demes*, geographically isolated populations, races, and even closely related species. Gene flow is more common among the adjacent demes of one species. The amount of migration between such demes is high, thus ensuring that their gene pools will be similar. This sort of gene flow contributes little to the evolutionary process, as it does little to alter gene frequencies or to contribute to variation within the species.

Much more significant for the evolutionary process is gene flow between two populations of a species that have not interbred for a prolonged period of time. Populations of a species separated by geographical barriers (as a result, for example, of seed dispersal to a distant locale) often develop very dissimilar gene combinations through the process of natural selection. In isolated populations, dissimilar alleles become fixed or are present in much different frequencies. When circumstances do permit gene flow to occur between populations, it results in the breakdown of gene complexes and the alteration of allele frequencies, thereby reducing genetic differences in both. The degree of this homogeniza-

tion process depends on the continuation of interbreeding among members of the two populations over extended periods of time.

Hybridization

The migration of a few individuals from one breeding population to another may, in some instances, also be a significant source of genetic variation in the host population. Such migration becomes more important in the evolutionary process in direct proportion to the differences in gene frequencies—for example, the differences between distinct species. Biologists call interbreeding between members of separate species *hybridization*. Hybridization usually does not lead to gene exchange or gene flow, because hybrids are not often well adapted for survival and because most are sterile. Nevertheless, hybrids are occasionally able to breed (and produce fertile offspring) with members of one or sometimes both the parent species, resulting in the exchange of a few genes or blocks of genes between two distinct species. Biologists refer to this process as *introgressive hybridization*. Usually, few genes are exchanged between species in this process, and it might be more properly referred to as "gene trickle" rather than gene flow.

Introgressive hybridization may, however, add new genes and new gene combinations, or even whole chromosomes, to the genetic architecture of some species. It may thus play a role in the evolutionary process, especially in plants. *Introgression* requires the production of hybrids, a rare occurrence among highly differentiated animal species but quite common among closely related plant species. Areas where hybridization takes place are known as *contact zones* or *hybrid zones*. These zones exist where populations overlap. In some cases of hybridization, the line between what constitutes different species and what constitutes different populations of the same species becomes difficult to draw. The significance of introgression and hybrid zones in the evolutionary process remains an area of some contention among life scientists.

Speciation

Biologists often explain, at least in part, the poorly understood phenomenon of *speciation* through migration and gene flow—or rather, by a lack thereof. If some members of a species become geographically isolated from the rest of the species, migration and gene flow cease. Such geographic isolation can occur, for example, when populations are separated by water (as occurs on different islands or other landmasses) or valleys (different hillsides). The isolated population will not share in any mutations, favorable or unfavorable, nor will any mutations that occur among its own members be transmitted to the general population of the species. Over long periods of time, this genetic isolation will result in the isolated population becoming so genetically different from the parent species that its members can no longer produce fertile progeny should one of them breed with a member of the parent population. The isolated members will have become a new species, and the differences between them and the parent species will continue to grow as time passes. Scientists, beginning with Darwin, have demonstrated that this sort of speciation has occurred on the various islands of the world's oceans and seas.

Paul Madden

See also: Adaptive radiation; Coevolution; Community-ecosystem interactions; Evolution: convergent and divergent; Genetic drift; Genetics: mutations; Hybrid zones; Hybridization; Population genetics; Reproductive isolating mechanisms; Species and speciation.

Sources for Further Study

Christiansen, Freddy B. *Population Genetics of Multiple Loci*. New York: Wiley, 2000. Multiple-loci population genetics considers the mixture or recombination of genes during reproduction and their transmission through the population. This important work details the author's unifying theory, developed over two decades, that provides an accessible and natural extension from classical population genetics to multiloci genetics.

Endler, John A. *Geographic Variation, Speciation, and Clines*. Princeton, N.J.: Princeton University Press, 1977. Includes an excellent chapter on gene flow and its influence on the evolutionary process. Endler sees evolution as a very slow and gradual process in which gene flow and small mutations cause massive change over long periods of time.

Foster, Susan A., and John A. Endler, eds. *Geographic Variation in Behavior: Perspectives on Evolutionary Mechanisms*. New York: Oxford University Press, 1999. Includes illustrations, maps, bibliographical references, index.

Leapman, Michael. *The Ingenious Mr. Fairchild: The Forgotten Father of the Flower Garden*. New York: St. Martin's Press, 2001. A biography of nurseryman Thomas Fairchild, who invented artificial hybridization and caused a scandal in the early eighteenth century.

Mousseau, Timothy A., Barry Sinervo, and John A. Endler, eds. *Adaptive Genetic Variation in the Wild*. New York: Oxford University Press, 2000. Featuring a superb selection of papers from leading authors, this book summarizes the state of current understanding about the extent of genetic variation within wild populations and the ways to monitor such variation. It is a valuable resource for professionals and graduate students in genetics, biology, ecology, and evolution.

Real, Leslie A., ed. *Ecological Genetics*. Princeton, N.J.: Princeton University Press, 1994. A collection of essays by five distinguished ecological geneticists combining fieldwork and laboratory experiments. Chapter 1 covers gene flow specifically.

Stuessy, Tod F., and Mikio Ono, eds. *Evolution and Speciation of Island Plants*. New York: Cambridge University Press, 1998. The proceedings of a congress on plant speciation on the islands of the Pacific. Covers a wide range of topics with case studies of specific islands and species.

GENE REGULATION

Categories: Cellular biology; genetics

Genetic regulation is the manner in which a cell carries out transcription of its DNA (by copying it to messenger RNA) and the production of corresponding protein, called translation.

When a gene is expressed, one strand of that gene's double helical deoxyribonucleic acid (DNA) is copied, in the process of *transcription*, by an enzyme called RNA (ribonucleic acid) polymerase to make a *messenger RNA* (mRNA). The mRNA then associates with a *ribosome* (formed by specific ribosomal RNAs and proteins), where the nucleic acid base sequence is read, to produce the corresponding protein in a process called *translation*. A section of three nucleotides, called a *codon*, codes for one amino acid. Small RNA molecules called *transfer RNAs* (tRNAs) carry specific amino acids to the ribosome during translation. That amino acid is added to the growing polypeptide chain when the anticodon part of the tRNA pairs with a codon of the mRNA being translated.

Prokaryotic vs. Eukaryotic Regulation

Prokaryotes (*Bacteria* and *Archaea*) and eukaryotes (all other life-forms, including plants and animals) carry out transcription and translation in very similar ways, but there are some differences. The RNA polymerases and ribosomes differ between the two types of cell. In a prokaryote, such as a bacterium, translation of an mRNA can begin as soon as the first part of the mRNA molecule has been made from the DNA template. This is said to be "coupled" transcription and translation. In the eukaryotic cell, mRNA is transcribed in the nucleus and crosses the nuclear envelope to go to the cytoplasm, where the mRNA is translated into proteins on ribosomes.

Promoters

Transcription begins at a *promoter*, which is a DNA sequence at the start of a gene that tells the RNA polymerase where to start transcribing the DNA to make mRNA. The promoter includes the site where transcription begins (the initiation site) as well as sequences upstream, which are not transcribed. In eukaryotes, RNA polymerase II recognizes a DNA sequence called the TATA box (because it contains many thymine (T) and adenine (A) bases) that is about thirty nucleotides upstream of the initiation site. Another element often found in promoters is the CAAT box (which includes cytosine (C) bases).

Other regulatory elements include *enhancer* sequences, which can be in either orientation in the chromosome and far from the coding region of the gene. Enhancers may be involved in regulating the specificity of expression of a gene in a particular tissue or organ. *Silencer sequences* are structurally similar to enhancers but function to decrease gene expression. Other proteins, called *transcription factors*, aid the RNA polymerase in locating and binding to the promoter. The DNA-binding domains of eukaryotic transcription factors may have one of several types of structural motifs, such as a helix-turn-helix structure, a zinc finger (a cysteine- and histidine-rich region of the protein that complexes zinc), or a leucine zipper. Once the polymerase binds, it separates the two strands of DNA at the initiation site, and transcription begins. Transcription ends at a termination site on the DNA. In eukaryotes, a common termination sequence is AATAAA.

Other Regulatory Elements

Genes with related functions often have similar regulatory elements. A number of environmentally induced genes in plants contain the G-box with the sequence 5'CCACGTGG3'. Genes that respond to light, ultraviolet radiation, cold, and drought have the G-box and at least one additional regulatory element. For example, genes that respond to the hormone abscisic acid also have an abscisic acid-responsive element.

Role of Chromatin

The chromosomes of eukaryotes consist of unique, single-copy genes among a complex pattern of repetitive DNAs. In addition, there are DNAs in organelles outside the nucleus, such as mitochondria and chloroplasts. Each eukaryotic chromosome contains a single long DNA molecule that is coiled, folded, and compacted by its interaction with chromosomal proteins called *histones*. This complex of DNA with chromosomal proteins and chromosomal RNAs is *chromatin*. The DNA of higher eukaryotes appears to be organized into looped domains of chromatin by attachment to a nuclear scaffold. The loops are anchored at matrix-attachment regions that function in the structural organization of the DNA and may increase transcription of certain genes by promoting the formation of a less-condensed chromatin.

Transcription and mRNA Processing in Eukaryotes

In the eukaryote, as mRNA makes its way from the nucleus to the ribosomes in the cytoplasm, the mRNA molecule is modified (RNA processing). Both ends of the mRNA are altered. The 5' end (the end that is first formed during transcription) is capped by the addition of a modified guanine (G) residue (7-methylguanosine). The 5' cap protects the mRNA from degradation by nucleases and serves as a signal for the attachment of small ribosomal subunits to the mRNA. While still in the nucleus, the 3' end of the mRNA is also modified. An enzyme adds thirty to two hundred adenine nucleotides (called a poly-A tail). The poly-A tail inhibits mRNA degradation and may aid in mRNA transport to the cytoplasm.

In addition, many eukaryotic mRNAs are spliced in a cut-and-paste process in which a large part of the RNA molecule initially synthesized is removed. The parts of the DNA that code for the RNA that will be discarded are the intervening sequences, or introns. The DNA that codes for the parts of the RNA that will be translated are exons.

Susan J. Karcher

See also: Bacteria; DNA in plants; DNA replication; Eukaryotic cells; Genetic code; Genetics: mutations; Nucleic acids; Nucleus; Plant cells: molecular level; Prokaryotes; Ribosomes.

Sources for Further Study

Buchanan, Bob B., Wilhalm Gruissem, and Russell L. Jones. *Biochemistry and Molecular Biology of Plants*. Rockville, Md.: American Society of Plant Physiologists, 2000. A comprehensive textbook. Includes illustrations, photos, sources, and index.

Calladine, C. R., and Horace R. Drew. *Understanding DNA: The Molecule and How It Works*. San Diego: Academic Press, 1997. Focuses on the structure of DNA. Includes figures, exercises, and index.

Griffiths, Anthony J. F., et al. *An Introduction to Genetic Analysis*. New York: W. H. Freeman, 2000. An upper-level college textbook. Includes CD-ROM, illustrations, problem sets with solutions, glossary, and index.

Lewin, Benjamin. *Genes VII*. New York: Oxford University Press, 2000. This standard college-level textbook contains individual chapters on messenger RNA, protein, synthesis, transcription, and chromosomes. Includes illustrations, glossary, and index.

Lodish, Harvey, et al. *Molecular Cell Biology*. New York: W. H. Freeman, 2000. Comprehensive text with an experimental emphasis. Includes CD, figures, and index.

GENETIC CODE

Categories: Genetics; reproduction

The genetic code defines each amino acid in a protein, or polypeptide, in terms of a specific sequence of three nucleotides, called codons, in the DNA. The genetic code is therefore the key to converting the information contained in genes into proteins.

The genetic code defines each amino acid in a protein, or *polypeptide*, in terms of a specific sequence of three *nucleotides*, called codons, in the deoxyribonucleic acid (DNA). Therefore, the genetic code is called a *triplet code*. The four different nucleotides in DNA can form sixty-four different triplet codons. Because there are only twenty amino acids found in proteins, some amino acids are encoded by more than one codon. Therefore, the genetic code is said to be redundant, or *degenerate*. Three of the triplet codons do not encode any amino acids. These are *stop codons*, which identify the end of the message (similar to the period at the end of a sentence) encoded in genes. The genetic code is nearly universal; that is, specific codons code for the same amino acids in nearly all organisms. However, a few exceptions have been found, primarily in *mitochondrial DNA* (mtDNA), but also in a few protozoa and a single-celled algae, such as *Acetabularia*. (DNA is found in three places in eukaryotic cells: the nucleus, plastids, and mitochondria.)

Role of RNA

There are two distinct steps in the conversion of DNA sequences into protein sequences: *transcription* and *translation*. Transcription is the process by which the nucleotide sequence of a gene (in the DNA of a chromosome) is used to make a complementary copy of RNA (*ribonucleic acid*). DNA is a double-stranded molecule, and genes are arranged along DNA on each strand. Wherever there is a gene on one strand, called the *coding strand*, the other strand opposite the gene contains a nucleotide sequence that is complementary. The opposite strand is called the *non-coding strand*. When transcription of a gene occurs, it is the coding strand that is transcribed into a complementary stand of RNA. The transcribed RNA is complementary in the sense that for each adenine (A) in the DNA, a uracil (U) is incorporated into the RNA. Likewise, for each uracil, guanine (G), and cytosine (C) in the DNA, an adenine, cytosine, and guanine are incorporated into the RNA, respectively. Thus, for the codon AGT in the DNA, it becomes UCA in the RNA.

In *prokaryotes* (bacteria), the RNA resulting from transcription is called *messenger RNA* (mRNA), and it is immediately ready for the next step, translation. In the subsequent translation process, the mRNA is translated into a protein sequence by a *ribosome*. Ribosomes are macromolecular assemblies composed of various proteins and a second type of RNA, *ribosomal RNA* (rRNA). Ribosomes bind to mRNA molecules near the 5′ end and scan along the nucleotides until they reach an *initiation* or *start codon*, which indicates where translation should begin and establishes the *reading frame*. The start codon is always AUG and is the first translated codon in all mRNAs. Translation ends when a stop codon is reached. Most organisms have three stop codons (UAA, UGA, and UAG), and each gene ends with one of these. A stop codon does not code for an amino acid but simply identifies the end of the gene. Because nucleotide bases are read three at a time along a continuous chain of nucleotides, shifting the reading frame by inserting or deleting a single nucleotide within a gene can dramatically alter the amino acid sequence of the protein it can produce.

Although ribosomes are essential for translating mRNAs, they are not directly responsible for interpreting the codons. *Transfer RNA* (tRNA) molecules are single-stranded RNA molecules that exhibit extensive intramolecular base-pairing such that the tRNA has a two-dimensional structure with a stem and three loops resembling a three-leaf clover. The middle loop contains a region, composed of three nucleotides, called the *anticodon*, which is complementary to a specific codon. The tRNA molecules directly decode the mRNA sequence or translate it into a correct amino acid sequence by their ability to bind to the right codon. Each tRNA carries a specific amino acid to the ribosome where protein synthesis occurs. The binding of amino acids to tRNAs occurs at a place on the tRNA called the *amino acid attachment site*. Amino acids are added to tRNAs by special enzymes called *aminoacyl-tRNA synthetases*. Each tRNA has a special region on the stem below the amino acid attachment site whose nucleotide sequence determines which amino acid needs to be attached to the tRNA. This code is often called the second genetic code, and geneticists have discovered that if the nucleotide region is changed in this region, the wrong amino acid gets attached.

There are at least one type of tRNA for each of the twenty amino acids. It is the pairing of the codons of the mRNA molecules and anticodons of tRNA molecules that determines the order of the amino acid sequence in the polypeptide chain. The third base of the anticodon does not always properly recognize the third base of the mRNA codon. The third base is called the *wobble* position because nonstandard base pairing can occur there. This phenomenon, together with the degeneracy of the genetic code means that cells do not have to have sixty-one different tRNA types (one for each codon that speci-

fies an amino acid). For example, only two tRNAs are needed to recognize four different codons for the amino acid glycine. Consequently, most organisms have about forty-five different tRNA types.

Polypeptide Sorting

A typical plant cell contains thousands of different kinds of proteins. For the cell to function properly, each of its numerous proteins must be localized to the correct cellular membrane, or cellular compartment. The process of protein sorting, or protein targeting, is critical to the organization and functioning of plant cells. Protein sorting relies upon the presence of special *signal sequences* at one end of the protein molecule. These signal sequences direct proteins to various sites. For example, some proteins synthesized on ribosomes in the cytoplasm are targeted to organelles, such as the mitochondria or chloroplasts. Other proteins, such as those found in the plant cell wall, are targeted to the cytoplasmic membrane for transport out of the cell to the cell wall. The signal sequences are frequently removed once the protein has reached its intended destination.

Lisa M. Sardinia

See also: DNA in plants; DNA replication; Endomembrane system and Golgi complex; Gene regulation; Genetics: mutations; Nucleolus; Nucleus; Proteins and amino acids; Ribosomes; RNA.

Sources for Further Study

Gonick, Larry, and Mark Wheelis. *The Cartoon Guide to Genetics*. New York: HarperPerennial, 1991. Slightly silly, nonintimidating explanation of classical and modern genetics. Cartoon drawings and amusing anecdotes assist in describing complex concepts. Suitable for high school students.

Lewin, Benjamin. *Genes VII*. New York: Oxford University Press, 1999. Thorough, comprehensive reference book. All sections, including the section on the genetic code, are detailed and include the history of discoveries in the area. Includes illustrations, bibliographic references, and glossary.

Paolella, Peter. *Introduction to Molecular Biology*. Boston: WCB/McGraw-Hill, 1998. One chapter is devoted to the genetic code, with other chapters on protein synthesis. Straightforward explanation with adequate illustrations.

GENETIC DRIFT

Categories: Ecology; genetics; reproduction and life cycles

Genetic drift refers to random changes in the genetic composition of a population. It is one of the evolutionary forces that cause biological evolution, the others being selection, mutation, and migration, or gene flow.

Drift occurs because the genetic variants, or *alleles*, present in a population are a random sample of the alleles that adults in the previous generation would have been predicted to pass on, where predictions are based on expected migration rates, expected mutation rates, and the direct effects of alleles on fitness. If this sample is small, then the genetic composition of the offspring population may deviate substantially from expectation, just by chance. This deviation is called genetic drift. Drift becomes increasingly important as population size decreases. The key feature of drift that distinguishes it from the other evolutionary forces is the unpredictable direction of evolutionary change.

Anything that generates *fitness variation* among individuals (that is, variation in the ability of individuals to survive and reproduce) will increase the magnitude of drift for all genes that do not themselves cause the fitness variation. Because of their indeterminate growth, plants often vary greatly in reproductive potential because of local environmental variation, and this magnifies genetic drift. For example, the magnitude of drift in most annual plants is more than doubled by size variation among adults. This makes sense if one considers that larger individuals contribute a larger number of offspring to the next generation, so any alleles they carry will tend to be overrepresented.

Fitness variation caused by selection will also increase the magnitude of drift at any gene not directly acted upon by the selection. If an individual has high fitness because it possesses one or more favorable alleles, then all other alleles it possesses will benefit. This is called *genetic hitchhiking*. This is a potent source of evolution because the direction of change at a hitchhiking gene will remain the same for multiple generations. However, it is not possible to predict in advance what that direction will be because where and when a favorable mutation will occur cannot be predicted.

The opportunities for drift to occur are greatly influenced by *gene flow*. Most terrestrial plants are characterized by highly localized dispersal. Thus, even in large, continuous populations, the pool of potential mates for an individual, and the pool of seeds that compete for establishment at a site, are all drawn from a small number of nearby individuals known as the *neighborhood*. If the neighborhood is sufficiently small, genetic drift will have a significant impact on its genetic composition.

For these and other reasons, population size alone is not sufficient to predict the magnitude of drift. The *effective size* of a population, N_e, is a number that is directly related to the magnitude of drift through a simple equation. Thus, N_e incorporates all characteristics of a population that influence drift.

Loss of Variability

The long-term consequence of drift is a loss of genetic variation. As alleles increase and decrease in frequency at random, some will be lost. In the absence of mutation and migration, such losses are permanent. Eventually, only one allele remains at each gene, which is said to be fixed. Thus, all else being equal, smaller populations are expected to harbor less genetic variation than larger populations.

An important way in which different plant populations are not equal is in their reproductive systems. With self-fertilization (selfing), or asexual reproduction, genetic hitchhiking becomes very important. In the extreme cases of 100 percent selfing or 100 percent asexual reproduction, hitchhiking will determine the fates of most alleles. Thus, as a new mutation spreads or is eliminated by selection, so too will most or all of the other alleles carried by the individual in which the mutation first arose. This is called a *selective sweep*, and the result is a significant reduction in genetic variation. Which alleles will be swept to fixation or elimination cannot be predicted in advance, so the loss of variation reflects a small N_e. Consistent with this expectation, most populations of flowering plants that reproduce partly or entirely by selfing contain significantly less genetic variation than populations of related species that do not self-fertilize.

Extinction

Mutations that decrease fitness greatly outnumber mutations that increase fitness. In a large population in which drift is weak, selection prevents most such mutations from becoming common. In very small populations, however, alleles that decrease fitness can drift to fixation, causing a decrease in average fitness. This is one manifestation of a phenomenon called *inbreeding depression*. In populations with very small N_e, this inbreeding depression can be significant enough to threaten the population with extinction. If a population remains small for many generations, mean fitness will continue to decline as new mutations become fixed by drift. When fitness declines to the point where offspring are no longer overproduced, population size will decrease further. Drift then becomes stronger, mutations are fixed faster, and the population heads down an accelerating trajectory toward extinction. This is called *mutational meltdown*.

Creative Potential

By itself, drift cannot lead to adaptation. However, drift can enhance the ability of selection to do so. Because of diploidy and sexual recombination, some types of mutations, either singly or in combinations, will increase fitness when common but not when rare. Genetic drift can cause such genetic variants to become sufficiently common for selection to promote their fixation. A likely example is the fixation of new structural arrangements of chromosomes that occurred frequently during the diversification of flowering plants. New chromosome arrangements are usually selected against when they are rare because they disrupt meiosis and reduce fertility. The initial spread of such a mutation can therefore only be caused by strong genetic drift, either in an isolated population of small effective size or in a larger population divided into small neighborhoods.

John S. Heywood

See also: Endangered species; Gene flow; Genetics: mutations; Population genetics; Selection.

Sources for Further Study

Denny, Mark, and Steven Gaines. *Chance in Biology: Using Probability to Explore Nature.* Princeton, N.J.: Princeton University Press, 2000. Explores randomness, using biological examples. Some math skills required. Includes problems with solutions, index.

Futuyma, Douglas J. *Evolutionary Biology.* 3d ed. Sunderland, Mass.: Sinauer Associates, 1998. A standard text with a lucid chapter on drift. Includes suggested readings, figures, tables, numerical examples, and index.

Hedrick, Philip W. *Genetics of Populations.* 2d ed. Boston: Jones & Bartlett. An introduction to mathematical models of evolution, with extensive coverage of genetic drift and effective population size. Includes figures, tables, numerical examples, bibliographical references, and index.

GENETIC EQUILIBRIUM: LINKAGE

Categories: Cellular biology; genetics

Genetic equilibrium is the tendency for genes located close together on the same chromosome to be inherited together. The farther apart the genes, the less likely it is that they will be passed along together.

The genetic complement of any organism is contained on one or more types of *chromosomes*. Whether there are only a few chromosomes or many (such as in a diploid organism), each type occurs as a set of two, called *homologs*. Each gene at a particular *locus*, or site, along the chromosome occurs twice in the same cell (except for some loci which only occur on one of the two sex chromosomes), one copy of each homolog. The particular information at each locus may be different because genes can exist in several forms.

An alternative form of a particular gene is called an *allele*. For example, one of the genes for flower color in pea plans can exist as a white allele or a purple allele. A given pea plant could have a white allele on one homologous chromosome and a purple allele on the other. Each homologous chromosome, therefore, contains an allele at a locus which may or may not be the same as the allele on its homolog.

During reproduction, this chromosomal material is copied, thereby duplicating the individual genes which lie along the chromosome. During *mitosis*, a copy of each chromosome is distributed to each of the two new nuclei. In *meiosis*, however, during gamete production, the chromosome copies are separated so that only a single chromosome from each pair of homologous chromosomes is distributed to each of four new nuclei. Before this happens, however, the homologs and their duplicates, called *sister chromatids*, become aligned. The arms of sister chromatids undergo crossover near the beginning of the first part of meiosis, which results in the exchange of homologous regions of these chromosomes.

The result of this *crossing over* is that alleles that were once on one homolog are now on the other. In mitosis, genes on the same chromosome exhibit linkage and tend to remain together and be inherited by the daughter cell together; in meiosis, these linked genes can become recombined in new associations so that linkage is partial. Individuals with chromosomes exhibiting these new combinations of alleles are called *recombinants*.

Discovery of Linkage

Mendelian genetics (named for Gregor Mendel) predicts a 3:1 phenotypic ratio in a monohybrid cross (a cross involving only one gene having two alleles, one dominant and one recessive) and a 9:3:3:1 phenotypic ratio in a dihybrid cross (a cross involving two genes on different chromosomes, each having two alleles, one dominant and one recessive). Early in the twentieth century, geneticists began to notice that not all crosses produced off-

spring in the proportions predicted by Mendel's *law of independent assortment*. Cytologists also discovered that occasionally homologous chromosomes did not look exactly alike. Geneticists used these differences in chromosomes as cytological markers and associated them with genetic markers or alleles with specific effects.

In 1911 T. H. Morgan concluded that during segregation of alleles at meiosis, certain genes tend to remain together because they lie near each other on the same chromosome. The closer genes are located to each other on the chromosome, the greater their tendency to remain linked.

In 1909 *chiasmata* had been described. Chiasmata represent the locations of exchanges (crossover) between maternal and paternal homologous chromosomes. Morgan hypothesized that partial linkage occurs when two genes on the same chromosome are separated physically from each other by crossover during meiosis. Crossover provides new combinations of genes, genes which did not exhibit the linkage relationship in the parents but which were recombined. In these kinds of crosses, the parental phenotypic classes are most frequent in the offspring, while the recombinant classes occur much less frequently.

Genetic recombination results from physical exchange between homologous chromosomes that have become tightly aligned during meiotic prophase. A chiasma is the site of crossing over and is where homologs have lined up touching each other where they are homologous. Crossing over itself is the exchange of parts of nonsister chromatids of homologous chromosomes by symmetrical breakage and crosswise rejoining. Two papers providing convincing evidence of this were published within weeks of each other in 1930.

Harriet Creighton and Barbara McClintock worked with corn (*Zea mays*). They studied individuals in which the two copies of chromosome 9 had a strikingly different appearance. They studied two loci on chromosome 9, one affecting seed color (colored and colorless, dominant and recessive, respectively) and the other affecting endosperm composition (waxy or starchy, dominant and recessive, respectively). One homolog was dominant for both traits (colored and waxy) and lacked the knob and the extension. Plants with these two homologs of chromosome 9 had colored, waxy seeds. Recombinant offspring would have either colored and starchy seeds or colorless and waxy seeds and their copies of chromosome extension but no knob. Offspring that were like the parents showed no change in chromosome structure. This provided visual evidence that crossover had occurred.

Genetic Maps

The frequency of crossover can be used to construct a *genetic map*. The more closely linked genes are, the less frequently crossing over will take place between them. The recombinants will occur much less frequently than when linked genes are more widely separated.

With widely separated genes, the chances of double crossovers increases, so that the recombination frequency may actually underestimate the crossover frequency and, hence, the *map distance*. The map distance is a relative distance based on the percent of recombination and is not a precise physical distance. The presence of a chiasma in one region often prevents the occurrence of a second chiasma nearby. This phenomenon is called *interference*.

In many large, randomly mating natural populations, the genotype frequencies at each locus will typically be found at a mathematically determined *equilibrium*. In a single generation of random mating, unlinked loci separately attain equilibrium of genotype frequencies. This is not true of linked loci. If loci are unlinked, equilibrium occurs very rapidly, but if the loci are on the same chromosome, the speed of approach to equilibrium is proportional to the map distance between them.

Once equilibrium is attained, repulsion and coupling gamete frequencies do not depend on the degree of linkage. Another way of saying this is that the characters produced by alleles at linked loci show no particular association in an equilibrium population. When characters happen to be associated in a population, the association may form because alleles at separate loci that are in genetic disequilibrium result from recent population immigrations. They may also be the result of selection for certain allelic combinations. Like dominance, linkage can be confirmed only in controlled breeding experiments.

Mutations

Mutations are the ultimate source of variation. In populations, mutant alleles may accumulate over time because they are recessive to the normal allele. Recessive lethal alleles, as well as beneficial alleles, persist in populations because recessive alleles are hidden when in the heterozygous state (that is, in individuals who have one normal, dominant allele and one mutant, recessive allele). It is only when the mutant becomes widely distributed in the population that they are revealed.

With ten loci and four alleles at each locus, ten billion different possible genotypes will occur with equal frequency if all the alleles occur with equal frequency and segregate independently. This describes a state of linkage equilibrium. In a natural population, however, these conditions are rarely met. The probability is that some genotypes will be more common than others, even if the allele frequencies are all the same.

Diploid organisms typically have tens of thousands of gene loci. Because they have only a small number of chromosomes, usually less than forty, many loci lie on the same chromosome. The genotypes are highly biased toward already existing combinations. This does not alter the theoretical possibilities of particular genotypes, only the probability of their occurrence. It does ensure that variation is present in the population for adaptability to changing conditions, while maintaining large numbers of individuals that are adapted to existing conditions.

Functions of Linkage

Linked genes may control very different functions. For example, enzymes vary depending on climate. Northern species may possess an enzyme which functions at a lower temperature than the variant of the southern species. Linked to the gene which controls this highly adaptive allele may be an allele of another gene whose adaptive value is lower or neutral. This less adapted allele hitchhikes on the chromosome with the adaptive allele.

Linkage disequilibrium is decreased by recombination. The maintenance of favorable allelic combinations in linkage disequilibrium is enhanced by reducing recombination between the loci involved. This is achieved by inversions and translocations that include the loci involved. The genes included in these in the translocated or inverted region are sometimes called supergenes, because of their strong tendency to be inherited as a large unit of many genes. Inversions and translocation do not completely inhibit crossover in the regions involved, but they do reduce it. They also reduce the occurrence of recombinant chromosomes, because if a crossover does occur in one of these regions, most of their resulting recombinant chromosomes are either lethal or cause varying levels of sterility. Whenever linkage disequilibrium is favored by natural selection, chromosomal rearrangements increasing linkage among loci will also be favored by natural selection.

Judith O. Rebach, updated by Bryan Ness

See also: Chromosomes; DNA in plants; DNA replication; Genetics: Mendelian; Genetics: mutations; Genetics: post-Mendelian; Mitosis and meiosis; RNA.

Sources for Further Study

Blixt, S. "Why Didn't Mendel Find Linkage?" *Nature* 256 (1975): 206. A short article describing the linkage patterns of genes not subjected to experiment by Gregor Mendel, even though they are noted in Mendel's records.

Comfort, Nathaniel. *The Tangled Field: Barbara McClintock's Search for the Patterns of Genetic Control.* Cambridge, Mass.: Harvard University Press, 2001. A biography of the geneticist whose work on corn led to the understanding of gene linkage that won her a Nobel Prize. Places McClintock's work within the context of the history of science.

Fedoroff, N. V. "Transposable Genetic Elements in Maize." *Scientific American* 250 (June, 1984): 84-98. The transposable genetic elements were first described by Barbara McClintock. In this lucid article, the elements are analyzed to show that by changing their position on a chromosome or by moving to another chromosome, they change the activity of nearby genes.

Russell, Peter J. *Genetics.* 5th ed. Menlo Park, Calif.: Benjamin/Cummings, 1998. A thorough undergraduate textbook on genetics. Chapter 12 covers replication and recombination of genetic material.

GENETICALLY MODIFIED BACTERIA

Categories: Bacteria; biotechnology; economic botany and plant uses; genetics

Bacteria may be genetically modified through the introduction of recombinant DNA molecules into their cells. Such bacteria may be used to produce human insulin or introduce disease-resistant genes into plants, as well as numerous other applications.

The ability to genetically engineer bacteria is the outcome of several independent discoveries. In 1944 Oswald Avery and his coworkers demonstrated gene transfer among bacteria using purified DNA (deoxyribonucleic acid), a process called *transformation*. In the 1960's the discovery of *restriction enzymes* permitted the creation of hybrid molecules of DNA. Such enzymes cut DNA molecules at specific sites, allowing fragments from different sources to be joined within the same piece of genetic machinery.

Restriction enzymes are not specific in choosing their target species. Therefore, DNA from any source, when treated with the same restriction enzyme, will generate identical cuts. The treated DNA molecules are allowed to bind with one another, while a second set of enzymes called *ligases* are used to fuse the fragments together. The recombinant molecules may then be introduced into bacteria cells through transformation. In this manner, the cell has acquired whatever genetic information is found in the DNA. Descendants of the transformed cells will be genetically identical, forming *clones* of the original.

Bacterial Plasmids

The most common forms of genetically altered DNA are bacterial *plasmids*, small circular molecules separate from the cell chromosome. Plasmids may be altered to serve as appropriate *vectors* (carriers of genetic material) for genetic engineering, usually containing an *antibiotic* resistance gene for selection of only those cells that have incorporated the DNA. Once the cell has incorporated the plasmid, it acquires the ability to produce any gene product encoded on the molecule.

The first such genetically altered bacteria used for medical purposes, *Escherichia coli*, contained the gene for the production of human insulin. Prior to creation of the insulin-producing bacterium, diabetics were dependent upon insulin purified from animals. In addition to being relatively expensive, insulin obtained from animals produced allergic reactions among some individuals. Insulin obtained from genetically altered bacteria is identical to that of human insulin. Subsequent experiments also engineered bacteria able to produce a variety of human proteins, including human growth hormone, interferon, and granulocyte colony-stimulating factor.

Use in Plants

Genetically modified bacteria may also serve as vectors for the introduction of genes into plants. The bacterium *Agrobacterium tumefaciens*, the cause of the plant disease called crown gall, contains a plasmid called *Ti*. Following infection of the plant cell by the bacterium, the plasmid is integrated into the host chromosome, becoming part of the plant's genetic material. Any genes that were part of the plasmid are integrated as well. Desired genes can be introduced into the plasmid, promoting pest or disease resistance.

In April of 1987 scientists in California sprayed strawberry plants with genetically altered bacteria to improve the plants' freeze resistance, marking the first deliberate release of genetically altered organisms in the United States to be sanctioned by the Environmental Protection Agency (EPA). The release of the bacteria climaxed more than a decade of public debate over what would happen when the first products of biotechnology became commercially available. Fears centered on the creation of bacteria that might radically alter the environment through elaboration of gene products not normally found in such cells. Some feared that so-called super bacteria might be created with unusual resistance to conventional medical treatment. Despite these concerns, approval for further releases of genetically altered bacteria soon followed, and the restrictions on release were greatly relaxed. By 2002, permits for field tests of hundreds of genetically altered plants and microorganisms had been granted.

Richard Adler

See also: Biotechnology; DNA: recombinant technology; Genetically modified foods; Plant biotechnology.

Sources for Further Study

Dale, Jeremy W. *Molecular Genetics of Bacteria*. 3d ed. New York: John Wiley, 1998. Uses a molecular approach to introduce students to bacterial genetics. Provides clear explanations of techniques and terminology.

Levin, Morris A., ed. *Engineered Organisms in Environmental Settings: Biotechnological and Agricultural Applications*. Boca Raton, Fla.: CRC Press, 1996. Overview of environmental applications of genetically modified organisms; describes their releases into the environments and their observed effects.

Miesfeld, Roger L. *Applied Molecular Genetics*. New York: John Wiley, 1999. Discusses key biochemical and cell biological principles behind commonly used applications of molecular genetics, using clear terms and well-illustrated flow schemes. Includes references and a list of Web sites.

GENETICALLY MODIFIED FOODS

Categories: Agriculture; biotechnology; economic botany and plant uses; food; genetics

Foods derived from living organisms that have been modified using gene-transfer technology are known as genetically modified foods. Many of these foods are plant crops.

Applications of genetic engineering in agriculture and the food industry could increase world food supplies, reduce environmental problems associated with food production, and enhance the nutritional values of certain foods. However, these benefits are countered by food-safety concerns, the potential for ecosystem disruption, and fears of unforeseen consequences resulting from altering natural selection. Humans rely on plants and animals as food sources and have long used microbes to produce foods such as cheese, bread, and fermented beverages. Conventional techniques such as cross-hybridization, production of mutants, and selective breeding have resulted in new varieties of crop plants or improved livestock with altered genetics. However, these methods are relatively slow and labor-intensive, are generally limited to intraspecies crosses, and involve a great deal of trial and error.

Transgenic Technology

Recombinant DNA techniques, which manipulate cells' deoxyribonucleic acid (DNA), developed in the 1970's enable researchers rapidly to make specific, predetermined genetic changes. Because the technology also allows for the transfer of genes across species and kingdom barriers, an infinite number of novel genetic combinations are possible. The first animals and plants containing genetic material from other organisms (*transgenics*) were developed in the early 1980's. By 1985 the first field trials of plants engineered to be pest-resistant were conducted. In 1990 the U.S. Food and Drug Administration (FDA) approved chymosin as the first substance produced

Genetically Modified Crop Plants Unregulated by the U.S. Department of Agriculture		
Crop	*Patent Holder*	*Genetically Engineered Trait(s)*
Canola	AgrEvo	herbicide tolerance
Corn	AgrEvo	herbicide tolerance
	Ciba-Geigy	insect resistance
	DeKalb	herbicide tolerance; insect resistance
	Monsanto	herbicide tolerance; insect resistance
	Northrup King	insect resistance
Cotton	Calgene	herbicide tolerance; insect resistance
	DuPont	herbicide tolerance
	Monsanto	herbicide tolerance; insect resistance
Papaya	Cornell	virus resistance
Potato	Monsanto	insect resistance
Squash	Asgrow	virus resistance
	Upjohn	virus resistance
Soybean	AgrEvo	herbicide tolerance
	DuPont	altered oil profile
	Monsanto	herbicide tolerance
Tomato	Agritope	altered fruit ripening
	Calgene	altered fruit ripening
	Monsanto	altered fruit ripening
	Zeneca	altered chemical content in fruit

Source: U.S. Department of Agriculture Animal and Plant Health Inspection Service (APHIS).

by modified organisms to be used in the food industry for dairy products such as cheese. That same year the first transgenic cow was developed to produce human milk proteins for infant formula. The well-publicized Flavr Savr tomato, modified to delay ripening and rotting, obtained FDA approval in 1994.

Goals and Uses

By the mid-1990's, more than one thousand genetically modified crop plants were approved for field trials. The goals for altering food crop plants by genetic engineering fall into three main categories: to create plants that can adapt to specific environmental conditions to make better use of agricultural land, increase yields, or reduce losses; to increase nutritional value or flavor; and to alter harvesting, transport, storage, or processing properties for the food industry. Many genetically modified crops are sources of ingredients for processed foods and animal feed.

Herbicide-resistant plants, such as the Roundup Ready soybean, can be grown in the presence of glyphosphate, an *herbicide* that normally destroys all plants with which it comes in contact. Beans from these plants were approved for food-industry use in several countries, but there has been widespread protest by activists such as Jeremy Rifkin and environmental organizations such as Greenpeace. Frost-resistant fruit containing a fish antifreeze gene, insect-resistant plants with a bacterial gene that encodes for a pesticidal protein (*Bacillus thuringiensis*), and a viral disease-resistant squash are examples of other genetically modified food crops that have undergone field trials.

Scientists have also created plants that produce healthier unsaturated fats and oils rather than saturated ones. Genetic engineering has yielded coffee plants whose beans are caffeine-free without processing and tomatoes with altered pulp content for improved canned products. Ge-

netically modified microbes are used for the production of food additives such as amino acid supplements, sweeteners, flavors, vitamins, and thickening agents. In some cases, these substances had to be obtained from slaughtered animals. Altered organisms are also used for improving fermentation processes in the food industry.

Food and Environmental Issues

Food safety and quality are at the center of the genetically engineered food controversy. Concerns include the possible introduction of new toxins or allergens into the diet and changes in the nutrient composition of foods. Proponents argue that food sources could be designed to have enhanced nutritional value.

A large percentage of crops worldwide are lost each year to drought, temperature extremes, and pests. Plants have already been engineered to exhibit frost, insect, disease, and drought resistance. Such alterations would increase yields and allow food to be grown in areas that are currently too dry

---

Seeds of Dissent

Creation of transgenic crops has been alleged to create the risk of new allergenic proteins. Despite these risks, as of 2001, both giant agribusiness and the United States Department of Agriculture resisted informational labeling of genetically modified (GM) food products, arguing that GM foods are as safe as conventional foods and that there is no evidence they cause allergic reactions. Because GM products are often mixed with conventional products, unlabeled foods make it hard to trace individuals' allergic reactions. Similarly, people with culturally based dietary restrictions do not know whether produce they eat contains proteins derived from beef, pork, fish, or other animals.

Other potential risks of transgenic crops include increases in toxins and decreases in nutritional value. Accidental crossbreeding with wild species of plants has caused critics of "genetic pollution" to raise the specter of a steady stream of animal and microbial genes entering the gene pools of plants in wild ecosystems.

Additionally, herbicide-resistant weeds remind activists that soybeans dubbed Roundup Ready Soybeans have been bred to tolerate glyphosphate, enabling that herbicide's wider use. Development of pesticide resistance among insects was feared when U.S.-grown transgenic corn was shipped, unlabeled, to Europe in 1996. Situations like this could lead to increased application of chemicals already in use by farmers or a switch to different, more toxic chemicals.

These and other controversies and fears have caused many European nations and Japan to prohibit agricultural production of GM crops. Some nations have banned imports of all GM products.

or infertile, positively impacting the world food supply.

Environmental problems such as deforestation, erosion, pollution, and loss of biodiversity have all resulted, in part, from conventional agricultural practices. Use of genetically modified crops could allow better use of existing farmland and lead to a decreased reliance on pesticides and fertilizers. Critics fear the creation of "superweeds"—either the engineered plants or new plant varieties formed by the transfer of recombinant genes conferring various types of resistance to wild species. These weeds, in turn, would compete with valuable plants and have the potential to destroy ecosystems and farmland unless stronger poisons were used for eradication. The transfer of genetic material to wild relatives (*outcrossing*, or "genetic pollution") might also lead to the development of new plant diseases. As with any new technology, there may be other unpredictable environmental consequences.

Diane White Husic

See also: Biotechnology; DNA: recombinant technology; Genetically modified bacteria; Plant biotechnology.

Sources for Further Study

American Chemical Society. *Genetically Modified Foods: Safety Issues*. Washington, D.C.: Author, 1995. A thorough overview of the technology, applications, and risks associated with genetically engineered foods.

Anderson, Luke. *Genetic Engineering, Food, and Our Environment*. White River Junction, Vt.: Chelsea Green, 1999. Anderson explains why genetic engineering has become such an important issue and provides an introduction to its social, environmental, and health implications.

Paredes-Lopez, Octavio, ed. *Molecular Biotechnology for Plant Food Production*. Lancaster, Pa.: Technomic, 1999. Provides a comprehensive set of reviews of current knowledge of molecular biotechnology, including genetic engineering, associated with the production of plant food crops.

Rissler, Jane. *The Ecological Risks of Engineered Crops*. Cambridge, Mass.: MIT Press, 1996. A carefully reasoned science and policy assessment showing that the commercialization and release of transgenic crops on millions of acres of farmland can pose serious environmental risks. The authors propose a practical, feasible method of conducting precommercialization evaluations that will balance the needs of ecological safety with those of agriculture and business and that will assist governments seeking to identify and protect against two of the most significant risks.

Shannon, Thomas A., ed. *Genetic Engineering: A Documentary History*. Westport, Conn.: Greenwood Press, 1999. A collection of articles and excerpts for students researching genetic engineering. Topics include the debates in genetic engineering, animal application, agriculture, the human genome project, issues in research, diagnostic applications of genetic information, ethical issues, and cloning.

Yount, Lisa. *Biotechnology and Genetic Engineering*. New York: Facts on File, 2000. Offers a chronology from the birth of agriculture to recent findings, information on court cases and significant legislation, and concise biographies of influential figures. Includes annotated bibliography and a list of organizations and agencies.

GENETICS: MENDELIAN

Categories: Disciplines; genetics; history of plant science

Mendelian genetics is the classical mechanistic explanation of heredity in sexually reproducing organisms. It explains how genetic information is passed from one generation to another.

In 1866 the Augustinian monk Gregor Mendel (1822-1884) published a paper titled "Versuch über Pflanzenhybriden" (*Experiments in Plant-Hybridisation*, 1910), describing the heredity of mutant characteristics of garden peas. Mendel founded the modern science of genetics with these experiments, because they led him to propose the existence of hereditary factors, now called *genes*, and rules describing their inheritance, now referred to as *Mendel's laws*. The importance of Mendel's work was not recognized until 1900, sixteen years after his death, when the movements of *chromosomes* during cell division were carefully studied. Since then, Mendel's laws have been shown to hold true throughout nature. The biochemical nature of genes has been discovered, the genetic code has been broken, and genetics has assumed a central role in modern biology, medicine, and agriculture.

Mendel's Experiments

Mendel was not the only researcher interested in genetics, or inheritance, as it was called in his day. The prevailing theories of his time, though, differed considerably from his final conclusions. It was believed by most biologists that inheritance involved the blending of some sort of element from each parent. The result is offspring that are intermediate between the parents. This process, as understood, was analogous to blending two colors of paint. Experiments by others made it difficult to challenge this theory, because they often studied numerous complex traits simultaneously, only analyzed offspring from a single generation, and worked with small numbers of organisms.

One of the reasons for Mendel's success was that he simplified the problem of heredity by analyzing a few simple, easily distinguishable, hereditary differences among a species that was easy to breed. He also initially studied one trait at a time and followed the inheritance of each trait for several generations, using large enough numbers to solidify his conclusions.

He began by selecting strains of garden peas that differed by a single trait from normal strains, such as wrinkled versus smooth, green peas versus yellow, and tall plants versus short. Once each strain bred true for a mutant variation, he crossed it with a different strain to see which trait was passed on to the offspring. Mendel observed that all the hybrid offspring of each individual cross resembled one of the parent types and not the other, rather than a mixture of the two types or an intermediate form (the conventional "mixed-paint" theory). Crosses between tall and short parent strains, for example, produced hybrids that were tall only. Mendel defined this phenomenon as *dominance* of one trait over the alternate trait, which he called the *recessive* trait.

Mendel then discovered that crossing two hybrids resulted in the reappearance of the recessive trait but only in one-fourth of the offspring. A cross between two tall hybrids, for example, produced about three-fourths tall plants and one-fourth short plants. Mendel proposed that hereditary factors (now called genes) existed for each of the traits with which he was working. He also proposed that hereditary factors exist in pairs, such that each individual inherits one from the pair carried by each parent. Mendel hypothesized that the pairs of factors would be separated, and one would be randomly included in each *gamete* (male or female germ cell—pollen or ovule, in the case of plants).

Testing the Theory

Mendel tested this theory with further crosses. (Because Mendel did not know of the existence of genes, he did not have a clear way to refer to the ge-

Mendel's Pea Plants

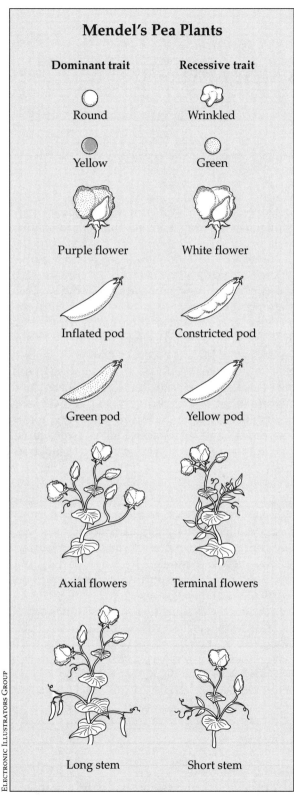

Dominant trait	Recessive trait
Round	Wrinkled
Yellow	Green
Purple flower	White flower
Inflated pod	Constricted pod
Green pod	Yellow pod
Axial flowers	Terminal flowers
Long stem	Short stem

ELECTRONIC ILLUSTRATORS GROUP

Mendel evaluated the transmission of seven paired traits in his studies of garden peas.

netic factors responsible for his results. The word "gene" is used in this discussion for convenience.)

He predicted, for example, that the original tall parent had two tall genes for height (symbolized as *TT*), and the original short parent had two short genes for height (*tt*). The hybrid would inherit one of each (*Tt*), but because the tall gene is dominant, the hybrid's appearance would be tall. Mendel predicted that crossing a *Tt* hybrid with one of the short (*tt*) plants should produce half tall (*Tt*) offspring and half short ones (*tt*). That is exactly what Mendel observed. He was also able to predict the outcomes of other crosses involving different traits. Mendel concluded that his theory worked: Paired hereditary factors must exist, and only one of the two, chosen at random, could be passed on to each offspring by each parent. Mendel labeled this phenomenon *segregation*, meaning that the parent's two hereditary factors are physically segregated into different cells during the production of gametes. This principle of segregation is now called Mendel's first law of inheritance.

Mendel's second law of inheritance describes the principle of *independent assortment*, which states that different hereditary factors segregate independently of one another. Mendel observed, for example, that if he crossed a tall and purple parent (*TT-PP*) with a short and white one (*tt-pp*), the hybrid offspring were tall and purple, as these genes are dominant. Then, when he crossed the tall and purple hybrid (*Tt-Pp*) with another, identical hybrid, the progeny showed an "assortment" of the two traits (tall and purple, tall and white, short and purple, short and white) in a 9:3:3:1 ratio, respectively. This is the ratio expected if each trait's genes segregate independently. Stated another way, whether a parent passes on a tall or short factor does not influence whether that parent also passes on the purple or white factor.

These two laws of heredity summarize Mendel's discovery of discrete genetic factors and their patterns of inheritance: Mendel had proposed that discrete genetic factors exist, had explained how they are passed on, and had supported his theories with experimental evidence. Mendel's discovery, however, was virtually ignored. He died in 1884 without receiving recognition for his work. Mendel's laws were independently rediscovered in the year 1900, and then their fundamental importance and general applicability were widely recognized.

Results of Mendel's Pea-Plant Experiments

Parental characteristics	First generation	Second generation	Second generation ratio
Round × wrinkled seeds	All round	5,474 round : 1,850 wrinkled	2.96 : 1
Yellow × green seeds	All yellow	6,022 yellow : 2,001 green	3.01 : 1
Gray × white seedcoats	All gray	705 gray : 224 white	3.15 : 1
Inflated × pinched pods	All inflated	882 inflated : 299 pinched	2.95 : 1
Green × yellow pods	All green	428 green : 152 yellow	2.82 : 1
Axial × terminal flowers	All axial	651 axial : 207 terminal	3.14 : 1
Long × short stems	All long	787 long : 277 short	2.84 : 1

Morgan's Contributions

Microscopic bodies in the nuclei of cells, called chromosomes, had been discovered by the end of the nineteenth century, and in 1901 it was proposed that chromosomes are the physical structures that contain Mendel's hereditary factors, or genes. Chromosomes were a likely structure for the location of genes because chromosomes occur in pairs, duplicate when the cell divides, and segregate into sperm and egg cells such that only one of the two chromosomes in each pair is passed on to any single offspring by each parent. The chromosomal theory of heredity made it easier for biologists to think of genes as physical objects of analysis, and studies of Mendelian patterns of inheritance and their chromosomal basis progressed rapidly.

A geneticist named Thomas Hunt Morgan at Columbia University made several key discoveries using fruit flies between 1910 and 1920. He and his colleagues discovered mutations in flies that showed different patterns of inheritance in males and females, which led to association of these genes with the sex-determining X and Y chromosomes. Traits affected by genes on these chromosomes show a sex-linked pattern of inheritance in which recessive traits appear more often in males than in females. Human sex-linked traits, for example, include hemophilia, color-blindness, and baldness.

Fruit flies have three pairs of chromosomes besides the sex chromosomes, and Morgan's laboratory team showed that traits could be grouped together in "linkage groups" corresponding to their four pairs of chromosomes. They realized that Mendel's second law describing the principle of independent assortment corresponded to the assortment of chromosomes being passed from parent to offspring. Any genes on different chromosomes would be passed on independently, while genes linked together on the same chromosome would be passed on together as a unit. The discovery of linkage groups supported the idea that chromosomes were made up of collections of a large number of genes linked together.

Sturtevant's Contributions

Morgan's laboratory group, however, also observed occasional exceptions to this pattern of linkage, when offspring showed unexpected new combinations of linked genes that did not exist in either parent. Alfred H. Sturtevant, a student in Morgan's laboratory, proposed that the paired chromosomes carrying different forms of the same genes (one carrying recessive forms, for example, a-b-c, versus the other, carrying dominant forms A-B-C) could undergo a reciprocal exchange of part of the chromosome. One chromosome pair could exchange, for example, C for c, resulting in new a-b-C and A-B-c combinations of the linked genes.

Sturtevant also discovered that such recombination events occur with different frequencies between different genes. Only 1 percent of the A and B genes might be switched in each cross, for example, but 20 percent of the A and C genes might recombine in the same cross. Sturtevant proposed that the genes are linked together in a linear sequence and that the frequency of recombination between them is a function of the physical distance separating them on the chromosome. Two genes that are far apart should recombine more frequently than two genes close together, since there would be a greater opportunity for the breakage and the exchange of different chromosomal material to occur between them.

Sturtevant proposed that differences in the frequency of recombination among linked genes on the same chromosome could be used to "map" the genes in a linear sequence that would reveal their order and relative positions on the chromosome. This principle turned out to be universal, and it allows genes to be mapped to specific locations on each chromosome in all organisms that can be systematically bred. Mendel's genes had, by the 1920's, been associated with chromosomes, and individual genes on each chromosome could be ordered and mapped using recombination analysis.

Mid-Twentieth Century Developments

The following two decades were marked by two important parallel developments in genetics. The first was a mathematical and experimental synthesis of Mendel's genetic theory with Charles Darwin's theory of natural selection. It was shown that the genetic mechanism described by Mendel provided the hereditary mechanism required for Darwin's theory of natural selection. The revision of Darwin's work that resulted is often referred to as the neo-Darwinian synthesis.

The second development was progress in identifying the biochemical nature of genes, primarily by the extension of genetic analysis to bacteria and viruses. These studies led to the identification of *deoxyribonucleic acid* (DNA) as the hereditary molecule and to the identification of its biochemical structure by James Watson and Francis Crick in 1953. Once the biochemical structure of genes was identified, an understanding of how DNA replicates and carries a genetic code that directs the synthesis of proteins followed rapidly.

One more revolutionary breakthrough that set the stage for the current era of genetics was the rapid development of *recombinant DNA technology* in the 1970's and its refinement and broad application in the 1980's. Recombinant DNA technology is a collection of methods that allows DNA sequences of one organism to be recombined with those of another. The application of these techniques is commonly referred to as "genetic engineering." The fact that the chemical structure of DNA and the genetic code for protein synthesis are virtually the same for all organisms makes recombinant DNA a powerful technology. Recombinant DNA techniques, together with an understanding of the genetic code and the ability to identify and map specific genes, have opened up a new era of biological investigation and applications to medicine and agriculture.

Bernard Possidente, Jr., updated by Bryan Ness

See also: Chromosomes; DNA: historical overview; DNA in plants; DNA: recombinant technology; DNA replication; Genetic equilibrium: linkage; Genetics: mutations; Genetics: post-Mendelian; Mitosis and meiosis.

Sources for Further Study

Henig, Robin Marantz. *The Monk in the Garden: The Lost and Found Genius of Gregor Mendel.* Boston: Houghton Mifflin, 2001. This biography of Mendel is very readable, even for those with a minimal genetics background.

Jacob, Francois. *The Logic of Life.* Princeton, N.J.: Princeton University Press, 1993. History of the study of heredity, from the ideas of the Greeks through the discovery of the genetic code, by a Nobel Prize-winning molecular geneticist. Presented as a serious intellectual history.

Orel, Vitezslav. *Mendel.* New York: Oxford University Press, 1996. A detailed account of the life and work of Gregor Mendel. Orel is the head of the Moravian Museum exhibit on the history of Mendelian genetics at the site of Mendel's monastery and original garden laboratory in what is now the city of Brno, Czechoslovakia. A fascinating biography.

Sturtevant, Alfred H. *A History of Genetics.* Cold Spring Harbor, N.Y.: Cold Spring Harbor Laboratory Press, 2001. Comprehensive history of genetics written by the discoverer of the technique of linkage mapping by recombination analysis. The book is clearly written and nontechnical and is especially interesting for its firsthand account of the discoveries made in Thomas Hunt Morgan's "fly lab" in the early 1900's.

GENETICS: MUTATIONS

Categories: Cellular biology; evolution; genetics

A mutation is a heritable, sudden change in the structure of a gene, which has no relation to the individual's ancestry. The change can occur spontaneously or can be the result of exposure to ultraviolet radiation or chemicals.

In 1901 Hugo de Vries coined the term *mutation* to describe changes in the hereditary material of evening primrose (*Oenothera*). "Mutation" is a derivative of the Latin verb *mutare*, meaning "to move or change." The word was first used to describe spontaneous, heritable changes in the phenotype of an organism.

In the modern era of genomics, mutations can be defined as changes in DNA (deoxyribonucleic acid) sequences, that is, changes in the structure of a gene. The changes can occur spontaneously or can be induced via ionizing radiation (ultraviolet radiation) or chemicals, such as aflatoxin B_1 and ethylmethane sulfonate. A common cause of spontaneous mutations is *deamination*, in which the amino group on the number 2 carbon of cytosine (C) is removed, converting C to uracil (U) in DNA. Another cause is copying errors during DNA replication: slippage or shifting of the translational reading frame. Spontaneous mutations also may be caused by *depurination*, in which the bond between deoxyribose sugar and a purine base, adenine (A) or guanine (G), is hydrolyzed, or by *depyrimidination*, the hydrolization of the bond between deoxyribose sugar and a pyrimidine base, either C or thymine (T). Depyrimidination is less common than depurination. The sites where a base is missing are called apurinic sites (when a purine base is missing) or apyrimidinic sites (when a pyrimidine base is missing) or simply *AP sites*.

An individual with a mutation is called a *mutant*. When a mutation occurs in the reproductive tissue of an individual plant, it can be transmitted to the next generation. When a mutation occurs in the somatic tissue, it will be limited only to that generation and affects only the cells in which it occurs.

Heredity vs. Genetic Combinations

Mutation is the only process that creates new genetic variation that results from a change or changes in the structure of a gene or genes. Genetic variability also can arise from new genetic combinations produced through the processes of crossing over, gene segregation, and chromosome assortment. In the case of recombination, all the genes are already present in an individual, and new variation simply results from the shuffling of those genes during gamete formation; there are no structural changes in genes.

Sites and Types of Mutation

Gene mutation involves a change in a single base pair or a deletion of a few base pairs. It usually affects the function of a single gene. The substitution of one base (or nucleotide) for another base (or nucleotide) is called a *point mutation* or a *substitution mutation*. The replacement of a pyrimidine (cytosine and thymine) with another pyrimidine or the replacement of a purine (adenine and guanine) with another purine is termed *transition*. The replacement of a pyrimidine with a purine or the replacement of a purine with a pyrimidine is termed *transversion*.

Base pair or nucleotide changes can produce one of the following types of mutation:

- *Missense mutation*, which results in a protein in which one amino acid is substituted for another amino acid.

- *Nonsense mutation*, in which a stop codon is substituted for an amino acid codon, which results in premature termination of a protein.

- *Frameshift mutation*, which causes a change in the reading frame. These mutations can introduce a different amino acid into the protein and have a much larger effect on protein structure. Small deletions also have effects similar to those of frameshift mutations. A nonfunctional protein may be produced, unless the frameshift is near the terminal end of a gene.

- *Chromosomal mutations* or *abnormalities*, which involve deletions or insertions of several contiguous genes, inversion of genes on a chromosome, or the exchange of large segments of DNA between nonhomologous chromosomes.

Effects of Mutations

Gene flow is the exchange of genes between different populations of the same species, produced by migrants and commonly resulting in changes in gene frequencies at many loci (locations) in the recipient gene pool. The effect of a mutation carried into a new population will depend upon the number of migrants that were mutants and the size of the recipient population.

Genetic drift is the random fluctuations of gene frequencies as a result of sampling errors. Drift occurs in all populations, but its effects are most striking in small populations. Due to periodic reductions in population size, genetic drift can affect gene frequencies. A large population may contract and expand again with an altered gene pool (called the *bottleneck effect*). A consequence of genetic drift is reduced variability. A mutation has a better chance of spreading faster in a smaller population than in a larger population.

The "founder effect" is a term first coined by Ernest Mayr in 1942, who referred to small groups of migrants that succeed in establishing populations in a new place as "founders." Two founders could carry only four alleles at each gene locus. If a rare (mutant) allele were included among them, its frequency would be considerably higher (0.25) than it was in the parental population.

Nonrandom mating, also called *assortative mating*, occurs when male and female plants are not crossed at random. If the two parents of each pair tend to be more (or less) alike than is to be expected by chance, then positive (or negative) assortative mating occurs. Positive assortative mating promotes *homozygosity*, whereas negative assortative mating tends to promote *heterozygosity*. Mating of similar homozygotes would increase their frequency at the expense of heterozygotes. A mutation should spread more quickly under assortative mating than under nonassortative mating.

Inbreeding is defined as the coming together, at fertilization, of two alleles that are identical by descent. This is the result of mating between closely related plants. A mutation has a better chance of establishing under mating systems in which close relatives are involved than under those where inbreeding is prevented. Assortative mating represents the mating of individuals with similar *phenotypes*, whereas inbreeding represents the mating of individuals of similar *genotypes*.

Autogamy, or self-fertilization, is the strictest form of inbreeding. A mutation would spread more quickly under self-pollination than under cross-pollination.

If a mutation occurs at a homozygous locus ($aa \times Aa$, or $AA \times Aa$), the result would be greater diversity. If a mutation occurs at a heterozygous locus ($Aa \times AA$ or aa), it would result in more uniformity. Depending on the size of the population, gene frequency will change.

Manjit S. Kang

See also: Genetic drift; Nonrandom mating; Population genetics.

Sources for Further Study

Doolittle, Donald P. *Population Genetics: Basic Principles*. New York: Springer-Verlag, 1986. From the advanced series in agricultural sciences. Includes bibliography, index.

Redei, G. P. *Genetics Manual*. New Jersey: World Scientific. 1998. This book is an excellent encyclopedic reference, with eighteen thousand entries related to genetics.

GENETICS: POST-MENDELIAN

Categories: Disciplines; genetics; history of plant science

Thirty years after the work of Gregor Mendel in the nineteenth century, several rediscoveries of his work in genetics brought his theories to the fore. At about the same time, the discovery of chromosomes, coupled with the earlier knowledge, took genetics in a new direction.

Gregor Mendel (1822-1884) is often considered the founder of the science of genetics. Though his experiments with pea plants became the basis for understanding genetics in all plants and animals, he died unknown. In 1900 three simultaneous "rediscoveries" of Mendel's studies put his name at the forefront of biology. With the reintroduction to the world of Mendel's genetic laws, biologists began to look more closely at genetic phenomena. These researchers used Mendel's laws as the basis for more in-depth studies of genetics, which led to the modern understanding of genes, chromosomes, and their inheritance.

Rediscovery of Mendel

In 1900, working independently of one another, biologists Erich Tschermak, Hugo de Vries, and Karl Erich Correns each published data that reasserted Mendel's historic principles of heredity. Each scientist came to this rediscovery from a slightly different perspective.

Tschermak, an Austrian botanist, coincidentally started pea plant breeding experiments in 1898. He performed these experiments for two years before he accidentally discovered a reference to Mendel's work from thirty years earlier. When he read Mendel's papers, Tschermak found that he had duplicated many of Mendel's breeding experiments, and, embarrassingly, his own work was not as thorough. The Austrian published his own findings and gave credit to Mendel for performing the original breeding work. Tschermak is known for applying the genetic principles he helped rediscover to developing wheat-rye hybrids and a disease-resistant oat hybrid.

Hugo de Vries, whose primary concern was understanding how evolution worked, was a professor at the University of Amsterdam. De Vries wanted to find a genetic basis for Charles Darwin's theory of natural selection to understand how species could change over time. De Vries studied the evening primrose and found that, after cultivating the plant for years, several varieties arose through abrupt, unexplained genetic changes. Based on these changes, he came up with a theory of *mutation* in which he hypothesized that rapid alterations in organisms could explain how evolution could quickly produce new species. For eight years, starting in 1892, de Vries conducted breeding experiments that led him to the same laws of heredity that Mendel had discovered. When he reported his own work, de Vries was very careful to attribute his concepts to Mendel.

Karl Correns was a German botanist at the University of Tübingen in the 1890's. By coincidence, Correns conducted breeding experiments with peas that reproduced Mendel's experiments. In a survey of the literature, Correns found Mendel's papers, published many years earlier. Much of Correns's life work was spent in developing additional evidence to support Mendel's hypotheses. Correns was the first researcher to suggest that if certain genes were physically close to each other, they might be "coupled" in some way and be consistently inherited in offspring. His concept explained why some traits did not seem to follow Mendel's *law of independent assortment*, which stated that all traits separated independently of one another when inherited by offspring.

Chromosomes

Chromosomes were not discovered until the end of the nineteenth century, so Mendel was never able to suggest any physical basis for his genetic theories. It was not until the science of cytology (the study of cells) was founded that scientists started to

AP/WIDE WORLD PHOTOS

Mutation creates new genetic variation that results from a change or changes in the structure of a gene or genes. Such genetic changes can influence the color of an organism; these clover leaves have mutated to exhibit a unique purple hue.

examine cells and their replication more closely. They discovered that *somatic cells* (that is, nonreproductive cells) consistently went through a pattern of division in which chromosomes were duplicated and separated between two new *daughter cells*.

Walter Sutton and Theodore Boveri, working with grasshopper cells, were the first scientists to notice that chromosomes in somatic cells occur in pairs. Sutton and Boveri suggested a connection between the pairs of chromosomes and Mendelian genetics. They believed that chromosomes carried the units of inheritance and that the way chromosomes divided accounted for how Mendel's laws functioned. Their work formed the basis for the *chromosomal theory of inheritance*.

The chromosomal theory of inheritance suggested that Mendel's genes reside on chromosomes and that when plants and animals reproduce, half their genetic material comes from each parent, forming sets of chromosomes. For example, barley

has fourteen chromosomes in each somatic cell. Seven of those chromosomes are contributed from the "mother" plant and seven from the "father" plant, to make a total of fourteen chromosomes in the offspring. Therefore, half of the genes from all organisms come from each parent to determine the progeny's genetic makeup.

Each chromosome is essentially one long, linear strand of deoxyribonucleic acid (DNA), wrapped up and compacted for easy duplication and transport by the cell. There are two copies of each chromosome (called homologous pairs or homologs) in every somatic cell of an organism, each with the same physical appearance.

Such a cell with two copies of each chromosome type is called a *diploid* cell. Reproductive cells, known as *gametes*, have half the number of chromosomes and are known as *haploid* cells. It is these haploid cells from each parent that comprise the new diploid cells of the offspring. Half the chromosomes in each diploid cell come from each parental haploid cell.

Copies of the same gene on each chromosome pair are found at the same location (called a *locus*) and control traits of the organism. The copies of the same gene at a locus are called *alleles*. For example, one copy of chromosome #1 might be from the male parent and have a *dominant allele* (symbolized by *A*), and the other copy of chromosome #1 might be from the female parent and have a *recessive allele* (symbolized by *a*). These two alleles together (*Aa*), each on a separate chromosome, would constitute the *genotype*, and the expression of these alleles produces the *phenotype* (physical traits of an organism).

Linkage

In 1905 William Bateson and Reginald Punnet were the first to show clear evidence that Correns's theory of "genetic coupling" was correct. They crossed sweet peas having purple flowers and long pollen grains with sweet peas having red flowers and round pollen grains. According to Mendel's rules, the offspring in the second sweet pea generation should have *segregated* (genetically separated) into four phenotype combinations (purple/long, purple/short, red/long, and red/short) in a ratio of 9:3:3:1, because the two genes controlling these traits should have separated independently of each other. Bateson and Punnet did not obtain a 9:3:3:1 ratio. Instead, the parental traits stayed together in the offspring more often than expected, and more offspring looked like the parents than expected: purple/long or red/short. Bateson and Punnet called this phenomenon *linkage*, and any genetic traits that followed this pattern were said to be linked to each other.

In 1910 American geneticist Thomas Hunt Morgan explained the physical basis for linkage. Through his experiments with fruit flies, Morgan found that alleles for different traits only followed Mendel's law of independent assortment if they were on different chromosomes or if they were very far apart when they were on the same chromosome. If the genes for two traits were on the same chromosome, they were often passed down to the next generation jointly and stayed together consistently from generation to generation. Morgan further found that the physically closer that two alleles were to each other on a chromosome, the more closely "linked" they were to each other, staying together in offspring a greater percentage of the time.

Incomplete Dominance

Linkage was one of the first phenomena to break the Mendelian laws, but there were many additional conditions that Mendel would have puzzled over, such as *incomplete dominance*. Usually in a heterozygous organism (one with a dominant allele and a recessive allele at the same locus), the phenotype is controlled by the dominant allele, and the trait from the recessive allele will be masked. When incomplete dominance occurs, the dominant trait is weakened, and the heterozygotes look as though they have a trait partway between the recessive and dominant traits. For example, if a red-flowered snapdragon, *RR*, is crossed with a white-flowered snapdragon, *rr*, all the first-generation offspring are heterozygous, *Rr*. If the trait were dominant, then all the flowers in the offspring would be red. However, the trait displays incomplete dominance, and all the flowers are pink.

Multiple Alleles

Although an individual can have only up to two alleles at a locus, more than two alleles can exist in a population. For example, some populations of red clover are estimated to have hundreds of alleles at a locus for self-sterility. As a result, most individuals have alleles that are different from those of other members of the population, thus preventing self-pollination and making out-crossing with other plants successful. Some plants can also have more than two alleles at a locus if they are *polyploid*.

A polyploid plant has more than two homologous chromosomes of each type. The most common type of polyploid is a tetraploid, which has four homologous chromosomes of each type. With four chromosomes of each type, a locus has four alleles instead of just two. Other levels of polyploidy exist in plants, even as much as cases with ten, twenty, or more homologous chromosomes of each type.

Gene Interactions

Gene interactions occur when two or more different loci (gene locations) affect the outcome of a single trait. The most common type of gene interaction is known as *epistasis*. Epistasis describes a situation where an allele at one locus masks the phenotypic effects of a different locus. The gene being masked is called the *hypostatic* gene, while the gene doing the masking is called the *epistatic* gene.

Bateson and Punnet discovered this phenomenon during their sweet pea breeding experiments.

They crossed purple-flowered plants with white-flowered plants. In the first generation, they got all purple-flowered offspring—so they concluded that purple was the dominant gene. In the next generation, they did not get a 3:1 ratio of purple-flowered to white-flowered plants, as would be expected if purple was dominant. They got a ratio of 9:7 purple to white-flowered plants. It turned out that there were two different loci involved in the control of petal color: at the first locus C (purple) was dominant to c (white), and at the second locus P (purple) was dominant to p (white). When either locus was homozygous recessive, either cc or pp, the flowers were white, regardless of the genotype of the other locus. The recessive alleles were epistatically affecting (or masking) the dominant alleles.

Polygenic Inheritance

Certain traits are too complex to be controlled by a single locus. These traits are controlled by a complex of two or more loci, a phenomenon known as *polygenic inheritance*. In humans, multiple loci control height, intelligence, and skin coloration. These multiple genes lead to *continuous variation*, meaning that one observes a wide range of phenotypic variation. The first experiment demonstrating continuous variation was conducted by Swedish scientist Herman Nilsson-Ehle in 1910. He studied the inheritance of the red pigment on the hulls of wheat. He found that red-hulled wheat crossed with white-hulled wheat for several generations gave him plants ranging in pigment from white, light-pink, and pink to medium, basic, and dark-red. Nilsson-Ehle found that three loci control this color variation in the wheat.

Pleiotropy

Pleiotropy also breaks Mendel's laws. Usually, one locus controls a single trait. A pleiotropic gene is a single locus that controls multiple traits. If there is a loss of function mutation in a pleiotropic gene, the organism is affected in multiple ways. One example of such a trait can be found in the plant *Arabidopsis thaliana*. This plant has a mutant allele known as tu8. This gene was originally isolated as a mutant in the biochemical pathway that makes glucosinolate, a chemical used against pathogens. This tu8 mutation also causes the plant to be dwarfed, late-flowering, and heat-sensitive. Genetic experiments by researchers James Campanella and Jutta Ludwig-Mueller have shown that all these traits are controlled by a mutation in a single gene.

Cytoplasmic Inheritance

Finally, *cytoplasmic inheritance*, often known as *maternal inheritance* or *extranuclear inheritance*, is a phenomenon referring to any genetic traits not inherited from nuclear genes. For example, both chloroplasts and mitochondria contain their own genetic information that is inherited by every generation. This information is inherited in a different fashion from that of nuclear genes. Nuclear genes, in the form of chromosomes, are donated equally by each parent. The genetic information from chloroplasts and mitochondria is not donated equally. Offspring come from the joining of the male and female gametes, but the size of these gametes differs drastically. Male gametes, both animal sperm and plant pollen, are often one-one hundredth the volume of an egg cell. Because of their small size, male gametes often have little cytoplasm. Because the chloroplast and mitochondria reside in the cytoplasm, it is usually the case that none of the organellar DNA of the male gamete is included in the offspring. The female parent alone donates the chloroplast and mitochondrial alleles. Although maternal inheritance is the rule in most plants, a few groups, such as some members of the evening primrose family (*Onagraceae*) have displayed biparental inheritance of organellar DNA.

James J. Campanella

See also: Chromosomes; DNA in plants; DNA replication; Gene regulation; Genetic code; Genetic equilibrium: linkage; Genetics: Mendelian; Genetics: mutations; Mitosis and meiosis; Nucleic acids; Nucleus; Reproduction in plants; RNA.

Sources for Further Study

Burris, Robert, et al. *Historical Perspectives in Plant Science*. Ames: Iowa State University Press, 1994. Examines the historical perspectives of genetics and plant science.

Griffiths, Anthony, et al. *Modern Genetic Analysis*. New York: W. H. Freeman, 1999. This genetics textbook offers detailed explanations of the exceptions to Mendel's laws.

Snedden, Robert. *The History of Genetics*. New York: Thomson Learning, 1995. The text, presented using nontechnical terms, is geared toward the young adult.

Tudge, Colin. *The Impact of the Gene: From Mendel's Peas to Designer Babies*. New York: Hill and Wang, 2001. A narrative on the development of genetics, from Mendel's pea experiments to the end of the twentieth century.

Verne, Grant. *Genetics of Flowering Plants*. New York: Columbia University Press, 1978. Focuses specifically on the history and genetics of plants.

GERMINATION AND SEEDLING DEVELOPMENT

Categories: Anatomy; physiology; reproduction and life cycles

With germination, the growth of a seedling, spore, or bud begins. Seedling development begins with the close of germination.

To germinate, seeds must be nondormant and in a suitable environment. Seeds germinate within a restricted range of temperatures, moisture, oxygen, light, and freedom from chemical inhibitors. Wild seeds display many *adaptations* that predispose germination within specific habitats and seasons. By contrast, seeds of crops and other cultured plants usually lack controls that prevent germination. The control system was lost because some seeds in the population lacked controls and were chosen when they germinated in the care of a culturist. For that reason, most cultivated plants that start from seeds show little or no germination control. Most of the information on germination control, therefore, covers wild species of plants.

Seeds and Dormancy

Seeds are the exclusive means of regeneration for the annual flowering plants. In other plants, seeds are an alternative strategy to regeneration by *buds*, *bulbs*, *rhizomes*, *stolons*, or *tubers*. In those plants, the primary roles of the seed are to disperse the population and to reinvigorate the genetic diversity of the germ line.

Seed *dormancy* occurs in most plants, and when a seed is dormant it will not germinate, even if it is in the right environmental conditions. The dormant state may begin with maturation of the seed embryo, or it may develop in climate extremes after the seed falls from its parent. Dormancy prevents immediate germination when the mature seed is in an inappropriate environment, and it is a programmed phase in the life cycle. Dormancy's function ends, and a *germination window* opens, at a time when the emerging seedling will have the optimum chance for survival.

After-ripening removes the dormancy and allows the seed to respond to germination stimuli. Seeds of summer annuals after-ripen when exposed to winter and early spring temperatures, a treatment called *stratification*. Exposure to cold temperatures can also promote dormancy in some seeds.

Not all nongerminating seeds are innately dormant. There are also nondormant and conditionally dormant seeds. Neither type may germinate when the seeds mature, simply because the parent prevents contact with the soil and absorption of water or because the temperature range is below that necessary for germination. Wild seeds may experience a deepening of dormancy as a result of exposure to the temperatures of the dormant season. Nondormant seeds may simultaneously experience biochemical reactions that deepen dormancy and cause them to after-ripen.

Dormancy is caused by one or more conditions of the seed. Physiological dormancy of the embryo is the most common. It may be caused by the presence of an *inhibitor molecule*, an inadequate level of a growth hormone, or some other internal factor. Examples of the latter include blockages in membrane function or in synthesis of an enzyme or its nucleic acid messenger.

Other causes of dormancy are a hard or impervious seed coat, an underdeveloped embryo, or some combination of those factors. Some hard-coated seeds require physical scraping, such as tumbling down a swift-flowing stream and being scraped on the streambed. Others may require exposure to a forest fire or passage through the gut of an animal to weaken the seed coat.

Germination

The seeds from the previous year's crop of summer annuals wait to germinate in spring. Seeds of many species will germinate when soil temperatures reach a threshold constant. Others require daily fluctuations of temperature, waiting until the daily fluctuation becomes sufficiently large.

Seeds of most species require a light stimulus to germinate. The light is absorbed by the pigment *phytochrome*, which is positioned in the cotyledons of the embryo. Phytochrome acts as a shade detec-

tor. White light and, especially, red light, convert the phytochrome molecule to an active form. The rearrangement of the molecule causes it to attach a different part of itself to a new location on a cell membrane within the cotyledons of the seed. Transformed phytochrome allows seeds to germinate. By contrast, far-red light, which is absorbed by the transformed phytochrome molecule, transforms the molecule back to its original shape—that is, it deactivates it. When sunlight is transmitted through green leaves to the forest floor, much of the visible light with wavelengths shorter than 700 nanometers is absorbed or scattered. This *shade light* is rich in far red, and it tends to deactivate phytochrome.

Not all seeds germinate at the beginning of the following growing season. Light-demanding seeds that have become buried or have fallen into the shade will be stressed by the absence of an activat-

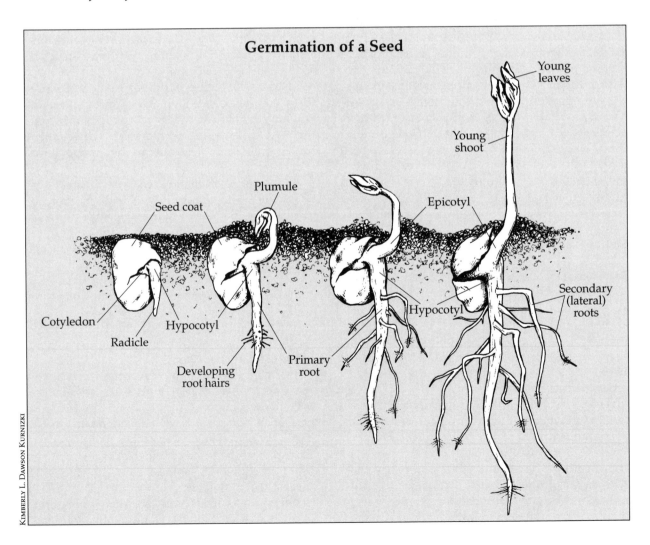

Germination of a Seed

Young leaves

Young shoot

Seed coat

Plumule

Epicotyl

Secondary (lateral) roots

Hypocotyl

Cotyledon

Radicle

Hypocotyl

Developing root hairs

Primary root

KIMBERLY L. DAWSON KURNIZKI

ing light signal, while the embryos experience an environment that is otherwise growth-promoting. Stressed seeds may enter a *secondary dormancy* and will need to undergo a second interval of after-ripening before again becoming nondormant. Seeds emerging from secondary dormancy may require a smaller light stimulus. Following primary dormancy, one or more complete light cycles may be necessary. By contrast, seeds may be fully activated by only a brief pulse of light when they emerge from secondary dormancy.

Many seeds become dry on the parent plant as a part of maturation. Drying is believed to end the seed-building phase and start the pregermination phase. Most domestic seeds require no after-ripening but will germinate if allowed to take up water. Water uptake is a first step in germination. Biochemical events that begin with water uptake in domestic seeds include metabolism along three separate pathways and an increased use of oxygen. Excessive moisture, even at temperatures too low for germination, may lower seed viability. Thus, seeds are best preserved under cold, dry conditions.

With proper storage, a reasonable percentage of seeds may live for many years. Record long-lived species include seeds of *Canna* (arrowroot, six hundred years), *Albizia* (mimosa and silk trees), and *Cassia* (about two hundred years). Seeds of *Verbascum* (mullein) have survived at 40 percent viability for one hundred years. Weed seeds in soil banks may need to wait for hundreds of years before the forest is cleared by catastrophe and the environment once again is favorable for such pioneer species. By contrast, recalcitrant seeds require their embryos to be kept moist, or viability is quickly lost; examples include trees with large seeds, such as walnut, oak, hazel, and chestnut. These seeds usually live less than one year.

Seedlings

Seedling development begins with the close of germination. Cells in the embryonic root (the *radicle*) begin to divide and grow. In some seeds, or in unusual environmental conditions, other parts of the embryo emerge before the radicle. Development of the seedling is marked by the growth and elongation of the embryo stem. Seeds are classified according to which part of the stem grows more rapidly. In the *epigeal* type, the *hypocotyl*, which is the basal part of the stem between the radicle and the embryonic leaves (the *cotyledons*), grows,

thereby thrusting the cotyledons above the soil. In the *hypogeal* type, the *epicotyl* or upper part of the stem elongates, and the cotyledons remain underground.

Exposed to light, phytochrome in the cotyledons calls for an end to subterranean elongation, called *etiolation*, and the beginning of plantlike growth. Among its functions, phytochrome triggers the synthesis of chlorophyll; photosynthesis soon turns the cotyledons into sugar factories. At the same time, the epicotyl region of the embryo above the cotyledons is extending the *plumule* to form the first true, or foliage, leaves. In hypogeal seeds the first leaves emerge from the plumule.

Food reserves that are stored in the endosperm, cotyledons, and embryo will nourish the early growth of the plant until it can synthesize the necessary machinery for making its own food. Foods are stored in seeds as starch and other complex carbohydrates, fats, and proteins. Cereal grains contain large amounts (65-80 percent) of carbohydrates, which are stored in the endosperm. Seeds of legumes are famous for their high protein contents, which reach 37 percent in soybeans. The peanut, a legume, stands out by having both high protein (30 percent) and high fat content (50 percent). Legumes store food reserves in the embryonic leaves (cotyledons).

Food is transferred to the growing sites, primarily as sucrose and amino acids. Starches and other carbohydrates and fats are first converted to simple sugars and then to sucrose for transport. Synthesis of active phytochrome, second messengers, and the plant hormone gibberellin are involved in determining the rate of mobilization and transport. They promote synthesis of enzymes, such as those that break down food reserves into simple sugars.

The embryo selects one part of the root-stem axis for rapid growth, changing the relationship of the other parts to the external environment. The cotyledons are also versatile: They may act as first leaves or may remain attached to another part of the embryo, such as the endosperm. There, they act as absorptive organs to transport mobilized food reserves to the growing parts of the seedling. In onions, a single cotyledon performs both functions. The exposed part carries out photosynthesis, while the buried part absorbs foods from the endosperm.

Ray Stross

See also: Angiosperm life cycle; Angiosperm plant formation; Dormancy; Hormones; Plant life spans.

Sources for Further Study

Bewley, J. Derek, and Michael Black. *Seeds: Physiology of Development and Germination.* 2d ed. New York: Plenum Press, 1994. An extensive examination of seed structure and biochemistry in relationship to its maturation and germination. Condensed from two earlier volumes. The topics force an emphasis on domestic seeds.

Fenner, Michael, ed. *Seeds: The Ecology of Regeneration in Plant Communities.* 2d ed. New York: Oxford University Press, 2001. An in-depth textbook on all aspects of seeds, from the evolution of optimal seed size to methods of dispersal, dormancy, and the various factors that end dormancy to the ecology of seedling regeneration.

Fosket, Donald E. *Plant Growth and Development: A Molecular Approach.* San Diego: Academic Press, 1994. An advanced undergraduate textbook on the molecular biology of plant growth. Chapter 8 covers embryogenesis, seed formation, and germination.

Meyerowitz, Elliot M., and Chris R. Somerville, eds. *Arabidopsis.* Plainview, N.Y.: Cold Spring Harbor Laboratory Press, 1994. A monograph on the model cruciferous weed plant *Arabidopsis thaliana.* Three chapters cover seed development, pattern formation, dormancy, and germination.

Raven, Peter H., Ray F. Evert, and Susan Eichhorn. *Biology of Plants.* 6th ed. New York: W. H. Freeman/Worth, 1999. College-level text refers to original experimental work. Accessible to high school readers as well.

GINKGOS

Categories: Evolution; gymnosperms; paleobotany; *Plantae;* taxonomic groups

The ginkgos, phylum Ginkgophyta, *constitute one of four phyla of the gymnosperms in the kingdom* Plantae. Ginkgo biloba, *the maidenhair tree, is the only living representative of the ginkgo family,* Ginkgoaceae, *a group of plants that have lived for millions of years and are identified by an abundant fossil record.*

The ginkgo is a hairless, deciduous tree with a straight trunk and pyramid-shaped foliage usually sparsely branched when young, becoming denser with age. Leaves are fan-shaped, 2 to 3 inches (5 to 7.5 centimeters) across, sometimes divided into two lobes. The ginkgo normally reaches heights of 80 to 100 feet (24 to 30 meters) and under favorable conditions grows to 125 feet (38 meters) or more. The bark is reddish-gray and corky, with irregular, wide fissures dividing rough plates. On old trees, the bark becomes gray, rough, and deeply furrowed.

Considerable diversity in branching habit occurs, sometimes with one side of the tree having erect branching and the other side spreading limbs. Young trees send out straight branches at a skyward angle and, until maturity, the sparse branching gives the tree an erratic appearance. Upon maturity, the branches round out and become widespread, yet retain an uneven crown.

As in many conifers, the long branches (shoots) and short, spurlike shoots of *Ginkgo biloba* are easily distinguished. The leaves are spirally arranged on both types but widely spaced on the long shoots, with leaves in crowded, rosettelike clusters on the short shoots. Branchlets (twigs) have a horizontal or drooping habit and are occupied with short, spurlike shoots. These shorter shoots increase in length only a fraction of an inch (2.5 centimeters) per year and may produce clusters of leaves annually for many years before abruptly lengthening out into long shoots bearing scattered leaves. The fan-shaped leaf has a marked resemblance to the fronds of the maidenhair fern, thus the common

Classification of Phylum *Ginkgophyta*

Class *Ginkgoopsida*
   Order *Ginkgoales*
      Family *Ginkgoaceae* (Ginkgo family)
         Sole living species: *Ginkgo biloba* (maidenhair tree)

name: maidenhair tree. However, in its native China it is commonly called ducks-foot tree, also based on leaf shape.

The leaves, which grow on slender stalks up to 3 inches (7.5 centimeters) long, have numerous veins radiating out from the base to an irregularly notched leaf margin. There is no central midrib vein on the somewhat leathery, textured leaf. Stomata (breathing pores in the leaves) occur on both the upper and lower surfaces of the leaves. The leaves emerge yellow-green in spring but turn green toward midsummer and become golden in autumn. Leaves on vigorous young trees can grow up to 6 to 8 inches (15 to 20 centimeters) in width. There is a morphological distinction between leaves of long branches and short shoots, with the leaves of long branches generally bilobate to four-lobed and those of short shoots only fan-shaped to bilobate.

Elliptical, naked seeds resembling a small plum appear on female trees in early spring. Seeds range from 0.75 to 1 inch long (1.9 to 2.5 centimeters) and are covered by a thin, yellowish-orange, fleshy outer wall enveloping a woody shell which contains an edible kernel in the shell interior. When falling to the ground in autumn, the seed covering begins to diminish in thickness over several months, giving off the vile odor of butanoic and hexanoic acids (butter and Romano cheese fatty acids), and is eventually lost from most seeds. *Ginkgo biloba* wood is light, brittle, yellowish in color, and of little value. It is used as a base wood in highly lacquered furniture and small carved items.

Reproduction

The ginkgo is dioecious: Male and female reproductive structures are borne on separate trees. The male reproductive structures appear in May and are inch-long, catkinlike structures bearing numerous paired, pollen-bearing organs. The pollen grains are similar to the elliptical grain of cycads. The pollen organs and ovules are confined to the short shoots of each ginkgo tree and arise in the leaf axils or inner bud scales.

The female reproductive structure consists of a long stalk, bearing on each side an erect, naked ovule, which is surrounded at the base with a collarlike rim. The paired ovules are borne in groups of two to ten. The three-layered ovules include a fleshy exterior layer (sarcotesta), an inner flesh, and a stony shell (sclerotesta) between the two. This three-layered structure is called the integument. The nucellus (the central cellular mass of the body of the ovule containing the embryro) is mostly free from the surrounding integument, except at its base, where it develops a pollen chamber at its apex.

Similar to the cycads, the gingko reproduces by means of flagellated sperm cells, which are carried by the wind-borne pollen to the female reproductive structures within the ovule. In the ginkgo, the vascular system is weakly developed and consists of a pair of braided bundles in the interior fleshy layer of the integument. Upon maturation of the microgametophyte (male gametophyte), pollen tubes are produced, as are large, motile sperm cells similar to those of the cycads. Megagametophyte (female gametophyte) development is similar to that in cycads as well.

Natural Regeneration

Studies into the seedling development of *Ginkgo biloba* reveal a unique mechanism of clonal regeneration that may help explain the species' long survival in the natural setting. The organ of clonal regeneration in the gingko is called the *basal chichi*. These organs are part of aggregates of suppressed shoot buds and are located in embryonic tissue of *Ginkgo biloba* seedlings. When damage occurs to the seedling axis, one of these subsurface buds grows down from the tree trunk to form a woody, stemlike basal chichi. Regeneration of *Ginkgo biloba* by basal chichi promotes survival of the tree in the forests of China today and may have been a factor in the protracted survival of the order since the Mesozoic era.

Habitat and Range

Ginkgo biloba, a distinctive tree suited for use in singular or in group plantings on lawns or along streets, is widely cultivated in all temperate zones. It prospers in moderately moist, fertile soil in humid, temperate regions. It is extremely resistant to disease and pests, and it is highly tolerate of smoke, dust, wind, and ice.

Ginkgo biloba is apparently native in eastern China, with documented semiwild trees growing on the west peak of Tian Mu Mountain in the Tian Mu Reserve, Zhejiang Province. *Ginkgo biloba* is planted in the eastern United States, Europe, and along the Pacific coast.

Fossil Record

Among the plants, *Ginkgo biloba* is probably the best-known example of a "living fossil." Although the ancestors to the order *Ginkgoales* date to the Paleozoic era, it was at the close of the Triassic period when they became a dominant part of the Mesozoic flora. During the Jurassic period, especially the middle Jurassic, *Ginkgoales* reached zenith numbers of species and its widest distribution.

Jurassic and Cretaceous fossil localities reveal circumpolar *Ginkgophytes* sites, including Alaska, Greenland, Zemlya Frantsa Iosifa (Franz Joseph Land), and Mongolia, with the Siberian locations especially productive. Southern Hemisphere *Ginkgoales* localities include Patagonia at the southern tip of South America, South Africa, India, Australia, and New Zealand. European fossil sites are known in England, Scotland, Germany, Italy, Hungary, Turkestan, and Afghanistan. Western Canada and the United States have *Ginkgoites* leaf remains from the Upper Mesozoic and Lower Tertiary deposits. The presence of *Ginkgophytes* in high northern latitudes during the Early Cretaceous period and its presence in southern latitudes, such as Argentina, during the Jurassic, suggests that the dispersal of the plant was from the southern to northern latitudes during the Upper Mesozoic era.

During the Tertiary period, the decline of the *Ginkgophytes* was evident from the presence of only two of nineteen species remaining in the fossil record. One of the remaining two species is the *Ginkgo adiantoides*, which declined sharply during the Oligocene period. This decline continued into the Miocene period in North America, with *Ginkgo adiantoides* disappearing from the fossil record at the end of the Miocene. *Ginkgo adiantoides* did continue into the Pliocene in Europe, however. Since the Pliocene, the fossil record indicates that *Ginkgoales* have been represented by the extant living fossil, *Ginkgo biloba*.

Researchers propose that the decline of *Ginkgophyta* was a result of competition from the angiosperms (flowering plants) for similar plant habitats. Also, the *Ginkgophytes* became more restricted to northern temperate forests in the Tertiary period. When glaciation occurred in these areas during the Pleistocene, these forests were destroyed by climate change.

Mariana Louise Rhoades

See also: Conifers; Cycads and palms; Evolution of plants; Fossil plants; Leaf arrangements; Paleobotany; Pollination; Reproduction in plants; Seeds; Shoots; Water and solute movement in plants.

Sources for Further Study

Cleal, Christopher J., and Barry A. Thomas. *Plant Fossils: The History of Land Vegetation.* Suffolk, England: Boydell Press, 1999. Relates the remarkable record of earth's floral evolution and land vegetation. Includes a generous selection of fossil plant portraits with explanatory notes.

Everett, Thomas, H. *The New York Botanical Garden Illustrated Encyclopedia of Horticulture.* New York: Garland, 1981. Older source, yet has good information on history and paleontology of gingko trees, physical description, and their botanical and horticultural background.

Moore, David, M., ed. *Marshall Cavendish Illustrated Encyclopedia of Plants and Earth Sciences.* New York: Marshall Cavendish, 1988. Covers biology of gingko tree, with readable charts on global ginkgo fossils and geological background on land plant development.

Raven, Peter H., Ray F. Evert, and Susan E. Eichhorn. *Biology of Plants.* 6th ed. New York: W. H. Freeman/Worth, 1999. Basic textbook covering living and extinct seed plants, including *Ginkgophyta*, with excellent graphs and glossary.

Steward, Wilson N., and Gar W. Rothwell. *Paleobotany and the Evolution of Plants.* New York: Cambridge University Press, 1993. Clearly written, reference-quality book concerning the origin and evolution of extant and extinct groups of plants as revealed by the fossil record.

GLYCOLYSIS AND FERMENTATION

Category: Cellular biology

Glycolysis is the beginning of the process of extracting usable energy from food. The disposal of the products of glycolysis when there is no oxygen available is the process of fermentation.

The simple sugar *glucose* is generally considered the starting point for looking at glycolysis and fermentation. Glucose is a simple *carbohydrate*, consisting of carbon, hydrogen, and oxygen. Most glucose is produced by plants; organisms that cannot photosynthesize must obtain glucose (or more complex carbohydrates) from their surroundings. Animals obtain food molecules by eating. Simpler forms of life, such as bacteria and yeast, simply absorb their food from their environment.

Breaking Chemical Bonds

The energy in glucose is locked up in the chemical bonds that hold the molecule together. The process of glycolysis breaks these chemical bonds in a series of carefully controlled chemical reactions. Each reaction can be greatly accelerated by the appropriate *enzyme*. Generally, cells have sufficient quantities of the necessary enzymes present at all times.

Each chemical step is regulated by either the amount of raw materials present or the amount of finished product. If the raw materials are in short supply, the rate of reaction will be slow. Also, if the finished products build up to a high concentration, the reaction will slow down. The energy of the chemical bonds in glucose must be released gradually. During most of the chemical steps, small amounts of energy are released. The amount of energy released is often not enough to perform significant biological work, in which case it is simply wasted as heat. The energy released during some steps, however, is captured in the special high-energy bond of *adenosine triphosphate* (ATP). ATP is one of the most important of the short-term energy storage molecules in cells and is a coenzyme for many important chemical reactions.

Adenosine Triphosphate

ATP belongs to a class of organic molecules known as *nucleotides*. It has an important role in the energy reactions in the cell. The term "triphosphate" indicates that there are three phosphate groups attached to the base molecule. The last two of these phosphates are held by a special kind of chemical bond known as a *high-energy bond*. It takes a greater amount of energy to form one of these bonds than to form the normal kinds of bonds that hold the atoms of other molecules together. When this bond is broken, a large amount of energy is released and is available to the cell to do work. Examples of such work are production of heat, synthesis of complex molecules, and movement of molecules across a membrane. When energy is required in a cell, the third phosphate of ATP is released. While the third phosphate group is routinely split off to release energy, the second one is rarely split off in cellular reactions. The cell must maintain a supply of ATP by means of the reverse reaction. The energy required for this reaction may come from fermentation when oxygen is unavailable. When oxygen is available, other components of cellular respiration are used, which include the *Krebs cycle* and *electron transport*.

Glycolysis

Energy from glycolysis is used to make ATP by two different processes. During glycolysis the glucose molecules are each split into two smaller molecules. The initial glucose molecules contain six carbon atoms each. Each molecule of glucose produces two molecules of pyruvic acid, and each pyruvic acid molecule contains three carbon atoms. During glycolysis, energy is released from the bonds of glucose molecules and is used to join free phosphate ions (also called *inorganic phosphate* or P_i) with molecules of adenosine diphosphate (ADP) to make ATP. This type of ATP synthesis is called substrate-level phosphorylation.

As a by-product, however, electrons are also stripped from glucose. These electrons are immediately trapped and held by another very important molecule, the electron carrier *nicotinamide adenine dinucleotide* (NAD). By convention, the empty electron carrier is denoted as NAD^+. When the molecule is carrying a pair of electrons, it is denoted as NADH, since the molecule also picks up a hydrogen nucleus, or proton. The electrons held by NADH represent potential energy. In the presence of oxygen, these electrons can be passed to the electron transport system to make ADP by *oxidative phosphorylation*, while at the same time regenerating NAD^+, which is required to maintain glycolysis. This second process for making ATP results in about eight times as much ATP per glucose molecule than from substrate-level phosphorylation in glycolysis. Because fermentation is carried out in the absence of oxygen, this process cannot be used. Instead, the NADH must be relieved of its electrons by an alternative process. The NAD^+ regeneration mechanism varies according to the type of organism.

Glucose molecules are relatively stable and do not split readily. For glucose molecules to split, they must be energized by the addition of two phosphate groups to each glucose molecule from two ATP molecules. The third phosphate from each ATP molecule is transferred, along with its high-energy bond. Therefore, the initial steps of glycolysis actually use ATP, depleting some of the cell's energy stores. Once glucose is energized, it readily splits under the influence of the appropriate enzyme. Each half of the glucose molecule then attaches another phosphate group from the cell's pool of P_i. In a series of reactions, each half of the glucose molecule generates two ATP molecules by substrate-level phosphorylation. Therefore, glycolysis results in a net gain of two molecules of ATP per molecule of glucose. At the end of glycolysis there are two three-carbon molecules of pyruvate left over for each original glucose molecule.

Fermentation

Under aerobic conditions, further energy from the chemical bonds of pyruvic acid is harvested by the Krebs cycle and electron transport system. When oxygen is not available (*anaerobic* conditions), however, the electrons must be removed from the NADH to regenerate NAD^+. While there are many ways of accomplishing this, the most common methods are alcoholic fermentation, as observed in yeast, where the end products are ethyl alcohol and carbon dioxide, and lactic acid fermen-

tation, as observed in the muscles of a mammal during strenuous physical exercise. In any event, no further energy is gained for the cell.

In yeast cells cultured in the absence of oxygen, a carbon atom and two oxygen atoms are first split from pyruvic acid, releasing a molecule of carbon dioxide (CO_2). This CO_2 gas is responsible for the bubbles that make bread rise and the carbonation in champagne. The remainder of each pyruvic acid molecule then receives a pair of electrons from NADH, producing a molecule of ethyl alcohol (ethanol). The alcohol evaporates from bread when it is baked but is retained for its mildly euphoric effect in alcoholic beverages. As far as the yeast is concerned, the alcohol is only produced as a way of regenerating NAD^+. It is not a desirable product and will eventually kill the yeast cells. Most yeast cells cannot tolerate an alcohol concentration greater than about 12 percent.

Cellular respiration is the process by which organisms harvest usable energy in the form of ATP molecules from food molecules. Fermentation is the form of respiration used when oxygen is not available. Fermentation is much less efficient than aerobic cellular respiration. Fermentation harvests only two molecules of ATP for every glucose molecule used. Aerobic respiration reaps a yield of more than thirty molecules of ATP. Additionally, the typical products of fermentation, alcohol or lactic acid, are toxic to the organism producing them. Most forms of life will resort to fermentation only when oxygen is absent or in short supply. These are described as *facultative anaerobes*.

While higher forms of life, such as animals, can obtain energy by fermentation for short periods, they enter an oxygen debt, which must eventually be repaid. The yield of two molecules of ATP for each glucose molecule used is simply not enough to sustain their high demand for energy. A few simple forms of life, mostly bacteria, rely solely on fermentation for their source of ATP. To some of these, oxygen is actually poisonous. These are described as *obligate anaerobes*, and they are only found under the completely anaerobic conditions of the deeper layers of mud in saltwater and freshwater marshes.

James Waddell

See also: ATP and other energetic molecules; Carbohydrates; Krebs cycle; Oxidative phosphorylation; Photorespiration; Photosynthesis; Respiration.

Sources for Further Study

Boulton, Chris. *Brewing Yeast and Fermentation*. Malden, Mass.: Blackwell Science, 2001. A text for brewers of all scales, describing the factors affecting brewing yeast fermentation performance and how they may be controlled. Commercially relevant and academically rigorous contributions have been blended in this text, providing the reader with a reference source and overview of the latest technological developments.

Campbell, Neil A., and Jane B. Reece. *Biology*. 6th ed. San Francisco: Benjamin Cummings, 2002. A very thorough and complete introductory biology text. The chapter 6 "Respiration: How Cells Harvest Chemical Energy" covers both aerobic and anaerobic respiration. The author points out the interconnectedness of many processes. Though aimed at college students, the text is easily read, and many of the diagrams stand by themselves.

Lehninger, Albert L. "How Cells Transform Energy." *Scientific American* 205 (September, 1961): 62-73. This old but classic review article covers glycolysis and other aspects of respiration as well as photosynthesis.

Stanbury, Peter F., Allan Whitaker, and Stephen J. Hall. *Principles of Fermentation Technology*. 2d ed. Tarrytown, N.Y.: Pergamon, 1995. An advanced textbook for graduate students of applied biology, biotechnology, microbiology, and biochemical and chemical engineering. Very thorough and up-to-date coverage of industrial fermentation procedures and applications. Photographs and diagrams illustrate the text.

GNETOPHYTES

Categories: Evolution; gymnosperms; *Plantae*; taxonomic groups

The gnetophytes are a small group of vascular seed plants composing the phylum Gnetophyta, *which is one of four phyla of gymnosperms that have living representatives.*

The *Gnetophyta* include only three genera—*Ephedra*, *Gnetum*, and *Welwitschia*—each of which belongs to a separate family, in a single order, the *Gnetales*. The gnetophytes have a number of features in common with the flowering plants (phylum *Anthophyta*, the angiosperms), which has sparked scientific interest in the evolutionary relationships between the two groups; they are the only gymnosperms, for example, in which vessels occur. There are about ninety species of gnetophytes. They are diverse in form and size, and their distribution varies widely, from moist, tropical environments to extremely dry deserts. Most gnetophytes are shrubs or woody vines. The leaves occur oppositely or in whorls of three.

Like most other gymnosperms, the gnetophytes bear their reproductive structures in strobili, or cones. The gnetophytes differ from other gymnosperms in that both the seed-producing (ovulate or female) cones and the pollen-producing (male) cones are compound; that is, they are, in turn, composed of cones. Both male and female cones contain oppositely arranged bracts, or modified leaves, which bear short, fertile shoots at the axil (the angle between the bract and the stem that bears it). Most gnetophytes are *dioecious*, meaning that they bear their pollen and ovulate cones on separate plants.

Angiosperm-like Features

The gnetophytes share with the angiosperms a number of structural and developmental characteristics. One of these is the presence of water-conducting tubes, called *vessels*, in the secondary xylem, or wood. Vessels, although present in angiosperms, do not occur in gymnosperms other than gnetophytes. Another similarity is that *archegonia*—structures that protect the egg—which are typical of gymnosperms but absent from angiosperms, are not found in either *Gnetum* or *Welwitschia* (although they are present in *Ephedra*). In addition, the cones

Classification of Gnetophytes

Class *Gnetopsida*
    Order *Ephedrales*
        Family *Ephedraceae* (Mormon tea family)
           Genus *Ephredra*
    Order *Gnetales*
        Family *Gnetaceae* (Gnetum family)
           Genus *Gnetum*
        Family *Welwitschiaceae*
           Genus *Welwitschia* (tumboa)

of gnetophytes bear some resemblance to angiosperm flower clusters, and the leaves of *Gnetum* are similar in form, structure, and venation to those of the *Eudicotyledones* of the angiosperms.

Another feature common to angiosperms and gnetophytes—but not found in gymnosperms other than gnetophytes—is double fertilization. In double fertilization, there is union of each of two sperm nuclei with a nucleus in the female gametophyte (the gamete-producing generation in plants), rather than just a union of a single sperm and egg nucleus. Further, in at least some species of all three gnetophyte genera, the reproductive structures produce nectar that attracts insects, as in many angiosperms. Insects play a role in the pollination of gnetophytes, in contrast to the typical gymnosperm's reliance on the wind.

The similarities between gnetophytes and angiosperms have led scientists, who have long thought that the ancestor of the angiosperms is a gymnosperm, to look closely at the gnetophytes. The living gnetophytes are considered too specialized to include the angiosperm ancestor. In addition, evidence suggests that such shared features as similar-appearing vessels, similar-appearing reproductive structures, and the absence of archegonia were derived independently in the two groups. Scientists have still not determined, however, whether the gnetophytes and angiosperms share a close, common ancestor. Unfortunately, the fossil record for gnetophytes is too sparse to shed much light on this question. Although the gnetophytes were once more diverse than they are today, there is no indication that the group was ever rich in genera or abundant in individuals. The earliest known gnetophyte fossils date back 140 million years, to the Early Cretaceous period of the Mesozoic era, which is about as far back in time as the angiosperm fossil record goes.

Researchers have turned to cladistics (phylogenetic analyses) to study fossil, structural, and molecular evidence in an attempt to determine evolutionary relationships. The results of these studies have been inconsistent, and additional research will be needed. There is strong evidence, however, that, within the gnetophytes, *Ephedra* is the closest to the common ancestor of the group, and that *Gnetum* and *Welwitschia* are derived sister clades.

Ephedra

The genus *Ephedra* includes about sixty species, most of them adapted to semiarid and desert conditions. *Ephedra* is the only genus of gnetophyte that occurs in the United States, with twelve species growing in the desert Southwest, some of them ranging into Mexico. Another twelve species occur over a wide area in South America. The rest grow in the Eastern Hemisphere, from central Asia westward across southwest Asia and into Mediterranean Europe and North Africa.

Most species of *Ephedra* are scraggly, profusely branched shrubs. Some are vinelike, commonly climbing over other vegetation. The leaves of most species are small, dry, brown scales. Their reduced size may be related to the plants' need to minimize evaporative water loss in their dry environments. Photosynthesis is carried on mostly in the branches, which remain green while young. The branches are jointed, giving rise to the genus's common name, joint fir. Both the branches and the leaves are arranged two or three to a node. With its jointed stems and small leaves, *Ephedra* superficially resembles the horsetail, *Equisetum*. The stems of *Ephedra* form secondary xylem, or wood, as do the stems of conifers and many angiosperms. *Ephedra* wood is extremely hard.

The cones of *Ephedra* are borne in the leaf axils and are very tiny—the smallest in the gnetophyte group. The female cones have at their tips one or two ovules borne on very short stalks. The male cones are in spikelike clusters. Male and female cones are produced on the same plant or on different ones, depending on the species. The mature female cones are fleshy and berrylike and often brightly colored.

Stem extracts of *Ephedra* have a long history of use as folk medicines. Many Eurasian *Ephedra* species, especially *Ephedra sinica*, contain ephedrine, an alkaloid chemical that the Chinese have used for more than five thousand years as a decongestant drug, called Ma-huang, to treat asthma and hay fever. Ephedrine alkaloids have not been found in New World species of *Ephedra*, but Western countries manufacture synthetic ephedrine, which is used in cough medicines. European settlers in the American Southwest and Mexico brewed a beverage known as Mormon tea or Mexican tea from stem fragments of *Ephedra*.

Gnetum

The genus *Gnetum* includes approximately thirty species, which grow throughout the moist tropics. Most of these are woody vines that climb on trees in the rain forests of central Africa, Asia, and northern South America and on some Pacific islands. The best-known species, *Gnetum gnemon*, however, is a tree native to Indonesia that grows to 10 meters. It is cultivated for its edible seeds and tender young leaves.

Gnetum stems characteristically bear two broad, leathery leaves at each node and produce secondary xylem, or wood. In all *Gnetum* species, male and female reproductive structures are borne on separate plants. The cones, like those of *Ephedra*, look like berries, and the seeds may be brightly colored.

Welwitschia

The genus *Welwitschia* includes a single species, *Welwitschia mirabilis*. This low-growing, perennial plant is restricted to a 150-kilometer-wide strip of coastal desert in Angola, Namibia, and South Africa. In this extremely arid environment, where there may be no precipitation for several years at a time, *Welwitschia* may survive, at least in part, by using dew and condensate from fog that rolls in off the ocean at night. Young plants seem to become established mainly during rare wet years. Some living *Welwitschia* plants have been dated at fifteen hundred years old.

Described by some as the strangest living plant, *Welwitschia* bears little resemblance to other gymnosperms. Most of the plant is taproot, buried in sand to a depth of 1 to 1.5 meters. At its tip, the taproot divides into smaller roots that probably absorb water unavailable to less deeply rooted plants. The exposed part of *Welwitschia* includes a massive, squat stem that emerges only a short distance above the ground. The stem widens with age, becoming up to a meter across, and may develop a crusty, barklike covering on its surface. This broad, woody, concave disk, having ceased elongating from the tip, produces only two leaves during the plant's lifetime. The wide, strap-shaped leaves continue to grow from their bases at a rate of 8 to 15 centimeters per year, for the life of the plant. Battered by wind and hot sand, the leaves break off at their tips and split lengthwise, giving older plants the appearance of having numerous leaves. With their worn tips, the leaves seldom exceed 2 meters in length, although they may reach 6 meters.

The only real branches that *Welwitschia* produces are the branch systems of the pollen and ovulate cones, which are borne on separate plants. These branch systems develop in the axils of the leaves, although they appear to grow from around the rim of the stem cup. The pollen cones, which are red, are produced in groups of two or three at the end of each branch. Ovulate cones are also red. Droplets of nectar lure flies to pollinate the plant. The seeds, generally only one to a cone, have winglike extensions that may aid dispersal by the wind.

Jane F. Hill

See also: Angiosperms; Angiosperm life cycle; Angiosperm plant formation; Cladistics; Eudicots; Evolution of plants; Gymnosperms; Medicinal plants.

Sources for Further Study

Mabberley, D. J. *The Plant Book: A Portable Dictionary of the Vascular Plants.* 2d ed. New York: Cambridge University Press, 1997. Reference work, with twenty thousand entries, including information on every family and genus of seed-bearing plant and fern.

Morin, Nancy R. *Flora of North America North of Mexico.* Vol. 2. New York: Oxford University Press, 1993. Covers twelve species of *Ephedra*. Includes descriptions of species and habitat, range, identification keys, and literature reviews.

Raven, Peter H., Ray F. Evert, and Susan E. Eichhorn. *Biology of Plants.* 6th ed. New York: W. H. Freeman/Worth, 1999. Comprehensive, introductory textbook. Includes a section on angiospermlike features of gnetophytes. Includes photos, glossary, index, and bibliography.

GRAINS

Categories: Agriculture; economic botany and plant uses; food

Grains are the fruits or seedlike fruits of plants, particularly members of the grass family, Poaceae. *Important cereal grain crops are all produced by annual grasses and are dry (desiccant) fruits with the ovary wall fused to the seed coat. Inside the fruit wall-seed coat covering (the bran) is a small embryo (germ) and a large amount of stored food (endosperm).*

Grains were the first domesticated crops and allowed the development of all of the great early civilizations. Several factors contribute to the importance of grains in agriculture: ease of growth, storage, and preparation; high yields; and high-energy, easily digestible content (starch). The wild relatives of cereal grains all disperse their seeds by the shattering, or breaking apart, of mature fruiting stalks. Harvesting these wild species is a problem because the seeds are flung everywhere when the fruiting stalk is disturbed. A first step in the domestication of all grains was the elimination of shattering so that inflorescences could be harvested. For grasses, such as wheat, that produce many stems, or tillers, arising from the base of the plant, selection led to synchrony in the production of the tillers so that all the inflorescences of a plant would set fruit at the same time. For grasses, such as corn, that had a thick main stem, selection led to the elimination of secondary branches and a concentration of seeds in one or a few large inflorescences. The second half of the twentieth century saw selection for shorter stature that allows grains to grow better in tropical regions. While thirty-five species of grasses have been domesticated, only five are major crops today: *wheat, rice, corn, sorghum,* and *barley.*

Wheat

Wheat is the most widely cultivated grain in the world and was among the earliest grains to be domesticated. Archaeological deposits from the Middle East, the native home of wheat, containing domesticated wheat seeds have been dated to ten thousand years ago. The first species of wheat domesticated was the diploid einkorn wheat (*Triticum monococcum*), soon followed by the tetraploid, free-threshing emmer wheat (*T. turgidum*), which made it easier for people to separate the fruits from the papery tissues, or chaff, in which they are enclosed. Today, emmer wheats are grown throughout the world and are especially well suited to making pasta and pastries. Bread wheat (*T. aestivum*) was the last to be domesticated. This species is a hexaploid, and its increased cell size has as an important secondary effect: the high production of proteins, known as glutens, that allow bread wheat to form an elastic dough that produces light, spongy bread.

While some wheat is eaten as a grain in such dishes as tabbouli, most wheat is used for flour: whole wheat if the bran and germ are ground along with the starchy endosperm and white if the bran and embryo are removed. Because white flour (even organic flour) lacks the vitamins present in whole wheat flour, federal law in the United States requires that it be enriched with five nutrients: riboflavin, niacin, folic acid, thiamin mononitrate, and iron.

Rice

The acreage devoted to rice (*Oryza sativa*) is less than that of wheat, but more rice is produced annually than wheat, and more people in the world depend on rice as their primary food. Rice was domesticated in the Yangtze River region of China, probably more than nine thousand years ago. In most of the world, rice is grown by germinating seeds and growing seedlings in a nursery. Seedlings are then planted by hand in fields covered with water. Rice does not need to grow in standing water, but it needs high rainfall if the fields are not irrigated. Because rice contains no gluten, it is not used for leavened bread, but it is well suited for cooking because the seeds retain their shape and have a soft, chewy consistency. The two major types of rice are long-grained *Indica*, preferred in India, and short-grained *Japonica*, preferred in China and Japan because of the sticky grains that adhere to one another upon cooking. Removing the bran produces white rice, which lacks the vitamins and fiber of brown rice. Consequently, rice is often enriched with vitamin B_1.

Corn

Corn (*Zea mays*) is the only major grain native to the New World. It was domesticated in southern Mexico about eight thousand years ago from an annual grass known as *teosinte*. Corn plants are monoecious (having both pistillate and staminate flowers on one plant), with male flowers forming the tassel on the top of a corn plant and the female flowers packed inside the ear. The silks of an ear are the styles, one leading to each kernel of corn. Much of the U.S. crop is used for animal feed, but a large portion is converted to cornstarch or corn syrup to be used in the brewing, paper making, and processed food industries. A by-product of the cornstarch industry is corn oil, extracted from the germs.

Sorghum

Sorghum (*Sorghum bicolor*) is native to sub-Saharan Africa, where it was domesticated by five thousand years ago. Grain sorghums are a major source of food for millions of people in Africa and India, but in the New World sweet sorghum (sorgo) is grown primarily for animal feed. The plants are robust, with modern cultivars having a single, thick stem bearing a mass of seeds at the apex. Sorghum is the most drought-tolerant of the major grains and is therefore widely grown in arid regions. In addition to the grain sorghums and sorgos, other varieties yield rough fiber for brushes and booms.

Barley

Barley (*Hordeum vulgare*) was probably the first grain to be domesticated; ancient cultivated fruits found in the Near East have been dated to 10,500 years ago. Initially, barley was preferred over wheat and was used to make flat breads, pastes, gruel, and beer. Barley became less popular after the domestication of emmer and then bread wheat. However, it has remained the grain of choice for brewing beer because of its superior flavor after malting. Malting consists of germinating the grain just enough for it to produce enzymes that break down the starch into simple sugars that yeast can then ferment.

Beryl B. Simpson

See also: Agricultural revolution; Agriculture: experimental; Agriculture: world food supplies; Corn; Plant domestication and breeding; Plants with potential; Rice; Vegetable crops; Wheat.

Sources for Further Study

Chapman, G. P., ed. *Grass Evolution and Domestication*. Cambridge, England: Cambridge University Press, 1992. An edited volume that treats grass evolution in general as well as the domestication of selected grains.

Iltis, H. H. "Homoeotic Sexual Translocations and the Origin of Maize (*Zea mays, Poaceae*): A New Look at an Old Problem." *Economic Botany* 54 (2000): 7-42. A discussion of a possible manner in which modern corn arose from its closest ancestor.

Pringle, H. "The Slow Birth of Agriculture." *Science* 282 (1998): 1446-1450. Covers the role of grains as agriculture emerged in various parts of the world.

Simpson, Beryl B., and Molly C. Ogorzaly. *Economic Botany: Plants in Our World*. 3d ed. Boston: McGraw-Hill, 2001. In this text, an entire chapter is devoted to grains. A lively and well-illustrated account of the major plant crops of the world.

Smartt, J. J., and N. W. Simmonds, eds. *Evolution of Crop Plants*. 2d ed. London: Longman Scientific & Technical, 1995. This volume provides an account of almost every crop species, with each article written by an authority on that species and its relatives.

Zohary, D., and M. Hopf. *Domestication of Plants in the Old World*. Oxford, England: Clarendon Press, 1993. Most cultivated plants were domesticated in the Old World, and Zohary and Hopf provide a summary of when, where, and how domestication of each occurred.

GRASSES AND BAMBOOS

Categories: Agriculture; angiosperms; economic botany and plant uses; food; gardening; *Plantae*

Grasses are monocotyledonous flowering plants (phylum Anthophyta, *the angiosperms) belonging to the family* Poaceae, *formerly* Gramineae. *The family is widespread and economically very valuable. All the grasses are herbaceous except for the bamboos, some of which are treelike.*

Grasses arose seventy million to eighty million years ago, in the Late Cretaceous period of the Mesozoic era. They succeeded partly because they concentrate their growth lower down on the leaf and stem than other plants do and thus can regenerate quickly when fire or herbivores remove the top part of the plant. This makes them ideal for human uses such as lawns and pasture for domestic livestock. Grasses are a very important food source for humans. Grasses provide all cereal grains—barley, corn, millet, oats, rice, rye, and wheat. Other grasses used for food include sorghum, sugarcane, and bamboo.

The grass family is important botanically and ecologically, too. With about 650 genera and 7,500 to 10,000 species, it is the fourth largest angiosperm family. In number of individual plants, it far outranks any other flowering plant family, composing the natural vegetation of about one-quarter of the earth's land surface. The family is especially abun-

PHOTODISC

Grasses are a very important food source for humans and provide all cereal grains, including wheat.

dant in semiarid climates and has the widest range of all the angiosperm families. Grasses provide a vital food source for many grazing, wild animals. By binding the soil with their roots, grasses protect it from erosion. In addition, grasses build up the soil when they die and decompose.

Stems, Roots, and Leaves

The aboveground stems of grasses are called *culms*. The roundness of culms distinguishes grasses from similar-looking plants, the sedges and rushes, which have differently shaped stems. Grass culms are divided into *nodes*, or joints, which are solid, and internodes (the regions between the nodes), which are usually hollow. Elongation of the culm occurs mainly at the bases of the internodes, rather than at the stem tip as in most plants. Grasses range in height from only a few centimeters, in annual bluegrass, to 30 or 40 meters, in treelike bamboos. Grasses may have underground stems, or *rhizomes*. The roots are slender and fibrous and form an extensive system that may compose a large proportion of the plant's total biomass. This large root system helps grasses obtain water in dry regions.

The leaves of grasses are generally arranged alternately on the stem and have no petiole, or stalk. The lower part of the leaf, the *sheath*, is wrapped around the culm like a split tube and is attached at its base to a node. The upper part of the leaf, the *blade*, diverges from the culm and is slender and elongated, tapering to a point at the tip.

Flowers and Fruits

Pollinated by the wind, grass flowers are simple and inconspicuous individually. They have stamens and pistils—the essential male and female reproductive structures—but lack petals and sepals, which most other angiosperms use to lure insect pollinators. Many grass flowers have three stamens and a single pistil that has two stigmas. Long, dangling anthers in the stamens and long, feathery stigmas help the wind transfer pollen efficiently from stamen to pistil.

Identification of grass species is based to a considerable extent on flower arrangement and characteristics of modified leaves, called bracts, that surround the flowers. Individual flowers, called florets, are usually arranged in clusters, called spikelets. Each floret and spikelet typically has at its base a pair of bracts. The spikelets may be crowded on an unbranched stalk, forming a spike, or borne at the ends of stalks having many branches, in a panicle.

Common Grasses of the United States

Common Name	Scientific Name
American beachgrass	*Ammophila breviligulata*
Baltic rush	*Juncus balticus*
Basin wildrye	*Leymus cinereus*
Basket rush	*Juncus textilis*
Beaked panicgrass	*Panicum anceps*
Bermudagrass	*Cynodon dactylon*
Big bluestem	*Andropogon gerardii*
Billion-dollar grass	*Echinochloa frumentacea*
Bitter panicgrass	*Panicum amarum*
Blue grama	*Bouteloua gracilis*
Bluebunch wheatgrass	*Pseudoroegneria spicata*
Broomsedge bluestem	*Andropogon virginicus*
Buffalograss	*Buchloe dactyloides*
Bushy bluestem	*Andropogon glomeratus*
California bulrush	*Schoenoplectus californicus*
Canada wildrye	*Elymus canadensis*
Caucasian bluestem	*Bothriochloa bladhii*
Cereal rye	*Secale cereale*
Cheatgrass	*Bromus tectorum*
Common rush	*Juncus effusus*
Common threesquare	*Schoenoplectus pungens*
Cosmopolitan bulrush	*Schoenoplectus maritimus*
Crested wheatgrass	*Agropyron cristatum*
Deergrass	*Muhlenbergia rigens*
Deertongue doc	*Dichanthelium clandestinum*
Desert wheatgrass	*Agropyron desertorum*
Eastern gamagrass doc	*Tripsacum dactyloides*
Fall panicgrass	*Panicum dichotomiflorum*
Field brome	*Bromus arvensis*
Florida paspalum	*Paspalum floridanum*
Hard fescue	*Festuca trachyphylla*
Indian ricegrass	*Achnatherum hymenoides*
Indian woodoats	*Chasmanthium latifolium*
Indiangrass	*Sorghastrum nutans*
Inland saltgrass	*Distichlis spicata*
Intermediate wheatgrass	*Thinopyrum intermedium*
Italian ryegrass	*Lolium perenne* ssp. *multiflorum*

(continued)

Common Grasses of the United States
(continued)

Common Name	Scientific Name
Kentucky bluegrass	*Poa pratensis*
Knotgrass	*Paspalum distichum*
Little bluestem	*Schizachyrium scoparium*
Maidencane	*Panicum hemitomon*
Nutgrass	*Cyperus rotundus*
Orchardgrass doc	*Dactylis glomerata*
Perennial ryegrass	*Lolium perenne*
Prairie cordgrass	*Spartina pectinata*
Purpletop tridens	*Tridens flavus*
Red fescue	*Festuca rubra*
Redtop	*Agrostis gigantea*
Reed canarygrass	*Phalaris arundinacea*
Salt rush	*Juncus lesueurii*
Saltmeadow cordgrass	*Spartina patens*
Sandberg bluegrass	*Poa secunda*
Santa Barbara sedge	*Carex barbarae*
Seashore paspalum	*Paspalum vaginatum*
Siberian wheatgrass	*Agropyron fragile*
Sideoats grama	*Bouteloua curtipendula*
Slender wheatgrass	*Elymus trachycaulus*
Slough sedge	*Carex obnupta*
Smooth brome	*Bromus inermis*
Smooth cordgrass	*Spartina alterniflora*
Snake River wheatgrass	*Elymus wawawaiensis*
Splitbeard bluestem	*Andropogon ternarius*
Streambank wheatgrass	*Elymus lanceolatus*
Switchgrass	*Panicum virgatum*
Tall fescue	*Lolium arundinaceum*
Tall flatsedge	*Cyperus eragrostis*
Tall oatgrass	*Arrhenatherum elatius*
Texas cupgrass	*Eriochloa sericea*
Timothy	*Phleum pratense*
Torpedo grass	*Panicum repens*
Tule	*Schoenoplectus acutus* var. *occidentalis*
Vanilla grass doc	*Hierochloe odorata*
Weeping lovegrass	*Eragrostis curvula*
Western wheatgrass	*Pascopyrum smithii*

Note: ssp. = subspecies; var. = variety

Source: U.S. Department of Agriculture, National Plant Data Center, *The PLANTS Database*, Version 3.1, http://plants.usda.gov. National Plant Data Center, Baton Rouge, LA 70874-4490 USA. Accessed April 11, 2002.

The grass fruit, called a caryopsis, or grain, consists of a single seed with its seed coat firmly adherent to the thin fruit wall. Wind plays a major role in dissemination of the fruits. Grasses may germinate, grow, set seed, and die in the same year (*annuals*) or live and produce seed for many years (*perennials*).

Bamboos

The bamboos are perennial, often tree-like, grasses belonging to the *Bambusoideae*, a subfamily that is thought to be an early offshoot in the grass family lineage. Bamboo taxonomy is poorly understood. One estimate holds that there are roughly 45 genera and 480 species.

Like other grasses, bamboos have jointed culms, which are hollow except at the nodes, where there are partitions. The culms, which originate from rhizomes, are often called *canes*. The canes are light and elastic. Their hardness is due not to secondary xylem, or wood, as in most trees and shrubs, but rather to scattered fibers in the outer walls of the cane internodes.

Canes of some bamboo species grow at rates as high as a meter a day. The upper nodes of fully elongated culms give rise to small, horizontal branches. Leaves are borne on these branches or on branches of these branches. The blades are narrow and often short. Although some bamboos flower every year, many bloom only at the end of their lifetimes, which may range from 10 to 120 years.

Bamboos are distributed mainly in tropical and subtropical regions, with large concentrations in Asia and South America. A few species reach mild temperate areas. In the United States, there is a single native species, *Arundinaria gigantea*, called cane. It forms canebrakes in southern bottomlands. Bamboos are grown as ornamentals in many parts of the world. Dense bamboo thickets are sometimes planted as living fences or barricades. In Asia, bamboos are very significant economically, providing materials for building, matting, and many other purposes. The young shoots are popular as food in eastern Asia.

Jane F. Hill

See also: Asian flora; Corn; Erosion and erosion control; European flora; Fruit: structure and types; Grains; Grasslands; North American flora; Pollination; Rice; Savannas and deciduous tropical forests; Wheat.

Sources for Further Study

Brown, Lauren. *Grasses: An Identification Guide*. Reprint. Boston: Houghton Mifflin, 1992. Emphasizes identification mainly of northeastern U.S. grasses. Uses nontechnical language. Includes line drawings.

Chapman, G. P. *The Biology of Grasses*. Wallingford, Oxon, England: CAB International, 1996. An advanced text, bridging the gap between introductory textbooks and technical research papers. Includes illustrations.

Clark, Lynn G., and Frederick W. Pohl. *Agnes Chase's First Book of Grasses: The Structure of Grasses Explained for Beginners*. 4th ed. Washington, D.C.: Smithsonian Institution Press, 1996. Updated edition of a classic introduction to the major groups of grasses in the Americas. Contains an added section on bamboos. Includes glossary, index, and line drawings.

Recht, Christine, and Max F. Wetterwald. *Bamboos*. Edited by David Crampton. Portland, Oreg.: Timber Press, 1992. Covers bamboos and their uses; structure; major genera and species; and cultivation and garden design. Includes index, bibliography, photographs, and line drawings.

GRASSLANDS

Category: Ecosystems

Grasslands are areas of intermittent rainfall which favor grass growth. The grass helps the soil become rich by facilitating the accumulation of nutrients and decaying plant material.

Grasslands once covered about a quarter of the world's land surface. Grasses' growth patterns help enrich the soil immensely. Because their soils become among the world's richest, grasslands are so intensely farmed and grazed that only small patches of natural grassland remain.

Climate and Geographic Location

Annual precipitation between 10 and 32 inches (25-80 centimeters), often with a dry period late in the growing season, supports grassland. Grassland temperature patterns vary. Fire and grazing favor grasses and often combine with climate to maintain grasslands.

Extensive grasslands generally are found in continental interiors. In North America, grasslands occur from the eastern foothills of the Rocky Mountains to the Mississippi River, from south central Canada to northeastern Mexico, in eastern Washington and Oregon, and in California's Central Valley. Grasslands on other continents include the steppes of Europe and Asia, areas fringing the major deserts of Africa and Australia, and the Pampas of South America.

Types of Grasslands

Extensive grasslands are often divided into *tall-grass*, *mixed-grass*, and *short-grass* regions. In prehuman settlement North American grasslands, the tall-grass prairie occurred in the moist eastern zone. Big bluestem, Indian grass, and switch grass grew 6-10 feet (2-3 meters) tall in this region. The short-grass prairie or plains occupied the drier western extreme. There, blue grama and buffalo grass seldom grew taller than 8 inches (20 centimeters). Mixed-grass prairie grew in between, with a mixture of tall, short, and middle-height grasses. Boundaries between regions were broad zones of gradual change.

Grasses and Grasslands

Grasses are well adapted to occupy regions with intermediate annual precipitation, fires, and grazing animals. Grasses have their main center of growth at or below the ground. Their slender, widespread roots compete intensely for nutrients and moisture, especially near the surface. The aboveground parts of the plants grow densely, and the entire aboveground plant dies every year, covering the ground with a dense mulch. These characteristics present difficulties for plants invading grasslands, as the grass roots usurp moisture and nutrients and the leaves and mulch intercept sunlight.

Under very dry conditions, when grasses cannot grow densely, shrubs and succulents (such as cacti) dominate, and deserts occur. With heavy rainfall and infrequent dry periods, trees compete well with grasses, and forests dominate the landscape. Grasslands are often bordered by forests at their moist edges and deserts at their dry boundaries. Under intermediate rainfall conditions, however, grasses are favored over all competitors.

Fire and grazing by animals tip the balance further in favor of the grasses. The late-season dry period typical of grasslands and the mulch built up after a year or more of growth are ideal conditions for the spread of fires. Whether started by lightning or by humans, fires spread quickly through the dried mulch. The tops of plants burn to the ground, but often little damage occurs underground. Because the primary growth center of most nongrass plants is above ground and that of grasses is below ground, fire is more harmful to woody plants and nonwoody, nongrass plants (*forbs*).

Because grazing removes the tops of plants, it does more damage to forbs and woody plants than to grasses. Many grasses actually increase growth after light grazing. Most extensive grasslands are occupied by large grazing animals, such as the bison and pronghorn of North American grasslands. These and other grazers played important roles in the maintenance of the native grasslands and in the lives of the people who lived there.

Grassland Soils

The presence of grasslands is determined by climate, fire, and grazing, but grasses impact their environment as well. In addition to their competitive role in excluding trees, shrubs, and forbs, grasses contribute to soil formation. All the aboveground parts of grass plants die each year, become mulch, and slowly decompose into the soil. Rainfall is generally insufficient to wash nutrients out of the reach of the grass roots, so the soil accumulates both nutrients and decaying plant material. The world's richest soils develop under these conditions.

Human Impact on Grasslands

Because of their soils, grasslands became agricultural centers. Domestic grasses became the predominant crops—corn in the tall-grass country and wheat in the mixed-grass region. The short-grass plains were too dry to support grain crops but became an important region for grazing domestic animals.

In the process of learning what activities the grasslands could and could not support, Americans changed the grasslands of the continent forever. Farming reduced native tall-grass prairie to one of the world's rarest habitats. Although grazing had less impact on the short-grass plains, vast areas have been *overgrazed* severely. Grasslands in other parts of the world have been similarly abused. Given the importance of grasslands to humanity, serious conservation measures must be taken to restore their productivity.

Carl W. Hoagstrom

See also: Biomes: types; Grasses and bamboos; Grazing and overgrazing; Rangeland.

Sources for Further Study

Brown, Lauren. *The Audubon Society Nature Guides: Grasslands*. New York: Knopf, 1985. A comprehensive field guide, fully illustrated with color photographs, to the trees, wildflowers, grasses, insects, birds, and other natural wonders of North America's prairies, fields, and meadows.

Collinson, Alan. *Ecology Watch: Grasslands*. New York: Dillon Press, 1992. Examines grasslands around the world and the life they support. Discusses how they are surviving changes brought about by people and climate. Includes index.

Joern, Anthony, and Kathleen H. Keeler, eds. *The Changing Prairie*. New York: Oxford University Press, 1995. Describes the ecology of the North American prairie and urges conservation measures to protect the remaining North American grasslands. Provides noneconomic arguments for the value of prairies, presents a synthesis of prairie ecology to facilitate the best possible management, and deftly summarizes conservation and management issues.

Sampson, Fred B., and Fritz L. Knopf, eds. *Prairie Conservation*. Washington, D.C.: Island Press, 1996. A comprehensive examination of the history, ecology, and current status of North American grasslands. Contains information on the historical, economic, and cultural significance of prairies, their threats, conservation, and restoration programs under way.

Steele, Philip. *Grasslands*. Minneapolis, Minn.: Carolrhoda Books, 1996. Describes grasslands of the world, including the plants, animals, and people found there. Suggests activities and questions for discussion. Presents case studies of specific grasslands.

GRAZING AND OVERGRAZING

Categories: Animal-plant interactions; economic botany and plant uses; environmental issues; soil

Animals that eat grass, or graze, can actually help the earth produce richer land cover and soil. When the land suffers ill effects because of too much grazing, overgrazing has occurred.

The effects of overgrazing occur where there are more grazing animals than the land and vegetation can support. Overgrazing has negatively affected regions of the United States, primarily in the Southwest. Areas that have been severely damaged by overgrazing typically show declining or endangered plant and animal species.

Herbivores are animals that feed on plant material, and grazers are herbivores that feed specifically on grass. Examples are horses, cows, antelope, rabbits, and grasshoppers. Overgrazing occurs when grazer populations exceed the *carrying capacity* of a specified area (the number of individual organisms the resources of a given area can support). In overgrazing conditions, there is insufficient food to support the animal population in question. Depending on the grazer's strategy, emigration or starvation will follow. Grasslands can handle, and even benefit from, normal grazing; only overgrazing adversely affects them.

Grasses' Defenses Against Grazing

Grasslands and grazers coevolved, so grasses can withstand grazing within the ecosystem's carrying capacity. All plants have a site of new cell growth called the meristem, where growth in height and girth occurs. Most plants have the meristem at the very top of the plant (the apical meristem). If a plant's apical meristem is removed, the plant dies.

If grasses had an apical meristem, grazers—and lawn mowers—would kill grasses. Grasses survive mowing and grazing because the meristem is located at the junction of the shoot and root, close to the ground. With the exception of sheep, grazers in North America do not disturb the meristem, and sheep do so only during overgrazing conditions. At proper levels of grazing, grazing actually stimulates grass to grow in height in an attempt to produce a flowering head for reproduction. Grazing also stimulates grass growth by removing older plant tissue at the top that is functioning at a lower photosynthetic rate.

Grazers

Mammalian grazers have high, crowned teeth with a great area for grinding to facilitate opening of plants' cell walls as a means to release nutrients. The cell wall is composed of cellulose, which is very difficult for grazers to digest. Two major digestive systems of grazing strategies have evolved to accommodate grazing. *Ruminants*, such as cows and sheep, evolved stomachs with four chambers to allow regurgitation in order to chew food twice to maximize cellulose breakdown. Intestinal bacteria digest the cellulose, releasing fatty acids that nourish the ruminants. Other grazers, such as rabbits and horses, house bacteria in the cecum, a pouch at the junction of the small and large intestines. These bacteria ferment the plant material ingested. The fermented products of the bacteria nourish these grazers.

Impacts in the Southwest

As previously mentioned, in the United States the negative effects of overgrazing are most intense in the Southwest. Some ecologists believe that one significant factor was the pattern of early European colonization of the area. Missions were abundant in the Southwest, and the missions owned cattle that were rarely slaughtered, except on big feast days. Because Catholic missionaries received some financial support from their religious orders in Europe, mission cattle were not restrained as strictly as were those owned by cattle ranchers, whose sole livelihood came from raising and selling cattle. Mission cattle roamed greater distances and began the pattern of overgrazing in the Southwest. The impact of overgrazing was particularly intense because much of the Southwest has desertlike conditions. Extreme environmental conditions result in particularly fragile ecosystems. Hence the Southwest was, and is, vulnerable to the effects of overgrazing.

Another possible—though disputed—contribution to overgrazing may stem from the fact that much of the land in the Southwest is public land under jurisdiction of the Bureau of Land Management. This federal agency leases land to private concerns for the purpose of grazing cattle or sheep. Some observers feel that the bureau has a conflict of interest in that its primary source of income is money obtained from leasing public land under its jurisdiction. They suspect that the bureau has granted, and fear that it may continue to grant, grazing leases in regions threatened with or suffering from overgrazing.

Effects of Overgrazing

Overgrazing can lead to a number of ecological problems. Depletion of land cover leads to soil *erosion* and can ultimately cause *desertification*. Other possible results are the endangering of some species of grass and the creation of *monocultures* in re-

Buffalo graze on prairie pasture at the Big Sky Buffalo Ranch in Granville, North Dakota. Grasslands can handle, and even benefit from, normal grazing; only overgrazing adversely affects them.

gions where certain species have been removed. Desertification is the intensification and expansion of deserts at the expense of neighboring grasslands. When overgrazing occurs along desert perimeters, plant removal leads to decreased shading. Decreased shading increases the local air temperature. When the temperature increases, the air may no longer cool enough to release moisture in the form of dew. Dew is the primary source of precipitation in deserts, so without it, desert conditions intensify. Even a slight decrease in desert precipitation is serious. The result is hotter and drier conditions, which lead to further plant loss and potentially to monocultures.

Overgrazing of grasslands, combined with the existence of nonnative species in an ecosystem, can result in the endangerment of species of native grasses. At one time, cattle in the Southwest fed exclusively on native grasses. Then nonnative plant species arrived in the New World in the guts of cows shipped from Europe. They began to compete with the native grasses. European grass species have seeds with prickles and burs; southwestern native grasses do not, making them softer and more desirable to the cattle. Hence European grasses experienced little, if any, grazing, while the much more palatable southwestern native grasses were grazed to the point of overgrazing. The result was drastic decline or loss of native grassland species. In such cases animals dependent on native grassland species must emigrate or risk extinction. For example, many ecologists conjecture that the Coachella Valley kit fox in California is threatened because of the loss of grassland habitat upon which it is dependent.

Solutions

Desertification is usually considered irreversible, but the elimination of grazing along desert perimeters can help to prevent further desertification. One kind of attempt to reestablish native grass species involves controlled-burn programs. Nonnative grassland species do not appear to be as fire-resistant as native grass species. Controlled burn programs are therefore being used in some overgrazed grassland areas to try to eliminate nonnatives and reestablish native grass species. If successful, such programs will improve the health of the ecosystem.

Jessica O. Ellison

See also: Biological invasions; Grasses and bamboos; Grasslands; Sustainable agriculture.

Sources for Further Study

Hodgson, J., and A. W. Illius, eds. *The Ecology and Management of Grazing Systems*. Wallingford, Oxon, England: CAB International, 1996. Synthesizes research from plant science, animal science, and ecology and looks at current issues in grazing across the world. Covers the principles of herbage growth and competition, animal nutrition and grazing behavior, and the interactions of plant and animal factors.

McBrien, Heather, et al. "A Case of Insect Grazing Affecting Plant Succession." *Ecology* 64, no. 5 (1983). An example of grazing research and many interrelated issues.

Sousa, Wayne P. "The Role of Disturbance in Natural Communities." *Annual Review Ecological Systems* 15 (1984). Discusses the importance of natural disturbances in maintaining an ecosystem's health.

Vogl, Richard J. "Some Basic Principles of Grassland Fire Management." *Environmental Management* 3, no. 1 (1979). A good introduction to controlled burns in California.

WallisDeVries, Michiel F., Jan P. Bakker, and Sipke E. Van Wieren, eds. *Grazing and Conservation Management*. Boston: Kluwer Academic, 1998. A comprehensive overview of the use of grazing as a tool in conservation management. Considers the ecological and historical background, the impact of grazing on community structure, management applications, and future prospects.

GREEN ALGAE

Categories: Algae; *Protista*; taxonomic groups; water-related life

The green algae are a diverse group of eukaryotic organisms classified in the phylum Chlorophyta. *They are considered eukaryotic because individual cells possess a prominent structural feature known as a nucleus, which houses the chemicals responsible for heredity and metabolic regulation. The phylum is one of several algal phyla in the kingdom* Protista, *where algae are grouped based upon pigmentation, carbohydrate storage reserves, and cell wall composition.*

Green algae are found in moist soils and freshwater and saltwater habitats; most are believed to be freshwater-dwelling. The phylum consists of at least eight thousand species. Some estimates place this number at seventeen thousand species. Several shared characteristics support the hypothesis that these organisms and terrestrial plants derived from a common ancestor.

General Characteristics

The green algae, or chlorophytes, may be unicellular, multicellular, colonial, or filamentous. Multicellular forms may demonstrate some tissue differentiation but not to the complexity displayed by terrestrial plants. Colonial algae tend to cluster in a pattern resembling a hollow sphere or disc. Some filamentous forms are coenocytic, meaning they have lost a portion or all of their cross walls.

The cell walls consist of cellulose. There are usually two layers of cellulose fortified by pectin. Some unicellular forms have a lorica (thin wall or cuticle), which is separated from the protoplast by a gelatinous matrix or water.

The phylum *Chlorophyta* is named for the prominent green chloroplast, a cell structure containing pigments that carry out photosynthesis, similar to that found in higher plants. The chloroplasts are green because the accessory pigments, which include xanthophylls and various carotenoids, do not mask the chlorophylls, the principal photosynthetic pigments, present. All classes contain chlorophylls *a* and *b*. Chlorophyll *c* has been found in a few species of the class *Prasinophyceae*. The chloroplasts are double-membraned structures with thylakoids (membranous folds) stacked in groups of from two to six.

The storage carbohydrate is starch. Starch grains can be found clustered around pyrenoids (protein bodies), if they are present. However, they are found generally scattered throughout the fluid portion of the chloroplast. Chlorophytes possess either two or four flagella (whiplike appendages for motility) at least once during their life cycle, although some forms have a single flagellum. In addition to providing motility, flagella may play a key role in the sexual process for some unicellular forms.

There is considerable debate over the classification of green algae. Most taxonomists currently classify *Chlorophyta* in the kingdom *Protista*. Because of the many similarities to terrestrial plants, many taxonomists feel that *Chlorophyta* should have its own kingdom. Living species of charophytes are grouped into three classes: *Chlorophyceae, Charophyceae,* and *Ulvophyceae.*

Chlorophyceae

The class *Chlorophyceae* is the largest in terms of the number of species listed. Members have two or more flagella; a diverse array of sexual and asexual reproductive strategies; production of a zygospore following sexual reproduction; and mitosis that involves phycoplasts (microtubules that separate daughter nuclei during division). Representative genera include *Chlamydomonas, Pandorina, Volvox,* and *Gonium.*

Chlamydomonas species are unicellular, with two apical flagella and a cup-shaped chloroplast. *Gonium* is a colonial species with four or more cells with no functional or morphological differentiation. *Pandorina* species form spherical colonies with limited differentiation and structural organization. Colonies of *Volvox* can consist of up to sixty thousand cells and demonstrate some structural specialization. Portions of the colony have cells with large flagella and stigmata. These cells appear to be specialized for colony motility. The posterior region consists of cells with small flagella and no stigmata. These seem to be responsible for reproduction.

Charophyceae

Charophyceae contains asymmetrical cells that may or may not be motile. Motile cells have two apical flagella. Sexual reproduction is characterized by the formation of a zygospore and zygotic meiosis. This class is similar to land plants in that nuclear envelopes dissolve during mitosis, which is not the case for the other two classes. The genus *Chara* includes members that resemble vascular plants. Chara species have a central axis and branchlike extensions. These organisms demonstrate apical growth that begins with an apical cell, which is analogous to the apical meristems of terrestrial plants. *Spirogyra* is a well-known filamentous genus that is distinguished by spiral chloroplasts. Sexual reproduction is characterized by the formation of a conjugation filament between two cells that allows for gamete transfer.

Ulvophyceae

Ulvophyceae is a diverse class of primarily marine organisms that can consist of small colonial forms, filamentous forms, thin sheets of cells, or coenocytic complexes. Reproduction is by alternation of generations, with meiosis occurring in spores. There may be two or more flagella, if flagella are present. The genus *Ulva*, also known as sea lettuce, displays a green sheet of cells that are found in intertidal waters. Reproduction involves an isomorphic alternation of generations. *Ulothrix* contains freshwater filamentous algae that can attach to surfaces via a holdfast. Ulothrix asexually generates zoospores and aplanospores. Species are able to reproduce sexually by formation of a heterothallic zygote/zygospore from isogamous gametes.

Reproductive Strategies

Chlorophytes reproduce sexually, which involves alternating haploid (organisms with half the complete chromosome set) and diploid stages. *Haplobiontic* haploid organisms consist of mature haploid forms that produce gametes by mitosis (division resulting in offspring cells identical to the parent form). Compatible gametes fuse and form a diploid zygote, which divides by meiosis (division resulting in four haploid offspring cells) to form four spores. A haplobiontic diploid organism consists of mature diploid forms that produce gametes by meiosis. *Diplobiontic* green algae are more complex, with a zygote undergoing mitosis. This results in the formation of a haploid and diploid thalli. The haploid thallus is referred to as the gametophyte, and the diploid thallus is referred to as the sporophyte. Gametophytes generate gametes, while sporophytes produce spores. This pattern is referred to as an alternation of generations.

The thalli may be identical (isomorphic) or different (heteromorphic). If a thallus produces both sperm and eggs, it is considered homothallic. If the egg and sperm are produced on separate thalli, the organism is heterothallic. Gametes may be isogamous (indistinguishable and motile) or heterogamous (two distinct types). Male gametes develop in gametangia known as antheridia. Female gametes develop in either oogonia (single-celled gametangia) or archegonia (multicelled gametangia). Zygotes often form thick-walled resting structures called zygospores.

The most common type of spore is the zoospore, which is a flagellated cell. Cells can form single zoospores or divide mitotically to produce many zoospores. Zoospores mature into vegetative cells within minutes or days, depending upon the species. Vegetative cells may or may not keep their flagella. Zoospores are typically formed in compartments called sporangia but may be formed following meiosis in a zygote. Most zoospores resemble members of the chlorophyte genus *Chlamydomonas*. Thick-walled, nonmotile spores called akinetes may be formed and can produce zoospores via mitosis or form filamentous structures. Some chlorophytes form aplanospores, which are nonmotile.

Ecology

Chlorophytes are found in diverse habitats all over the world. While most inhabit temperate, freshwater environments, marine and terrestrial forms also exist. Terrestrial forms include some living on moist soils, some on moist rocks, and some in snow-covered areas. Some terrestrial forms are specialized as *lichens*, a close association between an alga and a fungus, or living on animals such as turtles or sloths.

Because they are photoautotrophic, capable of making their own carbohydrates using sunlight energy, chlorophytes are critical to life on earth. Green algae are the planet's largest food source. They fix approximately 1,010 tons of carbon per year. As a result, they also contribute significantly to oxygen production.

Stephen S. Daggett

See also: Algae; Brown algae; *Charophyceae*; *Chlorophyceae*; Chrysophytes; Cryptomonads; Diatoms; Eutrophication; Evolution of plants; Lichens; Marine plants; Photosynthesis; Phytoplankton; *Protista*; Red algae; *Ulvophyceae*.

Sources for Further Study

Dillard, Gary E. *Common Freshwater Algae of the United States.* Berlin: J. Cramer, 1999. An up-to-date survey of freshwater organisms, excluding diatoms. Includes keys, illustrations, and bibliographical references.

Margulis, Lynn, et al. *Symbiosis in Cell Evolution.* New York: Freeman, 1993. An interesting look at the diversity and relatedness of organisms, including the green algae. Includes illustrations, bibliographic references, and index.

Raven, Peter H., Ray F. Evert, and Susan E. Eichhorn. *Biology of Plants.* 6th ed. New York: W. H. Freeman/Worth, 1999. A basic textbook that covers key algae groups in terms of cells, life cycles, genetics, and evolutionary relationships. Includes illustrations, glossary, and index.

Van Den Hoek, Christiaan, et al. *Algae: An Introduction to Phycology.* New York: Cambridge University Press, 1995. An excellent, illustrated survey of the algae. Includes glossary, bibliographic references, and index.

GREEN REVOLUTION

Categories: Agriculture; economic botany and plant uses; food; history of plant science

The Green Revolution implemented advances in agricultural science to raise food production levels, particularly in developing countries. These advances are associated with the spreading use of high-yield varieties (HYV) of wheat, rice, and corn developed through advanced methods of genetics and plant breeding.

Yield Increases

The Green Revolution can be traced back to a 1940 request from Mexico for technical assistance from the United States to increase wheat production. By 1944, with the financial support of the Rockefeller Foundation, a group of U.S. scientists were researching methods of adapting the new high-yield variety (HYV) wheat that had been successfully used on American farms in the 1930's to Mexico's varied environments. A major breakthrough is attributed to Norman Borlaug, who by the late 1940's was director of the research in Mexico. For his research and work in the global dissemination of the Mexican HYV wheat, Borlaug won the 1970 Nobel Peace Prize.

From wheat, research efforts shifted to rice production. Through the work of the newly created International Rice Research Institute in the Philippines, an HYV rice was developed. This so-called miracle rice was widely adopted in developing countries during the 1960's. Later research has sought to spread the success of the Green Revolution to other crops and to more countries.

Approximately one-half of the yield increases in food crops worldwide since the 1960's are attributable to the Green Revolution. Had there not been a Green Revolution, the amount of land used for agriculture would undoubtedly be higher today, as would the prices of wheat, rice, and corn, three crops that account for more than 50 percent of total human energy requirements. There is a concern, however, that the output benefits of the Green Revolution have had some negative equity and environmental effects.

Norman Borlaug was awarded the Nobel Peace Prize in 1970 for his research and work in the global dissemination of the Mexican high-yield varieties of wheat.

Equity and Environmental Issues

The Green Revolution has promoted input-intensive agriculture, which has, in turn, created several problems. In theory, a small-scale farmer will get benefits from planting the HYV seeds that are similar to those reaped by a large farm. In practice, however, small-scale farmers have had more difficulty in gaining access to Green Revolution technology. To use the new seeds, fields need adequate irrigation and the timely application of chemical fertilizers and pesticides. In many developing countries, small-scale farmers' limited access to credit makes the variety of complementary inputs difficult to obtain. Greater use of fertilizers is associated with rising nitrate levels in water supplies. Pesticides have been linked to community health problems. Long-term, intensive production has resulted in compaction and salinization of soil and other problems.

Because agriculture is increasingly dependent on fossil fuels, food prices have become more strongly linked to energy supplies of this type. This issue raises concerns about the sustainability of the new agriculture. Biotechnological approaches to generating higher yields, the expected future path of the Green Revolution, will raise an additional set of equity and environmental concerns.

Bruce Brunton

See also: Agriculture: modern problems; Agriculture: world food supplies; Corn; Genetically modified foods; Grains; High-yield crops; Hybridization; Rice; Sustainable agriculture; Wheat.

Sources for Further Study

Conway, Gordon. *The Doubly Green Revolution: Food for All in the Twenty-first Century.* Ithaca, N.Y.: Comstock, 1998. Argues that the next wave of agricultural revolution must emphasize conservation as well as productivity.

Hazell, Peter B. R., and C. Ramasamy, eds. *Green Revolution Reconsidered: The Impact of High-Yielding Rice Varieties in South India.* Baltimore: Johns Hopkins University Press, 1991. Case study uses data collected before and after high-yielding rice varieties were adopted in India. Discusses the impacts of increased income and employment as well as the distribution of welfare among rural households.

Manning, Richard. *Food's Frontier: The Next Green Revolution.* Berkeley: University of California Press, 2001. Survey of pioneering agricultural research projects under way in Africa, India, China, South America, and Mexico. Premise is that the Green Revolution created a number of serious problems.

Singh, Himmat. *Green Revolutions Reconsidered: The Rural World of Contemporary Punjab.* Oxford, England: Oxford University Press, 2001. Counters arguments against the Green Revolution. Maintains that in Punjab, agricultural intensification has rejuvenated rural society, both economically and socially.

GREENHOUSE EFFECT

Categories: Environmental issues; pollution

The greenhouse effect is a natural process of atmospheric warming in which solar energy that has been absorbed by the earth's surface is reradiated and then absorbed by particular atmospheric gases, primarily carbon dioxide and water vapor. Without this warming process, the atmosphere would be too cold to support life. Since 1880, however, the surface atmospheric temperature appears to be rising, paralleling a rise in the concentration of carbon dioxide and other gases produced by industrial activities.

Since 1880, carbon dioxide, along with several other gases—chlorofluorocarbons (CFCs), methane, hydrofluorocarbons (HFCs), perfluorocarbons (PFCs), sulfur hexafluoride, and nitrous oxide—have been increasing in concentration and have been identified as likely contributors to a rise in global surface temperature. These gases are called *greenhouse gases*. The temperature increase may lead to drastic changes in climate and food production as well as widespread coastal flooding. As a result, many scientists, organizations, and governments have called for curbs on the production of greenhouse gases. Since the predictions are not definite, however, debate continues about the costs of reducing the production of these gases without being sure of the benefits.

Global Warming and Human Interference

The greenhouse effect occurs because the gases in the atmosphere are able to absorb only particular wavelengths of energy. The atmosphere is largely transparent to short-wave solar radiation, so sunlight basically passes through the atmosphere to the earth's surface. Some is reflected or absorbed by clouds, some is reflected from the earth's surface, and some is absorbed by dust or the earth's surface. Only small amounts are actually absorbed by the atmosphere. Therefore, sunlight contributes very little to the direct heating of the atmosphere. On the other hand, the greenhouse gases are able to absorb long-wave, or infrared, radiation from the earth, thereby heating the earth's atmosphere.

Discussion of the greenhouse effect has been confused by terms that are imprecise and even inaccurate. For example, the atmosphere was believed to operate in a manner similar to a greenhouse, whose glass would let visible solar energy in but would also be a barrier preventing the heat energy from leaving. In actuality, the reason that the air remains warmer inside a greenhouse is probably because the glass prevents the warm air from mixing with the cooler outside air. Therefore the greenhouse effect could be more accurately called the "atmospheric effect," but the term greenhouse effect continues to be used.

Even though the greenhouse effect is necessary for life on earth, the term gained harmful connotations with the discovery of apparently increasing atmospheric temperatures and growing concentrations of greenhouse gases. The concern, however, is not with the greenhouse effect itself but rather with the intensification or enhancement of the greenhouse effect, presumably caused by increases in the level of gases in the atmosphere resulting from human activity, especially industrialization. Thus the term *global warming* is a more precise description of this presumed phenomenon.

A variety of other human activities appears to have contributed to global warming. Large areas of natural vegetation and forests have been cleared for agriculture. The crops may not be as efficient in absorbing carbon dioxide as the natural vegetation they replaced. Increased numbers of livestock have led to growing levels of methane. Several gases that appear to be intensifying global warming, including CFCs and nitrous oxides, also appear to be involved with *ozone depletion*. Stratospheric ozone shields the earth from solar ultraviolet (short-wave) radiation; therefore, if the concentration of these ozone-depleting gases continues to increase and the ozone shield is depleted, the amount of solar radiation reaching the earth's surface should increase. Thus, more solar energy would be intercepted by the earth's surface to be reradiated as long-wave radiation, which would presumably increase the temperature of the atmosphere.

However, whether there is a direct cause-and-effect relationship between increases in carbon dioxide and the other gases and surface temperature may be impossible to determine because the atmosphere's temperature has fluctuated widely over millions of years. Over the past 800,000 years, the earth has had several long periods of cold temperatures—during which thick ice sheets covered large portions of the earth—interspersed with shorter warm periods. Since the most recent retreat of the glaciers around ten thousand years ago, the earth has been relatively warm.

Problems of Prediction

How much the temperature of the earth might rise is not clear. So far, the temperature increase of around 1 degree Fahrenheit is within the range of normal (historic) trends. The possibility of global warming became a serious issue during the late twentieth century because the decades of the 1980's and the 1990's included some of the hottest years recorded for more than one century. On the other hand, warming has not been consistent since 1880,

and for many years cooling occurred. The cooling might have resulted from the increase of another product of fossil fuel combustion, sulfur dioxide aerosols, which reflect sunlight, thus lessening the amount of solar energy entering the atmosphere. Similarly, in the early 1990's temperatures declined, perhaps because of ash and sulfur dioxide produced by large volcanic explosions. In the late 1990's temperatures appeared to be rising again, thus indicating that products of volcanic explosions may have masked the process of global warming. The United States Environmental Protection Agency (EPA) states that the earth's average temperature will probably continue to increase because the greenhouse gases stay in the atmosphere longer than the aerosols.

Proper analysis of global warming is dependent on the collection of accurate temperature records from many locations around the world and over many years. Because human error is always possible, "official" temperature data may not be accurate. This possibility of inaccuracy compromises examination of past trends and predictions for the future. However, the use of satellites to monitor

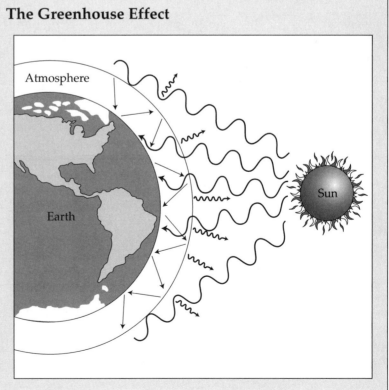

The Greenhouse Effect

Clouds and atmospheric gases such as water vapor, carbon dioxide, methane, and nitrous oxide absorb part of the infrared radiation emitted by the earth's surface and reradiate part of it back to the earth. This process effectively reduces the amount of energy escaping to space and is popularly called the "greenhouse effect" because of its role in warming the lower atmosphere. The greenhouse effect has drawn worldwide attention because increasing concentrations of carbon dioxide from the burning of fossil fuels may result in a global warming of the atmosphere.

Scientists know that the greenhouse analogy is incorrect. A greenhouse traps warm air within a glass building where it cannot mix with cooler air outside. In a real greenhouse, the trapping of air is more important in maintaining the temperature than is the trapping of infrared energy. In the atmosphere, air is free to mix and move about.

Atmosphere

Earth

Sun

U.S. Greenhouse Gas Emissions, 1990-1999

Type	1990	1994	1995	1996	1997	1998	1999
Carbon dioxide (carbon)[1]	1,350.5	1,422.5	1,434.7	1,484.1	1,505.2	1,507.4	1,526.8
Methane gas[1]	31.74	31.17	31.18	30.16	30.11	29.29	28.77
Nitrous oxide[2]	1,168	1,310	1,257	1,246	1,226	1,223	1,224
Chloroflurocarbons[2]	202	109	102	67	51	49	41
Halons[2]	2.8	2.7	2.9	3.0	3.0	3.0	3.0
Hydrofluorocarbons[2]							
HFC-23	3.0	3.0	2.0	3.0	3.0	3.4	2.6
HFC-125	(Z)	0.3	0.5	0.7	0.9	1.1	1.3
HFC-134a	1.0	6.3	14.3	19.0	23.5	26.9	30.3
HFC-143a	(Z)	0.1	0.1	0.2	0.3	0.5	0.7
Perfluorocarbons[2]							
CF-4	3	2	2	2	2	2	2
C-2F-6	1	—	1	1	1	1	1
C-4F-10	(Z)	(Z)	(Z)	(Z)	(Z)	(Z)	(Z)
Sulfur hexafluoride[2]	1	1	2	2	2	2	1

Note: Emission estimates were mandated by Congress through Section 1605(a) of the Energy Policy Act of 1992 (Title XVI). Gases that contain carbon can be measured either in terms of the full molecular weight of the gas or just in terms of their carbon content.

1. In millions of metric tons.
2. In thousands of metric tons.
Z. Less than 500 metric tons.
— Represents or rounds to zero.

Source: Abridged from U.S. Energy Information Administration, *Emissions of Greenhouse Gases in the United States,* annual. From *Statistical Abstract of the United States: 2001* (Washington, D.C.: U.S. Census Bureau, 2001).

temperatures has probably increased the reliability of the data.

Predictions for the future are hampered in various ways, including lack of knowledge about all the components affecting atmospheric temperature. Therefore, computer programs cannot be sufficiently precise to make accurate predictions. A prime example is the relationship between ocean temperature and the atmosphere. As the temperature of the atmosphere increases, the oceans would absorb much of that heat. Therefore, the atmosphere might not warm as quickly as predicted. However, the carbon dioxide absorption capacity of oceans declines with temperature. Therefore, the oceans would be unable to absorb as much carbon dioxide as before, but exactly how much is unknown. Increased ocean temperatures might also lead to more plant growth, including phytoplankton. These plants would probably absorb carbon dioxide through photosynthesis. A warmer atmosphere could hold more water vapor, resulting in the potential for more clouds and more precipitation. Whether that precipitation would fall as snow or rain and where it would fall could also affect air temperatures. Air temperature could lower as more clouds might reflect more sunlight, or more clouds might absorb more infrared radiation.

To complicate matters, any change in temperature would probably not be uniform over the globe. Because land heats up more quickly than water, the Northern Hemisphere, with its much larger landmasses, would probably have greater temperature increases than the Southern Hemisphere. Similarly, ocean currents might change in both direction and temperature. These changes would affect air temperatures as well. In reflection of these complications, computer models of temperature change range widely in their estimates. Predicted increases range from 1.5 to 11 degrees Celsius (3 to 20 degrees Fahrenheit) over the early decades of the twenty-first century.

Mitigation Attempts

International conferences have been held, and international organizations have been established to research and minimize potential detriments of global warming. In 1988 the United Nations Environment Programme and the World Meteorological

Organization established the International Panel on Climate Change (IPCC). The IPCC has conducted much research on climate change and is now considered an official advisory body on the climate change issue. In June, 1992, the United Nations Conference on Environment and Development, or Earth Summit, was held in Brazil. Participants devised the Framework Convention on Climate Change, considered the landmark international treaty. It required signatories to reduce and monitor their greenhouse gas emissions.

A more advanced agreement, the Kyoto Accords, was developed in December, 1997, by the United Nations Framework Convention on Climate Change. It set binding emission levels for all six greenhouse gases over a five-year period for the developed world. Developing countries do not have any emission targets. It also allows afforestation to be used to offset emissions targets. The Kyoto agreement includes the economic incentive of trading emissions targets. Some countries, because they have met their targets, would have excess permits, which they might be willing to sell to other countries that have not met their targets.

Margaret F. Boorstein

See also: Acid precipitation; Air pollution; Carbon cycle; Climate and resources; Ozone layer and ozone hole debate; Photorespiration; Rain forests and the atmosphere; Respiration.

Sources for Further Study

Berger, John J. *Beating the Heat: Why and How We Must Combat Global Warming*. Berkeley, Calif.: Berkeley Hills Books, 2000. Written for the general reader. Discusses causes, dangers, and remedies associated with climate change. Rebuts arguments from climate change skeptics.

Graedel, Thomas E. *Atmosphere, Climate, and Change*. New York: Scientific American Library/W. H. Freeman, 1997. Overview of how the earth's weather systems function and how human-caused events such as the increase in greenhouse gases affect the chemical composition of the atmosphere.

Houghton, John T. *Global Warming: The Complete Briefing*. 2d ed. New York: Cambridge University Press, 1997. Examines the scientific basis of global warming, likely impacts of climate change, the political aspects of it, and actions that could be taken by governments, industries, and individuals to mitigate the effects.

McKibben, Bill. *The End of Nature*. 10th anniversary ed. New York: Anchor Books, 1999. Discusses the extent to which humankind has become the single most important and decisive element in climatic calculus that determines the weather.

Rosenzweig, Cynthia. *Climate Change and the Global Harvest: Potential Impacts of the Greenhouse Effect on Agriculture*. New York: Oxford University Press, 1998. Analyzes the nature of predictable changes to world agriculture caused by global warming. Seeks to educate undergraduate students about how climactic factors affecting agriculture may be modified in the future.

GROWTH AND GROWTH CONTROL

Category: Physiology

The processes of primary and secondary growth take a plant from early cell division to its adult form. Growth control factors regulate these processes.

Plant growth is of two distinct types: *primary growth* and *secondary growth*. Primary growth results in increased length of stems or roots. Secondary growth increases the width of the plant and allows differentiation of cells into various distinct tissue types. Both types of growth occur in plant tissues called *meristems*. A meristem consists of tissue where extensive cell division takes place, and thus plant growth. There are two general types of meristems. Primary growth occurs at the *apical meristems*, and secondary growth occurs at the *lateral meristems*, which are known as the *vascular cambium* and the *cork cambium*.

Meristems

Apical meristems are located at the growing tips of the plant; there are apical meristems in the roots and in the buds on shoots of the aboveground part of the plant. New cells produced at the meristems are initially undifferentiated. They enlarge in the adjacent *zone of elongation*, mostly by increasing their water content. These cells eventually differentiate into the plant's primary tissue types: *dermal*, *vascular tissue*, and *ground tissues*.

Other meristematic tissue occurs along the stem, and lateral buds are capable of producing branches with their own meristems, but most elongation occurs from the apical meristem. Apical dominance prevents excessive branching and in some plants prevents all branching. If the apical meristem is removed by removing the end of the stem, the lateral buds will be released from apical dominance, and greater branching results. Eventually, one of the lateral meristems will grow more than the others and will impose apical dominance, becoming the new apical meristem.

Secondary growth increases the girth (diameter) of the stems and roots of woody plants. Lateral growth of the vascular cambium produces new vascular tissue, called secondary *xylem* and *phloem*. In trees and shrubs, this continual lateral growth produces wood. Cork cambium produces cells at the outer edges of roots and stems. At maturity these cells are dead and form the bark, their primary function being structural support and protection. The walls of cork cells contain a protective waxy substance called *suberin*.

Gametophytes and Sporophytes

All vascular plants, as well as mosses and liverworts and many algae, display a type of life cycle referred to as *alternation of generations*, which involves two distinct life-forms, the *gametophyte* and *sporophyte*. The sporophyte generation is genetically diploid and, as the name implies, produces spores by meiosis. Spores germinate and develop into gametophytes, which are genetically haploid and produce gametes (eggs and sperm) by mitosis. Sporophytes are larger than gametophytes and represent the dominant, or more noticeable, generation and typically live much longer than gametophytes. Trees, shrubs, herbs, and ferns all represent sporophytes. Gametophytes are extremely small by comparison and are therefore unknown to most nonscientists. They are free-living in ferns as tiny, heart-shaped masses of cells that grow on or just under the surface of the soil and are often smaller than the letters on this page. In flowering plants and gymnosperms, gametophytes are enclosed in the reproductive tissues of flowers and cones on the sporophytes that later develop into seeds. In mosses and liverworts, gametophytes are larger and represent the dominant generation, and the sporophyte grows from the structure on the end. Algae are more diverse. In some groups both gametophytes and sporophytes are indistinguishable, whereas in other groups one or the other is larger.

Both the gametophyte and the sporophyte develop from single cells—the *spores* and the *zygotes*, respectively. In seed plants, the first cell division of the zygote often defines the root cell line (or its equivalent) and the stem cell line (or its equivalent). The body produced by this early development is initially linear in many cases, laying out the primary axis of the plant body. The embryo grows from the zygote and as it matures is included in integuments that develop into the seed coat. A primary root and primary stem grow from a root apical meristem and a shoot apical meristem, respectively.

Primary and Secondary Tissue

In many species, the new cells in the sporophyte are produced primarily by the division of apical meristems, thus consisting almost exclusively of primary tissues. There are, however, some plants in which the sporophytes grow in girth. Some of these—such as the *calamites* (giant horsetails), the *Lepidodendron* (tree lycopods), and the seed ferns—are known only from the fossil record. Others are the trees and shrubs, so-called woody plants, that characterize the modern forests.

The wood of woody plants is composed almost entirely of secondary xylem—xylem that is not derived from the apical meristems but instead grows from the vascular cambium, a cylindrical meristem located under the bark. The bark of woody plants, also a secondary tissue, is composed of phloem and corky layers. The corky layers develop from a second cylindrical meristem, the cork cambium.

In addition to the secondary tissues, many plants as they grow produce secondary organs: branch stems and branch roots. These secondary organs are not derived from the original axis of the plant. In the early stages of their growth they are composed of primary tissues that are essentially identical to the primary tissues of primary organs. Cambial growth will produce secondary tissues in these branch stems and roots. The patterns of secondary tissue formation determine the form of the wood and bark of woody species. The patterns of secondary organ formation determine the architecture of the plant: the shape of the crown and the root system. This architecture plays an important role in the ability of the plant to compete for sunlight, water, and soil nutrients.

Growth Control

The patterns of secondary organ formation are controlled both by genetic factors and by environmental conditions. Horticulturists use plants as a source of dwarfing stocks that have a genetic predisposition to form branches early. In many cases, the dwarfing results from a failure of the stem to elongate in the *internodes* (the regions between the nodes, where leaves and lateral branches originate). Dwarfing appears to be particularly influenced by plant hormones called gibberellins, which stimulate internodal elongation in *dicots*. The effect of gibberellins is also influenced by the concentration of the other hormones within the plant.

Likewise, the architecture of columnar plants is under genetic and hormonal control. The Lombardy poplar, for example, has greatly reduced branching compared to the European poplar. This elongation

DIGITAL STOCK

Many plants as they grow produce secondary organs: branch stems and branch roots. These secondary organs are not derived from the original axis of the plant. The patterns of secondary organ formation determine the architecture of the plant: the shape of the crown and the root system. This architecture plays an important role in the ability of the plant to compete for sunlight, water, and soil nutrients.

of the principal axis is similar to that found in forest trees growing in the shade of the surrounding forest. The shaded environment stimulates the growth of the main axis of many tree species while inhibiting growth of the secondary stems. As a result, the stem reaches above the surrounding trees and is better able to compete for light.

The inhibition of secondary stem formation seems to be influenced primarily by hormones called auxins. High auxin concentrations inhibit the development of secondary stems, while low auxin levels stimulate the formation of branches. In some species, high cytokinin levels also stimulate secondary stem growth. Because cytokinins are produced in large quantities in the root tips, and auxins are produced in large quantities in the stem tip, the relationship between these two chemicals reflects the balance between the root system and the stem system.

Less is known about the mechanisms of control of branching in the root system. Some species, especially *monocots*, have many secondary roots of approximately equal size. Others have dominant primary roots, called *taproots*, with little development of secondary roots. Carrots carry this pattern to an extreme.

The pattern of secondary tissue formation is determined by an interplay between genetic factors and environmental conditions. Many plants complete their life cycles in a single year. This quick passage from seed to seed is under genetic control. Annuals rarely develop woody tissue; perennials survive many seasons and show increases in stem girth throughout their lives.

The annual rings seen in the cross-section of a tree are a result of seasonal variations in the production of secondary xylem. Variations in the thickness of annual rings are the result of genetic controls and environmental factors such as mean temperature, damage from insect pests or other pathogens, and water and nutrient availability. Even wind can have significant effects. Strong prevailing winds cause an effect called wind pruning, which results in reduced branching and shorter distances between the annual rings on the windward side.

Secondary tissues are extraordinarily complex. The patterns of cell division, apparently under genetic control, are influenced by a whole concert of hormones. Hormonal gradients and seasonal gradients of sugars and amino acids may play a role in these patterns of secondary tissue formation. The activity of mature phloem tissues and the concentrations of auxins, gibberellins, cytokinins, and perhaps of the gaseous hormone ethylene may all be important in regulating the activity of the cambia.

Craig R. Landgren, updated by Bryan Ness

See also: Angiosperm life cycle; Germination and seedling development; Hormones; Plant life spans; Plant tissues; Roots; Seeds; Shoots; Stems.

Sources for Further Study

Bryant, John A., and Donato Chiatante, eds. *Plant Cell Proliferation and Its Regulation in Growth and Development.* New York: Wiley, 1998. Concise and accessible overview of current research in plant cell proliferation, giving detailed examples of how it is regulated during the plant's growth and development. Documents the major advances that have taken place in this field over the last ten years.

Davies, Peter J., ed. *Plant Hormones: Physiology, Biochemistry, and Molecular Biology.* 2d ed. Boston: Kluwer Academic, 1995. A careful description and review of the hormones controlling plant reproduction and growth. Written for advanced students and professionals. Illustrated.

Fosket, Donald E. *Plant Growth and Development: A Molecular Approach.* San Diego: Academic Press, 1994. An upper-level undergraduate textbook that assumes no previous knowledge of plant biology. Includes chapter reviews and recommendations for further reading.

Kozlowski, Theodore T., and Stephen G. Pallardy. *Growth Control in Woody Plants.* San Diego: Academic Press, 1996. A combination textbook and reference work on the processes and mechanisms of woody plant growth. Covers reproductive systems, vegetative organs, physiological and environmental regulation of growth, and biotechnology. Includes an appendix of common and scientific plant names. Illustrated.

Raven, Peter H., and George B. Johnson. *Biology.* 5th ed. Boston: McGraw-Hill, 1999. Beautifully illustrated with many color drawings and photographs, this college-level biology text places plant growth in context of the biology of plants and places plants in context with the living and nonliving world. Chapter 31, "Vascular Plant Structure," provides a simple review of plant cell types, tissues, and organs. Chapter 32, "Plant Development," deals clearly with many questions concerning growth and growth controls.

GROWTH HABITS

Categories: Angiosperms; gardening; physiology

Plant habit, also known as plant life form, is the characteristic shape, appearance, or growth form of a plant species. It develops from specific genetic patterns of growth in combination with environmental factors and is part of the organization of every plant.

Growth of Plants

Development of a plant body is accomplished through growth, defined as increase in number of cells and size of a species. Rates of growth in plants are achieved in two ways: first, by *geometric increase*, in which all cells of the organism divide simultaneously, especially in a young embryonic plant; second, by *arithmetic increase*, in which only one cell undergoes division, especially in mature plants with localized growth in a region at the root and shoot apices. Generally, plants grow by a combination of both kinds of cell division to produce variations of form that finally develop a specific *habit* that is unique to a particular plant species.

Evolution of Growth Habits

The primary purpose for the evolution of different growth habits in plants is adaptation for permanent survival and reproduction of new individuals, typically under changing climatic conditions. Water availability, especially during the growing season, is the single most important environmental factor that limits plant distribution and productivity on a global basis. Competition in the past among plants for available water, nutrients, space, and light enhanced the evolution of adaptive growth forms in plants.

Some plants developed wood as a mechanism to counteract the destructive effects of wind, ice, mechanical damage, and fire. *Erect* and *dense* growth habits evolved to resist wind effects and other mechanical damages. Plants without wood adapted *prostrate*, *mat-forming*, *spreading*, *creeping*, or *climbing* habits.

As animals interacted with plants and in the past, both evolved simultaneously. Various plants developed prostrate and mat-forming habits in

Plants displaying the stemless growth habit have no visible stem aboveground and are composed mainly of leaves or leaflike structures, such as this Aloe vera. Stemless plants characterize aquatic and wetland vegetation, deserts, some grasslands, cultivated land, and wasteland.

AP/WIDE WORLD PHOTOS

order to endure intense grazing and trampling, or erect and tall growth forms to escape browsing and grazing.

In the past, individual plants that were able to adapt, survive, and produce more offspring were selected naturally for success. Different plants with varied growth habits colonized different habitats, becoming the dominant plants (largest or most abundant), and thereby the principal contributors in characterizing and sustaining different biomes. The various kinds of growth habits which evolved result in a variety of forms. It is not uncommon for one species of plant to exhibit growth habits among its different varieties.

Climbing Plants

Climbing plants are also called *vines*. The stems trail along or coil around other plants or structures as they grow upward. Examples include cucumber (*Cucumis sativus*), morning glory (*Ipomea* species), and grape vine (*Vitis* species). Climbers characterize moist forests and woodlands.

Clump-Forming Plants

Clump-forming or *tussocky plants* exhibit an aggregate of several shoots growing in a bunch from a common base, especially in grasses. Examples include the bunch grasses *Andropogon* and *Aristida* mosses (such as *Polytrichum* species) and sedges (*Carex stricta*). They characterize grasslands and are common in the prairies of the United States. They also grow in sandy locations, wetlands, and disturbed habitats.

Dense Plants

Dense plants grow many small, woody canes or stems very close together in an upright fashion. The majority are shrubs. Examples include *Ephedra*, southern arrowwood (*Viburnum dentatum*), mountain laurel (*Kalmia latifolia*), and creosote bush (*Larrea tridentata*). They characterize woodlands, grasslands, coastal vegetation, and deserts.

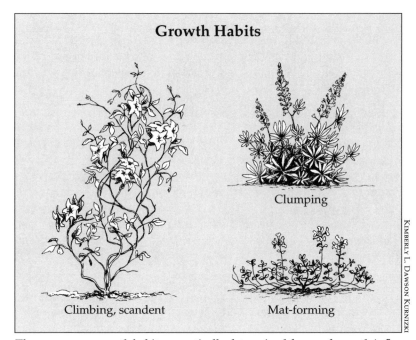

Growth Habits

Climbing, scandent

Clumping

Mat-forming

KIMBERLY L. DAWSON KURNIZKI

Three common growth habits, genetically determined forms of growth influenced by availability of light and water, wind exposure, soil conditions, and other environmental factors. There are many more such forms, including those shown above as well as erect, dense, prostrate, spreading, stemless, mounding, trailing, and other forms.

Erect Plants

In *erect plants*, one main stem grows in an upright position clearly above ground level. This is common in trees. Examples include banana (*Musa*), oak (*Quercus*), pine (*Pinus*), maple (*Acer*), and palm. They mainly characterize forests and woodlands and some grasslands.

Mat-Forming Plants

Mat-forming plants have many *stolons* (creeping stems) that grow in a trail along soil or water surfaces and spread out to produce a matlike cover. Examples include the grasses *Cynodon* and *Digitaria*, Kentucky bluegrass (*Poa pratensis*), crab grass (*Digitaria sanguinalis*), the aquatic ferns *Salvinia* and *Azolla*, and mosses, such as *Sphagnum*. They characterize grasslands, bogs, wetlands, secondary forest floors, and cultivated habitats.

Mound-Forming Plants

Mound-forming plants grow to form a rounded shape resembling a mound or swollen bump. Examples include the barrel cactus (*Ferocactus* and *Echinocactus*), several other species of cacti (such as *Gymnocalycium*), and *Euphorbia gymnocalycioides*. They characterize deserts, grasslands, and the tundra.

Open Plants

Upright, woody stems or canes growing in an erect fashion characterize *open plants*. Their growth resembles a dense habit but has fewer stems and an open, airier structure. Examples include some bamboos (*Bambusa*), black willow (*Salix nigra*), smooth alder (*Alnus serrulata*), and meadowsweet (*Spirea*). This is characteristic of some shrubs and small trees of forests, woodlands, wetlands, and grasslands.

Prostrate Plants

The stems of prostrate plants grow flat on the soil surface or almost touching (hugging) the ground but not trailing. Examples include the herbaceous milk-purslane weed (*Euphorbia supina*), common mullein (*Verbascum thapsus*), and some species of juniper. They are common in the tundra, grasslands, wetlands, and disturbed habitats.

Scandent Plants

Scandent plants have prominent stems in a leaning position. Examples include sugarcane (*Saccharum officinarum*), coconut palm (*Cocos nucifera*), bearberry (*Arctostaphylos uva-ursi*), blackberry (*Rubus fruticosus*), and some bamboos (*Bambusa*). They characterize the dwarfed, woody trees in the timberline

of the tundra, savanna, forest undergrowth, and coastal habitats.

Spreading Plants

Spreading plants exhibit a sprawling type of growth, resulting from profuse lateral branching in mostly woody or succulent stems. Examples include common juniper (*Juniperus communis*), blueberries (*Vaccinium*), prickly-pear cactus (*Opuntia*), Sumacs (*Rhus*), and ferns (such as *Adiantum*). They characterize forest undergrowth, grasslands, sandy coastal areas, deserts, cultivated lands, and some areas of the tundra.

Stemless Plants

Stemless plants have no visible stem aboveground and are composed mainly of leaves or leaflike structures. Examples include common dandelion (*Taraxacum officinale*), *Aloe vera*, sisal (*Agave*), onion (*Allium cepa*), and liverworts (such as *Marchantia polymorpha*). They characterize aquatic and wetland vegetation, deserts, some grasslands, cultivated land, and wasteland.

Samuel V. A. Kisseadoo

See also: Angiosperm cells and tissues; Angiosperm evolution; Angiosperm plant formation; Angiosperms; Arctic tundra; Biomes: types; Bryophytes; Community-ecosystem interactions; Conifers; Deserts; Savannas and deciduous tropical forests; Shoots; Species and speciation; Taiga; Wood.

Sources for Further Study

Bold, H. C., et al. *Morphology of Plants and Fungi*. New York: Harper and Row, 1987. Provides extensive detail on growth forms in all families of plants. Includes illustrations, bibliographical references, and index.

Mauseth, James D. *Botany: An Introduction to Plant Biology*. Sudbury, Mass.: Jones and Bartlett, 1998. Discusses the broad variety of plant life-forms that colonize different biomes of the world. Includes good pictures, illustrations, and index.

Pearson, Lorenz C. *The Diversity and Evolution of Plants*. Boca Raton, Fla.: CRC Press, 1995. The author discusses the origins of the wide variety of plants and the different life-forms that exist.

Rost, Thomas L., et al. *Plant Biology*. New York: Wadsworth, 1998. Outlines the essentials of plant growth and organization and relates the facts to good descriptions of biomes and plant habits.

GYMNOSPERMS

Categories: Gymnosperms; *Plantae*; taxonomic groups

Pine trees are a familiar example of gymnosperms, a series of evolutionary lines of woody vascular seed plants that produce seeds not encased in an ovary.

Two kinds of higher plants—the gymnosperms and *angiosperms*—have developed to become the dominant type of land plant. With the exception of a few aquatic angiosperms, they do not require water for pollen transfer and are thus free to live in a wide variety of habitats. Gymnosperms and angiosperms differ primarily in the amount of protection they provide their *ovules* (the part that, after fertilization, becomes a weed), with gymnosperms usually providing less than angiosperms.

Progymnosperms

The first group of gymnosperms to appear was the *progymnosperms*. These plants evolved from the trimerophytes about 365 million years ago. *Archaeopteris*, the best-known progymnosperm, was described in 1871 by Sir William Dawson. Dawson believed that *Archaeopteris* was an ancient *fern*. He reached this conclusion because the large, leafy branch systems of *Archaeopteris* resembled a fern frond. In 1960 Charles Beck showed that these branch systems were borne on a stem having typical gymnospermous wood. This discovery led to the recognition of the progymnosperms as a distinct plant group which completely altered biologists' view of gymnosperm evolution.

Archaeopteris reached an estimated height of about 18 meters (59 feet). The main axis of the plant gave rise to a series of lateral branch systems, bearing primary branches in a single plane. The flattened branch system resembled a fern frond. The primary branches were covered with spirally arranged leaves. Some leaves bore *eusporangia* (spore-bearing structures that originate from a few cells). The earliest progymnosperms were *homosporous*, meaning that all their spores looked alike. Later progymnosperms were *heterosporous*, producing two types of spores, large megaspores and smaller microspores. Seeds have not been found attached to any progymnosperm.

Classification of Gymnosperms

Division *Cycadophyta*
  Class *Cycadopsida*
    Order *Cycadales*
      Families:
        Cycadaceae (cycads)
        Zamiaceae (sago palms)

Division *Ginkgophyta*
  Class *Ginkgoopsida*
    Order *Ginkgoales*
      Family:
        Ginkgoaceae (ginkgoes)
Division *Coniferophyta* (conifers)
  Class *Pinopsida*
    Order *Pinales*
      Families:
        Araucariaceae (araucarias)
        Cephalotaxaceae (plum yews)
        Cupressaceae (cypress)
        Pinaceae (pines)
        Podocarpaceae (podocarps)
        Taxodiaceae (redwoods)
    Order *Taxales*
      Family:
        Taxaceae (yews)
Division *Gnetophyta*
  Class *Gnetopsida*
    Order *Ephedrales*
      Family:
        Ephedraceae (Mormon teas)
    Order *Gnetales*
      Families:
        Gnetaceae (gnetums)
        Welwitschiaceae

Source: U.S. Department of Agriculture, National Plant Data Center, *The PLANTS Database*, Version 3.1, http://plants.usda.gov. National Plant Data Center, Baton Rouge, LA 70874-4490 USA.

Two major lines of gymnospermous evolution arose from the progymnosperms—the cycadophyte and the coniferophyte lines. Two plant groups make up the cycadophyte line: the pteridosperms, or seed ferns, and the cycads. These plants have large, compound, frondlike leaves. The cordaites (known only from the fossil record) and the conifers make up the coniferophyte line, all of which have simple leaves.

Pteridosperms and Cordaites

The *pteridosperms* and the *cordaites* appeared first. These plants were common in the wet tropical and subtropical coal swamps that covered much of the central United States between 345 million and 225 million years ago. One of the best-known pteridosperms is *Medullosa*. *Medullosa* had an upright stem between 3 and 8 meters (10 and 26 feet) high. The lower portion of the stem was covered with adventitious roots (roots that develop from stems or leaves, rather than from other roots). A number of large compound leaves arose from the stem tip. Ovules and pollen organs occurred singly on the leaves and not in cones. Pteridosperm pollen organs consisted of a number of elongate eusporangia that were commonly fused to form a ring. The seeds of *Medullosa* were quite large. Some reached lengths of up to 11 centimeters. Unlike other pteridosperms, *Medullosa* had multiple vascular bundles in the stem. Other gymnosperms have only a single conductive strand in their stems.

The cordaites were derived from Archaeopteris-like progymnosperms. Some species of *Cordaites* were trees, others were shrubs, and some were similar to modern mangroves. *Cordaites* was common in swamp, floodplain, and upland environments. Long strap-shaped leaves up to 1 meter in length occurred at its branch tips. It resembled modern mangroves in having stilt roots.

Cones developed between the upper surface of some leaves and the stem of cordaites. Four rows of *bracts* were borne on the cone axis. Above each bract was a dwarf shoot that terminated in either male or female reproductive structures. Swedish botanist Rudolf Florin believed that the woody seed-bearing scale of modern pine could be derived from the dwarf shoot of cordaites through a series of extinct coniferalean intermediates. His interpretation has been adopted in many textbooks. It has also been shown that the conifers did not evolve directly from the cordaites, although both groups undoubtedly shared a common ancestor.

Cycadeoids, Cycads, and Conifers

When the coal swamps dried up, the pteridosperms and the cordaites were replaced by the *cycads*, *cycadeoids*, and *conifers*. The cycads and cycadeoids evolved from the medullosan seed ferns. The cycads and cycadeoids were among the dominant plants during the age of the dinosaurs. The conifers are related to the cordaites.

Some cycadeoids had slender, branching trunks, while others were short and stumpy. Both types had compound leaves. Cycadeoid cones contained both male and female reproductive structures. Earlier researchers thought that the cones of the beehivelike cycadeoids resembled primitive angiosperm flowers, but detailed reinvestigation of the cones showed that this was not true. The cycadeoids became extinct about sixty-five million years ago.

The cycads were more abundant in the past than they are now. Eleven genera and 160 species exist worldwide. They are dispersed in the modern tropics—in Africa, Cuba, Mexico, Australia, India, China, and Japan. *Zamia floridana* (coontie) is the only cycad native to the United States. Some cycads

---

Welwitschia: The Strangest Gymnosperm

There are many unusual plants in Africa, but one of the most unusual is *Welwitschia mirabilis*, a resident of the Namib Desert. It has a short, swollen stem only about 4 inches (10 centimeters) high, which terminates in a disc-like structure. Coming off the top of the stem are two straplike leaves. These two leaves last for the lifetime of the plant and continue to grow very slowly. As they grow, they twist and become shredded, so that an individual plant appears to have many leaves. The reproductive structures rise from the center of the stem, and, instead of flowers, *Welwitschia* has small cones.

Welwitschia is such a successful survivor of the Namib that it easily lives for hundreds of years. Some specimens have been dated to about two thousand years old. Older plants can reach tremendous sizes, with the top of the stem sometimes reaching 5 feet (1.5 meters) in diameter. Specimens of *Welwitschia* are extremely difficult to grow in cultivation, requiring special desert conditions and room for the deep taproot.

Pinaceae *(the pine family) is the largest family of living gymnosperms, containing about two hundred species in ten genera. Unlike a few conifers, such as the bald cypress and dawn redwood, which shed their leaves in the fall, pines are evergreens.*

51 genera. Conifers are most abundant in temperate areas, such as the northern parts of North America, Europe, and Asia, and New Zealand and southern Australia. The conifers are divided into seven families—the *Araucariaceae* (examples are the kauri pine and Norfolk Island pine), *Podocarpaceae* (typified by the yellow woods), *Pinaceae* (pine, spruce, hemlock, and fir), *Cupressaceae* (juniper and arborvitae), *Taxaceae* (yew), *Cephalotaxaceae* (cultivated in the United States as plum-yew or cow's-tail pine, it also has at least one Asian genus), and *Taxodiaceae* (bald cypress, redwoods). Some researchers separate the *Taxaceae* from the conifers on the basis of their arillate ovule (an ovule that forms an additional seed coat to the normal one). Several extinct conifer families are also known.

Ovule and Seed Complexes

The most distinctive features of the gymnosperms are their ovule and seed complexes. In the center of the ovule is the female *gametophyte*. Surrounding the female gametophyte are layers of tissue called the *nucellus* and the *integument*. The nucellus is a nutritive tissue from which the developing female gametophyte draws its nourishment. The integument surrounds the nucellus and is the outermost layer of the ovule. After fertilization, the integument develops into the seed coat.

The gymnosperms have exploited land habitats more successfully than the lower plants and pterophytes because gymnosperms do not require water for pollen transfer. The male gametophyte (pollen grain) is carried to the female gametophyte (ovule) through the air. The ovule exudes a sticky fluid (the pollination drop), which traps the pollen grains. As the sticky fluid dries, the pollen grains are drawn through the *micropyle* into the ovule. The *archegonia* (the organs that contain the eggs) are located directly below the micropyle. When the sperm are released, they fertilize the egg; a fertilized ovule is called a seed.

Modern Gymnosperm Types

Pines are the standard example of gymnosperms. *Pinaceae* (the pine family) is the largest family of living gymnosperms. It contains about two

are small, unbranched trees that grow to 18 meters (59 feet) tall and resemble palm trees. Others have subterranean stems, and only their leaves and cones show above the ground. All the cycads possess stiff, leathery, compound leaves, often with very sharp tips on each leaflet. The male and female reproductive structures are borne at the end of the stem in separate cones on different plants. Cycad cones are the largest cones that have ever been produced. Cycad ovules are also very large. Some reach lengths of 6 centimeters. Their ripe seeds are often brightly colored.

The dominant group of living gymnosperms is the conifers. About 550 species are divided among hundred species in ten genera. Pine trees are typically conical and represent the sporophyte generation. The leaves are needle-shaped and confined to short, lateral shoots. Unlike a few conifers, such as the bald cypress and dawn redwood, which shed their leaves in the fall, pines and most conifers are evergreens.

Sexual reproduction typically takes three years to complete. In the first year, the pollen (male) and seed (female) cones are formed. The cones are borne on different parts of the same plant. The small, upright male cones are borne in clusters on the lower branches of the tree. The mature seed cones are very large and hang down from the upper branches. In the spring, the male cones shed millions of pollen grains into the air. Only a few grains reach an ovule. Once there, the male gametophyte must wait from twelve to fourteen months before the female gametophyte matures and fertilization occurs. The seeds mature in the fall of the third year and are shed.

Included within the gymnosperms are four plants whose affinities are uncertain—ginkgo, or the maidenhair tree (*Ginkgo biloba*), and the gnetophytes (families *Gnetum*, *Ephedra*, and *Welwitschia*). Ginkgo is a large tree covered with fan-shaped leaves that turn golden yellow before being shed in the fall. Although a common ornamental in the United States, ginkgo is native to southeastern China. *Gnetum* is a broad-leaved vine found in the tropical rain forests of South America, western Africa, and southeastern Asia. *Ephedra* is found worldwide in cool, arid regions. Sixteen species of *Ephedra* grow in the western United States. Most species are highly branched shrubs with scalelike leaves. *Welwitschia mirabilis* is the most unusual plant species within the four families. The exposed portion of the stem gives rise to two strap-shaped leaves that are never shed and never stop growing during the life of the plant. *Welwitschia* is found only in coastal desert and inland savanna regions in southwestern Africa.

Gary E. Dolph

See also: Conifers; Cycads and palms; Ginkgos; Gnetophytes; Seedless vascular plants.

Sources for Further Study

Beck, C. B. *Origin and Evolution of Gymnosperms*. New York: Columbia University Press, 1988. This text deals solely with the gymnosperms and is intended for specialists. A number of detailed, theoretical points are covered that could not be included in more general texts.

Elias, T. S. *The Complete Trees of North America: Field Guide and Natural History*. New York: Chapman and Hall, 2000. This reference book discusses the characteristics and uses of all native North American gymnosperms.

Gifford, E. M., and A. S. Foster. *Morphology and Evolution of Vascular Plants*. 3d ed. New York: W. H. Freeman, 1989. This book provides one of the most comprehensive treatments of the living gymnosperms available (chapters 14 to 18). Of particular merit are its numerous labeled drawings and photographs. The pine life cycle is covered in exacting detail (chapter 17).

Kramer, K. U., P. S. Green, and E. Geotz, eds. *Pteridophytes and Gymnosperms*. New York: Springer-Verlag, 1991. The first volume of an encyclopedia of *Families and Genera of Vascular Plants*. Contains a revised classification, including a complete inventory of genera along with their diagnostic features, keys for identification, and references.

Stewart, W. N. *Paleobotany and the Evolution of Plants*. 2d ed. New York: Cambridge University Press, 1993. More than one-third of this book is devoted to a discussion of fossil gymnosperms (chapters 19 to 26). The line drawings are excellent, and they complement the photographs of the actual specimens nicely. Stewart treats the basic evolutionary questions paleobotanists are asking about the gymnosperms in a very readable fashion. His treatment of the origin of seeds (chapter 20) could serve as a model for the cautious and careful blending of paleobotanical knowledge with the inferred paths of evolution.

HALOPHYTES

Categories: Angiosperms; economic botany and plant uses; *Plantae*; soil; water-related life

Halophytes are salt-resistant or salt-tolerant plants that thrive and complete their life cycles in soils or waters containing high salt concentrations. Despite high salt content in the tissues of halophytes, they can be grown and harvested as food or animal fodder.

The Salinity-Plant Relationship

Salts are ionic molecules that typically dissolve in water and split into cations (positively charged ions) and anions (negatively charged ions). Rain dissolves salts from minerals in rocks and soil. These dissolved salts then either enter surface waters or percolate down to an aquifer. In arid regions where there is low rainfall, water evaporates quickly, leaving its salt load behind, leading to the development of saline soils, salt lakes, and brackish groundwater. Even in areas that are only moderately arid, excessive irrigation can cause soil salinity to rise with time, which can lead to loss of agricultural lands, because most crop plants tolerate very little salt in the soil. White, powdery, or crystalline residues coating the ground are indications of especially heavy salt accumulation.

Most plants can tolerate salts at very low levels,

Halophytic (salt-tolerant) trees, such as mangroves, can be harvested for wood and fuel. Mangrove roots serve as critical habitat and nutrient filters for many other species.

but many cannot live in soils or water with high concentrations of salts. The salts in saline areas are often dominated by sodium chloride, forming sodium and chloride ions in water. High concentrations of these ions prevent the growth of *glycophytes*, salinity-intolerant plants. Glycophytes grow only in areas with fresh water containing less than 125 parts per million of soluble salts. Halophytes can be *true halophytes* or *semi-halophytes*; both can thrive in areas with slightly to highly saline water containing soluble salts of 125 to 5,000 parts per million. Only the true halophytes can survive in areas with excessively saline water, containing greater than 5,000 parts per million soluble salts. The concentration of soluble salts is about 850 parts per million in Colorado River water and about 35,000 parts per million in seawater.

Water Potential

Movement of water between soil and plants occurs by osmosis. Water moves from an area of low solute concentration (low salt concentration) to an area of higher solute concentration. A common measure of the tendency of aqueous solutions to "attract" water is called *osmotic potential*. The osmotic potential of pure water is zero, which means that when coming into contact with water containing solutes, pure water will move into it. The osmotic potential of water with solutes is always expressed as a negative number, and the higher the concentration of solutes, the more negative the osmotic potential. Therefore, the osmotic potential in plants and growth media (water and soil) is always more negative than that of pure water. The osmotic potential in plants must always be more negative than the osmotic potential in soils, so that water can move from soil into the cells of plant roots. If large concentrations of soluble salts are present, as in saline or excessively fertilized soils, osmotic potential in the soil becomes more negative than in plants, causing water to move from roots to soils. As a result, glycophytes wilt and die, while halophytes possess adaptations that enable them to survive.

Growth and Survival Mechanisms

Many halophytes are considered *facultative halophytes*, indicating that they can grow in saline and nonsaline habitats. Others are *obligate halophytes* which can survive only in a saline environment. Some *Salicornia* species (pickleweed, for example) cannot survive under freshwater conditions.

Although some halophytes can grow normally in nonsaline soils, they are less abundant than glycophytes, suggesting that they may compete poorly with glycophytes in nonsaline environments.

Some halophytes require fresh water for germination, whereas others readily germinate in saline environments. Some germinate only during the rainy season, to take advantage of the relatively low salt concentration. Seed germination in the goosefoot family (*Chenopodiaceae*) is unique in that the hypocotyl (the stem between the cotyledons and the radicle), rather than the radicle (or embryonic root), emerges first. After the lower tip of the hypocotyl touches the soil, the radicle begins to emerge. The roots are ready to tolerate the salts, storing the absorbed salts in the hypocotyl tissues. In saltbush (*Atriplex*), germination is regulated by storing salts in the seed coat, and the developing embryo is desalinized in the seed.

In mangroves (*Avicennia* and *Rhizophora* species), seeds mature without undergoing dormancy and germinate while the fruit is still attached to the mother plant, which provides water and nutrients during germination. A seedling on the mother tree produces a long hypocotyl. This heavy hypocotyl causes the mature seedlings to fall on the mud, where they continue to mature.

Resistance of halophytes to salt stress involves two different adaptations: salt tolerance, which involves accumulating salts in the plant's cells, and salt avoidance, which involves adaptations to minimize the concentrations of salt in the cells or adaptations to bar salts from entering through plant roots. These two adaptations have led to the designations "salt accumulators" and "salt excluders," respectively.

Salt Accumulators

Salt accumulators absorb salts throughout the growing season, resulting in an increase in salt concentration in the cells and thus maintaining a water potential that is more negative than that of the soil. The difference in osmotic potential between plant cells and soil water forces the water to enter the cells by passing through the cell membrane via osmosis. Water in leaf cells escapes as vapor through pores, or stomata, a process called evapotranspiration. Evapotranspiration also helps move water from the roots up the stem to the leaves.

Salt accumulators, such as saltbush (*Atriplex*), smooth cordgrass (*Spartina alterniflora*), saltgrass

(*Distichlis spicata*), and tamarisk (*Tamarix petandra*), have specialized cells called salt glands located on the surfaces of their leaves, used for storing excess sodium chloride. As the glands fill with salt they eventually burst, releasing salts that form a crystalline coating on the leaves. The crystals fall or are dissolved by rain, which returns the salt to the soil. Other halophytes can accumulate and concentrate the salts in their cells up to a certain toxic concentration, at which point the salts cause the plant to die.

Some salt accumulators avoid salt stress by minimizing salt concentration in the cytosol of their leaf cells. Leaf cells regulate cytosolic salt levels by transporting sodium and chloride ions into the central vacuole. A high salt concentration in the vacuole causes it to take up more water and swell. As the water-filled vacuole pushes the cytosol toward the cell membrane and cell wall, the cell maintains its turgidity, typical of *succulent* halophytes, such as pickleweed (*Salicornia virginica*).

Salt Excluders

Some *salt excluders* avoid salt stress by defoliation or abscission (leaf release). When cytosolic salt concentrations approach toxic levels, excess sodium chloride accumulates in petioles, stems that connect leaves to the main stem). The petiole, including the leaf, dies and then detaches from the stem, dropping to the ground. The removal of salt-concentrated parts desalinates the plant, thus preventing buildup of sodium and chloride ions to toxic levels.

Other salt excluders avoid salt toxicity at the roots. The root epidermis (outer layer of cells) of some halophytes may not allow the passage of sodium and chloride ions through the cell membrane. Roots also have an endodermis (inner layer of cells) that contains waxy strips surrounding each cell, to obstruct the entry of sodium and chloride ions. Although this mechanism is less common, selective permeability of the cell membrane (the transport of other ions rather than sodium and chloride ions)

into the roots of halophytes is possible. In some plants, root cells are capable of actively pumping excess sodium and chloride ions out into the surrounding soil.

Halophytes as Crops

In arid regions, farmers often use groundwater to irrigate their crops. Even when salt concentrations in the groundwater are relatively low, salt can gradually accumulate in the soil as a result of high evaporation rates in these areas. Eventually, such soils become so salinized that most traditional crop plants cannot survive. Growing halophytes as crops could be a way to use salinized soils. Also, as the human population increases and fresh water becomes an expensive commodity, irrigating halophytes with salt water could be a way of conserving fresh water.

Some halophytes growing in deserts, estuaries, and seashores are potential sources of food, fuel, and forage. The seeds of eelgrass (*Zostera marina*) and saltgrass could be ground to make flour. Seeds of *Salicornia* species are a potential source of vegetable oil. Halophytic trees, such as mangrove, could be harvested for wood and fuel. Eelgrass and saltbush could be used as forage for livestock. Common reed (*Phragmites australis*) is a marsh plant used since ancient times for roofing and making baskets. Coconut palms (*Cocos nucifera*), found in occasionally inundated shorelines, yield oil from the nuts. Traditional uses of the coconut palm include weaving leaflets into baskets and mats, converting husks into charcoal, and using fibers to make brushes and mats. Other halophytes could be sources of waxes, gums, and pharmaceuticals.

Domingo M. Jariel

See also: Active transport; Adaptations; Aquatic plants; Irrigation; Osmosis, free diffusion and facilitated diffusion; Plants with potential; Root uptake systems; Soil salinization; Water and solute movement in plants; Wetlands.

Sources for Further Study
Choukr-Allah, Redouane, Atef Hamdy, and Clive V. Malcolm. *Halophytes and Saline Agriculture*. New York: Marcel Dekker, 1995. Focuses on halophytes to be grown in salt-affected lands and introduces the application of salt water for irrigating halophytes.

National Research Center. *Saline Agriculture: Salt-Tolerant Plants for Developing Countries*. Washington, D.C.: National Academy Press, 1990. Discusses uses of halophytes and other salt-tolerant plants.

Squires, Victor R., and Ali T. Ayoub. *Halophytes as a Resource for Livestock and for Rehabilitation* of Degraded Lands: Proceedings of the International Workshop on Halophytes for Reclamation of Saline Wetlands and as a Resource for Livestock, Problems and Prospects. New York: Kluwer Academic, 1994.

Ungar, Irwin A. *Ecophysiology of Vascular Halophytes*. Boca Raton, Fla.: CRC Press, 1991. Discusses seed germination, growth, and photosynthesis of halophytes, including water status and ionic contents in plants grown in saline environments.

HAPTOPHYTES

Categories: Algae; *Protista*; taxonomic groups; water-related life

The algal phylum Prymnesiophyta, *or* Haptophyta, *is a monophyletic taxon that contains two hundred to three hundred extant species that are 4-40 microns in size.*

The phylum *Haptophyta* is divided into two subclasses, the *Prymnesiophycidae* and *Pavlovophycidae*, that differ significantly from one another (see below). The phylum is usually classified within the kingdom *Chromista* with other algae containing chlorophyll *a* and *c* (excluding dinoflagellates), but their exact relationship to the heterokont algae remains unclear.

Although a few freshwater species are known, most are marine, and species diversity within the phylum is greatest in nutrient-poor waters of the tropical and subtropical open oceans. Most are motile biflagellate single cells or nonmotile coccoid (walled unicells) cells. Colonial, filamentous, and palmelloid forms are also known.

Cell Structure

Prymnesiophyte cells are typically eukaryotic. The nucleus is positioned centrally or nearer the posterior end of the cell and lies between the chloroplasts. Most cells probably possess a single highly branched mitochondrion, although it is possible that multiple mitochondria may be present in some species. An exceptionally large Golgi body (dictyosome) with many cisternae is located anteriorly, just beneath the flagellar basal region.

The motile cells of most prymnesiophytes bear two equal or slightly subequal heterodynamic flagella that lack mastigonemes (hairlike appendages) and arise from a shallow apical depression. However, the longer flagellum in members of the *Pavlovophycidae* is adorned with fibrous hairs and knobscales. Further differences between the *Prym-*

nesiophycidae and *Pavlovophycidae* are evident in the flagellar and mitotic apparatuses. In the *Prymnesiophycidae* the flagellar root system originating near the two basal bodies consists of three to four microtubular roots, the nuclear envelope breaks down prior to division, the mitotic spindle axis is straight, and a fibrous microtubular organizing center (MTOC) is absent. In contrast, the *Pavlovophycidae* flagellar root system includes only two roots, the nuclear envelope remains intact during division, the mitotic spindle is V-shaped, and a fibrous MTOC is present.

A *haptonema*—a thicadlike structure that, along with two flagella, extends from the cell wall—is found in many, but not all, species and gives this phylum its name. However, the haptonema is unique to the *Prymnesiophyta*. In many coccolithophorids the haptonema is reduced (vestigial) or absent. The haptonema, which may have evolved by duplication of a portion of the flagellar apparatus, arises between the flagellar basal bodies and at its base is composed of three to eight singlet microtubules. The microtubules are arranged in an arclike or crescentlike fashion and are surrounded by a layer of the endoplasmic reticulum. The movement of the haptonema (which does not beat) and flagella are highly coordinated. When present, the haptonema may be long or short and may or may not be capable of coiling. Coiling is induced by a rapid uptake of calcium from the environment. The haptonema can be used for attachment (in fact, the Greek word *haptein* means "to fasten") and, at least in some species, is used to capture prey (such as bacteria and smaller eukaryotes). In phagotrophic species a single membranebound food vacuole is posteriorly located, and the haptonema delivers captured prey to this organelle. The haptonema is also involved in tactile responses; for example, upon contact with an obstacle it may stimulate a change in the direction of swimming.

Chloroplasts and Nutrition

Most prymnesiophytes possess two chloroplasts that are surrounded by four membranes; in most species investigated, the outer membrane is studded with ribosomes and is confluent with the nuclear envelope. A girdle lamella, a thylakoid encircling the periphery of the plastid that is found in some other algae containing chlorophyll *a* and *c*, is absent. In members of the *Prymnesiophycidae*, eyespots are absent. Eyespots are present beneath an invagination of the plasmalemma in members of the *Pavlovophycidae* but usually are not associated with a flagellar swelling as in many other algal taxa.

The photosynthetic pigments of prymnesiophytes are diverse. All contain chlorophylls *a*, c_1, c_2, beta carotene, diadinoxanthin, and diatoxanthin, but chlorophyll c_3, fucoxanthin, 19'-hexanoyloxyfucoxanthin and 19'-butanoyloxyfucoxanthin are present or absent in different combinations in other species.

Pyrenoids are present and may be immersed (embedded within the chloroplast) or bulging, in which case they protrude from the periphery of the chloroplast and into the surrounding cytosol. Pyrenoids may be traversed by one or a few thylakoid membranes.

Most prymnesiophytes are photoautotrophs and, in addition to using photosynthesis for nutrition, probably are also capable of directly obtaining inorganic or organic nutrients dissolved in the surrounding water. A number of species, particularly those possessing a relatively long and flexible haptonema, are phagotrophic and ingest bacteria and smaller eukaryotes. Thus, mixotrophic species (those that combine photoautotrophic and heterotrophic means of obtaining food) are common among prymnesiophytes.

Cell Covering

Members of the *Pavolophycidae* lack body scales. However, cells of species placed in the *Prymnesiophycidae* are covered by organic base plate scales, mineralized calcium carbonate scales, or a combina-

tion of both that are external to the plasmalemma. Nearly 70 percent of all known species of prymnesiophytes are known as coccolithophorids. The external covering of these cells is composed of mineralized calcium carbonate (calcite) scales termed *coccoliths*. Individual coccoliths are intricately arranged around the cell to form a coccosphere. Both organic scales and coccoliths are produced within the Golgi complex and are released onto the cell surface near the point of flagellar insertion. An organic base plate scale serves as a matrix for coccolith calcification that may occur inside or outside the cell; thus, organic scales and mineralized scales are homologous structures.

Organic base plate scales are microfibrillar and composed of proteins, celluloselike and pectinlike substances. The microfibrils on the proximal side of the scale are arranged radially, whereas those on the distal surface are arranged spirally. In species bearing only organic scales (such as *Chrysochromulina*, *Phaeocystis*, and *Prymnesium*), one or more layers of organic scales may be present.

In coccolithophores the coccoliths are external to the organic scales (when present). *Pleurochrysis* is an example of a coccolithophorid that possesses organic scales as well as coccoliths, whereas the widely distributed coccolithophorids *Emiliana* and *Gephyrocapsa* bear mineralized scales only. There is tremendous diversity in coccolith morphology, ranging from those that are platelike to highly ornamented forms with rims and spines. The taxonomy of species within the group is based primarily upon structural differences among coccoliths. However, some species may bear coccospheres composed of morphologically different coccoliths, and transitions between coccolith types are now known to be associated with different life history phases. An accurate account of the biodiversity of prymnesiophytes species remains uncertain because different life history phases bearing different scales are often considered separate species.

The function of coccoliths is not known with certainty. It is likely that they serve multiple functional roles in some species, whereas in other species more specific functions may be attributed to morphologically different coccoliths. Coccoliths are effective at deterring only smaller grazers (such as protozoans), and coccolithophorids are readily eaten by other organisms. It has been suggested that coccoliths protect the delicate plasmalemma from osmotic, chemical, and physical disruption or invasive bacteria and viruses. Coccoliths, which may be shed or produced when needed, may also play a role in buoyancy control. The long spines on the coccoliths of *Rhabdosphaera* and other coccolithophores also reduce sinking rates. Calcification and photosynthesis in most coccolithophorids appears to be physiologically linked. It is possible that the carbon dioxide released during calcification may be used in the dark reactions of photosynthesis and that coccoliths increase the amount of surface area available for light capture.

There are more than one thousand different types of fossilized coccoliths, and these are among the most commonly used microfossils for stratigraphic analyses in the petroleum industry. In addition, because some coccolithophorids are restricted to water masses defined by a particular temperature range, fossil coccoliths are frequently used as paleoclimatic indicators. Because their calcium carbonate scales are birefringent, satellite imagery can be used to deduce the relative abundance and position of prymnesiophyte blooms.

Life History

Most prymnesiophytes are biflagellate motile single cells that reproduce asexually via binary fission. Life history stages may include transitions between nonmotile and motile forms and also the production of different scales (organic versus mineralized) and the production of mineralized scales of different morphologies. Thus, during their life histories flagellate cells may be morphologically transformed into amoeboid, coccoid, colonial, palmelloid (walled cells embedded in a mucilaginous envelope), or filamentous life forms. For example, *Pleurochrysis* possesses a nonmotile benthic colonial or filamentous haploid stage that alternates with a diploid motile coccolith bearing stage. Some life histories include alternations between two motile cell stages that bear completely different types of coccoliths. In *Hymenomonas* and *Ochrosphaera*, diploid coccolith-bearing cells alternate with cells possessing only organic scales.

Phylogeny and Fossil Record

Although they may have originated earlier, the coccoliths of prymnesiophytes first appear in the fossil of the Late Triassic, approximately 220 million years ago (mya). The abundance of coccolith fossils reached its peak during the Late Cretaceous (95-63 mya). Fossil records indicate that perhaps 80 percent of all coccolithophorids went extinct during the Cretaceous-Tertiary (K-T) event at the end of the Cretaceous period. Today's prymnesiophytes represent a radiation of the minority of species that survived the K-T extinction event, which also brought about the demise of dinosaurs.

According to most scholars, the most primitive prymnesiophytes are those lacking body scales and possessing a haptonema. Flagellates bearing organic scales are believed to have diverged next. The absence of a haptonema and the presence of coccoliths are considered derived features for the group, and these characteristics are found in most coccolithophorids.

Environmental Importance

Prymnesiophytes, particularly coccolithophorids, play important roles in coastal and open ocean environments. For example, they are integral contributors in global carbon and sulfur cycles.

Coccoliths that are shed, derived from dead cells, are ingested and expelled are transported to the sea floor. In some areas the accumulation of coccoliths has led to the formation of enormous deposits of chalk or limestone, a notable example being the White Cliffs of Dover in England. By this sedimentary process, calcium and carbon are cycled from the oceans back into the lithosphere. The ocean is the largest long-term sink of inorganic carbon on the planet, and carbonate deposits cover one-half of the sea floor, an area equal to one-third the surface of the earth. Coccoliths account for approximately 25 percent of the total yearly vertical transport of carbon to the deep sea. Blooms of coccolithophorids (such as *Emiliania*, *Pheocystis*) release dimethylsulfide that becomes aerosolized and subsequently acts as a nucleating agent for water droplets in the atmosphere, ultimately producing acid rain. Because coccoliths reflect light, large blooms may also have a cooling effect on the local climate.

Other species of prymnesiophytes, including some species of *Chrysochromulina*, *Prymnesium*, and *Phaeocystis*, are known to form blooms that are toxic to other marine organisms or that interfere with marine fisheries. On the other hand, *Pavolova* and *Isochrysis* are widely used as food in the aquaculture industry.

J. Craig Bailey

See also: Algae; Heterokonts; *Protista*.

Sources for Further Study

Graham, Linda E., and Lee W. Wilcox. "Haptophytes." In *Algae*. Upper Saddle River, N.J.: Prentice Hall, 2000. An introduction to the biology of the prymnesiophytes.

Green, J. C., and B. S. C. Leadbeater, eds. *The Haptophyte Algae*. Oxford, England: Clarendon Press, 1994. An overview of the biology of prymnesiophytes with excellent transmission electron micrographs of key subcellular features.

Winter, A., and W. G. Siesser, eds. *Coccolithophores*. New York: Cambridge University Press, 1994. A collection of articles on the biology of prymnesiophytes. Includes a beautiful collection of scanning electron micrographs of many coccolithophorids.

HARDY-WEINBERG THEOREM

Categories: Evolution; genetics

The Hardy-Weinberg theorem is the principal that, in the absence of external pressures for change, the genetic makeup of an ideal population of randomly mating, sexually reproducing diploid organisms will remain the same, at what is called Hardy-Weinberg equilibrium.

Population genetics is the branch of genetics that studies the behavior of genes in populations. The two main subfields of population genetics are theoretical (or mathematical) population genetics, which uses formal analysis of the properties of ideal populations, and experimental population genetics, which examines the behavior of real genes in natural or laboratory populations.

Population genetics began as an attempt to extend Gregor Mendel's laws of inheritance to populations. In 1908 Godfrey H. Hardy, an English mathematician, and Wilhelm Weinberg, a German physician, each independently derived a description of the behavior of allele and genotype frequencies in an ideal population of randomly mating, sexually reproducing diploid organisms. Their results, now termed the Hardy-Weinberg theorem, showed that the pattern of allele and genotype frequencies in such a population followed simple rules. They also showed that, in the absence of external pressures for change, the genetic makeup of a population will remain at an equilibrium.

Because evolution is change in a population over time, such a population is not evolving. Modern evolutionary theory is an outgrowth of the "New Synthesis" of R. A. Fisher, J. B. S. Haldane, and Sewell Wright, which was developed in the 1930's. They examined the significance of various factors that cause evolution by examining the degree to which they cause deviations from the predictions of the Hardy-Weinberg theorem.

Predictions

The predictions of the Hardy-Weinberg theorem hold if the following assumptions are true:

(1) The population is infinitely large.

(2) There is no gene flow (movement of genes into or out of the population by migration of gametes or individuals).

(3) There is no mutation (no new alleles are added to the population by mutation).

(4) There is random mating (all genotypes have an equal chance of mating with all other genotypes).

(5) All genotypes are equally fit (have an equal chance of surviving to reproduce).

Under this very restricted set of assumptions, the following two predictions are true:

(1) Allele frequencies will not change from one generation to the next.

(2) Genotype frequencies can be determined by a simple equation and will not change from one generation to the next.

The predictions of the Hardy-Weinberg theorem represent the working through of a simple set of algebraic equations and can be easily extended to more than two alleles of a gene. In fact, the results were so self-evident to the mathematician Hardy that he, at first, did not think the work was worth publishing.

If there are two alleles (A, a) for a gene present in the gene pool (all of the genes in all of the individuals of a population), let p = the frequency of the A allele and q = the frequency of the a allele. As an example, if $p = 0.4$ (40 percent) and $q = 0.6$ (60 percent), then $p + q = 1$, since the two alleles are the only ones present, and the sum of the frequencies (or proportions) of all the alleles in a gene pool must equal one (or 100 percent). The Hardy-Weinberg theorem states that at equilibrium the frequency of AA individuals will be p^2 (equal to 0.16 in this example), the frequency of Aa individuals will be $2pq$, or 0.48, and the frequency of aa individuals will be q^2, or 0.36.

The basis of this equilibrium is that the individuals of one generation give rise to the next generation. Each *diploid* individual produces *haploid* gametes. An individual of genotype AA can make only a single type of gamete, carrying the A allele. Similarly, an individual of genotype aa can make only a gametes. An Aa individual, however, can make two types of gametes, A and a, with equal probability. Each individual makes an equal contribution of gametes, as all individuals are equally fit, and there is random mating. Each AA individual will contribute twice as many A gametes as each Aa individual. The frequency of A gametes is equal to the frequency of A alleles in the gene pool of the parents.

The next generation is formed by gametes pairing at random (independent of the allele they carry). The likelihood of an egg joining with a sperm is the frequency of one multiplied by the frequency of the other. AA individuals are formed when an A sperm joins an A egg; the likelihood of this occurrence is $p \times p = p^2$ (that is, $0.4 \times 0.4 = 0.16$ in the first example). In the same fashion, the likelihood of forming an aa individual is $q^2 = 0.36$. The likelihood of an A egg joining an a sperm is pq, as is the likelihood of an a egg joining an A sperm; therefore, the total likelihood of forming an Aa individual is $2pq = 0.48$. If one now calculates the allele frequencies (and hence the frequencies of the gamete types) for this generation, they are the same as before: The frequency of the A allele is $p = (2p^2 + 2pq)/2$ (in the ex-

ample $(0.32 + 0.48) \div 2 = 0.4$), and the frequency of the a allele is $q = (1 - p) = 0.6$. The population remains at equilibrium, and neither allele nor genotype frequencies change from one generation to the next.

The Real World

The Hardy-Weinberg theorem is a mathematical model of the behavior of ideal organisms in an ideal world. The real world, however, does not approximate these conditions very well. It is important to examine each of the five assumptions made in the model to understand their consequences and how closely they approximate the real world.

The first assumption is infinitely large population size, which can never be true in the real world, as all real populations are finite. In a small population, chance effects on mating success over many generations can alter allele frequencies. This effect is called *genetic drift*. If the number of breeding adults is small enough, some genotypes will not get a chance to mate with one another, even if mate choice does not depend on genotype. As a result, the genotype ratios of the offspring would be different from those of the parents. In this case, however, the gene pool of the next generation is determined by those genotypes, and the change in allele frequencies is perpetuated. If it goes on long enough, it is likely that some alleles will be lost from the population, because a rare allele has a greater chance of not being included. Once an allele is lost, it cannot be regained.

How long this process takes is a function of population size. In general, the number of generations it would take to lose an allele by drift is about equal to the number of individuals in the population. Many natural populations are quite large (thousands of individuals), so that the effects of drift are not significant. Some populations, however, especially of endangered species, are very small: A number of plant species are so rare that they consist of a single population with less than one hundred individuals.

The second assumption is that there is no gene flow, or movement of genotypes into or out of the population. Individuals that leave a population do not contribute to the next generation. If one genotype leaves more frequently than another, the allele frequencies will not equal those of the previous generation. If incoming individuals come from a population with different allele frequencies, they also alter the allele frequencies of the gene pool.

The third assumption concerns mutations. A mutation is a change in the deoxyribonucleic acid (DNA) sequence of a gene—that is, the creation of a new allele. This process occurs in all natural populations, but new mutations for a particular gene occur in about 1 of 10,000 to 100,000 individuals per generation. Therefore, mutations do not, in themselves, play much part in determining allele or genotype frequencies. Mutation, however, is the ultimate source of all new alleles and provides the variability on which evolution depends.

The fourth assumption is that there is *random mating* among all individuals. A common limitation on random mating in plants is *inbreeding*, the tendency to mate with a relative. Because plants have a limited ability to move, and pollinators may not carry pollen very far, the plants in a population tend to mate with nearby individuals, which are often relatives. Such individuals tend to share alleles more often than the population at large.

The final assumption is that all genotypes are equally *fit*. Considerable debate has focused on the question of whether two alleles or genotypes are ever equally fit. Many alleles do confer differences in fitness; it is through these variations in fitness that natural selection operates. Newer techniques of molecular biology have revealed many differences in DNA sequences that appear to have no discernible effects on fitness.

As the cornerstone of population genetics, the Hardy-Weinberg theorem pervades evolutionary thinking. The advent of techniques to examine genetic variation in natural populations has been responsible for a great resurgence of interest in evolutionary questions. One can now test directly many of the central aspects of evolutionary theory. In some cases, notably the discovery of the large amount of genetic variation in most natural populations, evolutionary biologists have been forced to reassess the significance of natural selection compared with other forces for evolutionary change.

Richard Beckwitt

See also: Population genetics; Species and speciation.

Sources for Further Study

Audesirk, Gerald, and Teresa Audesirk. *Biology: Life on Earth*. 6th ed. Upper Saddle River, N.J.: Prentice Hall, 2001. An introductory college textbook designed for nonscience majors. Chapter 14, "The Processes and Results of Evolution," includes a complete explanation of basic population genetics, presented in a nontechnical way. The chapter is well illustrated and includes a glossary and suggestions for further reading.

Avers, Charlotte J. *Process and Pattern in Evolution*. New York: Oxford University Press, 1989. A text that introduces modern evolutionary theory to students who already have a background in genetics and organic chemistry. Chapter 7, "Gene Frequencies in Populations," covers basic population genetics. The book introduces most of the techniques used in the study of evolution and includes references to original research as well as other suggested readings.

Futuyma, Douglas J. *Evolutionary Biology*. 3d ed. Sunderland, Mass.: Sinauer Associates, 1998. Advanced text in evolution for students with previous exposure to calculus and a strong biology background, including genetics and various courses in physiology and ecology. The great strength of this book is in the presentation of areas of current research and argument in evolution, rather than a cut-and-dried array of "facts." There are numerous references to original research and a glossary.

Hartl, Daniel L. *A Primer of Population Genetics*. 3d ed. Sunderland, Mass.: Sinauer Associates, 2000. Intended for students with a college-level knowledge of biology; does not require prior exposure to genetics, statistics, or higher mathematics. Includes examples of the significance of population genetics ideas in many areas of biology and medicine. Each chapter has problem sets with answers. There are numerous references to original research.

Nagylaki, Thomas. *Introduction to Theoretical Population Genetics*. New York: Springer-Verlag, 1992. Thorough coverage of mathematical aspects of theoretical population genetics, organized to be covered in a one-year college course.

Starr, Cecie, and Ralph Taggart. *Biology: The Unity and Diversity of Life.* 9th ed. Pacific Grove, Calif.: Brooks/Cole, 2001. A textbook for an introductory college biology course. Chapter 36, "Population Genetics, Natural Selection, and Speciation," covers population genetics and mechanisms of evolution. The book is well provided with examples and many striking photographs.

Svirezhev, Yuri M., and Vladimir P. Passekov. *Fundamentals of Mathematical Evolutionary Genetics.* Translated by Alexey A. Voinov and Dmitrii O. Logofet. Boston: Kluwer Academic, 1990. A very clear exposition of the mathematical models of population genetics, placing the material in historical context and explaining mathematical concepts with lucidity and little fuss.

HELIOTROPISM

Categories: Movement; physiology

Heliotropism is a growth movement in plants that is induced by sunlight. It is sometimes called solar tracking, a directional response to the sun. Because plants react in a similar way toward artificial sources of light, heliotropism is sometimes termed phototropism, a growth movement induced by any light stimulus.

Plants that orient their leaves to receive maximum sunlight are called diaheliotropic. *Diaheliotropism* is the tendency of leaves or other organs of a plant to track the sun by turning their surfaces toward it. Tracking the sun maximizes the amount of direct solar radiation received. Diaheliotropic movements can increase radiation interception, enhance photosynthesis, and increase growth rates of plants.

Plants that move their leaves to avoid sunlight are called paraheliotropic. *Paraheliotropism* is a plant response to minimize surface exposure to the sun. By orienting leaves and other plant organs parallel to the sun's rays, light absorption is minimized. It is a process that some plants use to reduce dehydration by reducing leaf temperatures and water loss during times of drought.

Solar Tracking

Alfalfa, cotton, soybean, bean, and some wild species of the mallow family *Malvaceae* are plant species that exhibit *solar tracking*. Heliotropic plants change the orientation of their leaves toward the sun. This solar tracking mechanism can occur as a continuous adjustment throughout the day so that the leaf blades are always oriented perpendicular to the sun's rays. The leaves are in a nearly vertical position, facing the eastern horizon as the sun rises.

AP/Wide World Photos

Heliotropic plants, such as these sunflowers, are continuously moving their leaves, leaflets, and pinnules to readjust to prevailing light conditions. Movements occur every fifteen to sixty seconds, which is just slow enough to be imperceptible to most humans.

During the morning and later afternoon, when solar radiation is not at its most intense, the leaves move to a horizontal orientation. When there is increased solar radiation near midday, the leaves move to become more vertical so that they are not damaged by overheating. At sunset, the leaves are nearly vertical, facing the west when the sun sets. During the night, the leaves assume a horizontal position and reorient just before dawn, to start the solar tracking cycle over again. Leaves only track the sun on clear days. They stop when clouds block the sun.

Mechanisms

In many cases, the leaves of solar tracking plants are controlled by a specialized organ called the *pulvinus*. This organ is a swollen part of the petiole that may occur where it joins the stem, the leaf blade, or both. It contains motor cells that generate mechanical forces that control the orientation of the petiole and thus the leaf blade. The forces are produced by changes in the turgor in the pulvinus. The cells of this organ have highly elastic cell walls that allow them readily to change size and shape. The cells of the upper pulvinus have the capability of increasing their turgidity with water uptake, while the lower pulvinus can lose water very easily. The net effect is a force that moves the petiole.

Another mechanism producing heliotropism is produced by small mechanical changes along the length of the petiole and by movements of the younger parts of the stem. Heliotropic plants are continuously moving their leaves, leaflets, and pinnules to readjust to prevailing light conditions. Movements occur rather rapidly, every fifteen to sixty seconds, which is just slow enough to be imperceptible to most humans.

Alvin K. Benson

See also: Growth habits; Leaf arrangements; Leaf shapes; Nastic movements; Photoperiodism; Photosynthesis; Thigmomorphogenesis.

Sources for Further Study

Hart, James Watnell. *Plant Tropisms and Other Growth Movements*. Boston: Unwin Hyman, 1990. This text focuses specifically on the adaptations of plants to effect movement in response to various environmental stimuli. Bibliography, indexes, illustrations.

Haupt, W., and M. E. Feinleib, eds. *The Physiology of Movements*. New York: Springer-Verlag, 1979. This college-level textbook provides much detailed information about tropisms. Includes an index, illustrations, and extensive bibliography.

Nobel, Park S. *Physicochemical and Environmental Plant Physiology*. 2d ed. San Diego: Academic Press, 1999. Discusses plant chemistry, physiology, cells and tissues, and tropisms. Includes bibliographical references, appendices, and index.

Satter, R. L., and A. W. Galston. "Mechanisms of Control of Leaf Movements." *Annual Review of Plant Physiology* 32 (1981): 83-103. An excellent description of how leaves move. Summarizes much research and includes an extensive bibliography. Includes few illustrations.

HERBICIDES

Categories: Agriculture; pests and pest control

Herbicides are a class of pesticide used to kill or otherwise control weeds and other unwanted vegetation.

Herbicides are used to control grasses, weeds, and other plant pests. These chemical compounds kill plants or inhibit their normal growth. In general, herbicides work by interfering with photosynthesis, so that a plant dies from lack of energy or by a combination of *defoliation* (leaf removal) and *systemic* herbicidal action.

Herbicides are used to clear rights-of-way beneath power lines and along railways and roads. In agriculture and forest management, they are used

A plane sprays chemical herbicides on a cotton field near Leland, Mississippi. In agriculture, herbicides are used to control weeds or to remove the leaves from some crop plants to facilitate harvesting.

to control weeds or to remove the leaves from some crop plants to facilitate harvesting. While herbicides may be employed in lieu of tillage, their use is more often in conjunction with tillage and other agronomic practices. During wartime, defoliants and other herbicides have been used to destroy plants that an enemy uses for cover during battle or for food.

Types of Herbicides

Herbicides may be *selective* or *nonselective*. Selective herbicides, such as amitrole, atrazine, monuron, pyridine, 2,4-dichlorophenoxyacetic acid (2,4-D), and 2,4,5-trichlorophenoxyacetic acid (2,4,5-T), target a particular plant and will kill or stunt weeds among crop plants without injuring the crop. For example, 2,4-D targets soft-stemmed plants, while 2,4,5-T is effective against woody plants. Cereals (grains) are crops particularly suited for treatment with 2,4-D, because the compound does not harm narrow-leafed plants but kills broad-leaved weeds. Selective toxicity minimizes the environmental impact of an herbicide.

Nonselective herbicides (also called broad-spectrum or general-usage herbicides) are toxic to all plants. Examples include dinoseb, diquat, paraquat, and arsenic trioxide. Nonselective compounds are best suited for areas where all plant growth is to be suppressed, such as along railroad rights-of-way.

Some compounds, known as *contact herbicides*, kill only those plant parts to which they are directly applied. Others, called systemic herbicides, are absorbed through the plant's foliage or roots and carried to other parts of the plant. When mixed with the soil, some herbicides kill germinating seeds and small seedlings.

Popular *inorganic* herbicides include ammonium sulfate, sodium chlorate, sulfuric acid solutions, and borate formulations. Among the *organic* herbicides are the organic arsenicals, substituted amides and ureas, nitrogen heterocyclic acids, and phenol derivatives. Phenoxyaliphatic acids and their derivatives, a major group of organic herbicides, are selective poisons that readily travel from one part of a plant to another.

History

Agricultural societies have used simple chemical herbicides, such as ashes and common salts, for centuries. In 1896 a fungicidal compound known as Bordeaux mixture (a combination of copper sulfate, lime, and water) was found to be effective against some weeds. Subsequently, copper sulfate was employed as a selective weed killer in cereal crops. By the early 1900's sodium arsenate solutions and other selective inorganic herbicidal mixtures had been developed.

In the early 1940's a new generation of herbicidal compounds emerged. In an attempt to mimic natural plant hormones, the defoliant 2,4-D was created. At low concentrations, 2,4-D promotes retention of fruit and leaves; at higher concentrations, it overstimulates plant metabolism, causing leaves to drop off. A related chemical, 2,4,5-T, came into general use in 1948. The years after World War II saw the first large-scale application of herbicides in agriculture and other areas. The new defoliants rapidly gained acceptance because of their effectiveness against broad-leaved weeds in corn, sorghum, small grains, and grass pastures.

A few years after their development, these defoliants were employed as chemical weapons. During its conflict with communist guerrillas in Malaya during the late 1940's and early 1950's, Britain sprayed 2,4,5-T on crops and jungle foliage to deprive the guerrillas of food and cover. The United States conducted a similar antifood and antifoliage campaign in South Vietnam during the 1960's. In this campaign, dubbed Operation Ranch Hand, massive quantities of herbicidal mixtures were sprayed from aircraft onto Vietcong food plantations, infiltration routes, staging areas, and bases. The quantity and frequency of the spraying greatly exceeded recommended levels; in addition, mechanical problems or military need often forced aircraft to dump their herbicide loads all at once, drenching the jungle below. Soldiers, civilians, and the environment were subjected to unusually high concentrations of defoliants.

One of the herbicides used in this campaign was Agent Orange, a mixture that included 2,4-D and 2,4,5-T. Commercial preparations of 2,4,5-T contain varying amounts of *dioxin*, a highly toxic contaminant. Agent Orange has been implicated in the increased incidence of stillbirths and birth defects among the Vietnamese living in the areas sprayed, in illnesses suffered by American and Australian soldiers who were involved in the operation, and in birth defects among the children of these veterans. In 1970 the United States placed severe restrictions on domestic and agricultural use of 2,4,5-T, at about the same time the defoliation campaign was halted.

U.S. Regulation of Herbicides

In 1947 the Federal Insecticide, Fungicide, and Rodenticide Act (FIFRA) authorized the U.S. Department of Agriculture (USDA) to oversee registration of herbicides and other pesticides and to determine their safety and effectiveness. In December, 1970, the newly formed Environmental Protection Agency (EPA) assumed statutory authority from the USDA over pesticide regulations. Under the Federal Environmental Pesticide Control Act of 1972, an amendment to FIFRA, manufacturers must register all marketed pesticides with the EPA before the product is released. Before registration, the chemicals must undergo exhaustive trials to assess their potential impact on the environment and human health. An EPA decision to grant registration is based on the determination that unreasonable adverse effects on human health or the environment are not anticipated within the constraints of approved usage. Since October, 1977, the EPA has classified all pesticides to which it has granted registration as either restricted-usage (to be applied only by certified pest control operators) or unclassified (general-usage) pesticides.

Karen N. Kähler

See also: Agriculture: modern problems; Biopesticides; Genetically modified foods; Pesticides.

Sources for Further Study

Carson, Rachel. *Silent Spring*. 1962. Reprint. Boston: Houghton Mifflin, 1994. A hard look at the effects of insecticides and pesticides on songbird populations throughout the United States, whose declining numbers yielded the silence to which the title attests.

Cremlyn, R. J. *Agrochemicals*. New York: Wiley, 1991. Discusses the growth in sophistication and application of chemical pesticides or agrochemicals since World War II. Physiochemical factors and biochemical reactions are discussed as an introduction to subsequent chapters dealing with major chemical groups used to control different kinds of pests.

Crone, Hugh D. *Chemicals and Society*. New York: Cambridge University Press, 1986. Includes an extended discussion on herbicides, toxicity, public perception, and the use of Agent Orange.

Hansen, Gary W., Floyd E. Oliver, and N. E. Otto. *Herbicide Manual*. Washington, D.C.: U.S. Department of the Interior, Bureau of Reclamation, 1983. Provides information on herbicides and their composition, mode of action, and use.

Rao, V. S. *Principles of Weed Science*. 2d ed. Enfield, N.H.: Science Publishers, 2000. Rao discusses the application of technology to meet the challenges of the emergence of weeds resistant to herbicides, shifts in weed flora, and concerns about the safety of herbicides.

Ware, George W. *Complete Guide to Pest Control*. 3d ed. Fresno, Calif.: Thomson Publications, 1996. Provides information on herbicides and their composition, mode of action, and use.

Zimdahl, Robert L. *Fundamentals of Weed Science*. 2d ed. San Diego: Academic Press, 1999. Covers new developments in weed science as well as relevant aspects of the discipline's historical development. The focus is on weed biology and ecology, but coverage of herbicides and chemical weed control is also included.

HERBS

Categories: Angiosperms; economic botany and plant uses; food; medicine and health

The term "herb" has a variety of meanings but is most frequently used in one of three ways. To a botanist, an herb is a plant with a soft, flexible stem and a life cycle that is completed in one growing season. A person interested in medicinal plants would use the word "herb" to describe any plant with medicinal properties. In a culinary situation, an herb is a plant used to impart flavor to food.

Herb or Spice?

Herbs and spices are both used in cooking to modify the taste and smell of food. A clear distinction between herbs and spices is difficult to draw, but there are some broad differences that are useful to know. Plants referred to as herbs, such as basil (*Ocimum basilicum*) or rosemary (*Rosmarinus officinalis*), typically have been used in temperate regions throughout much of recorded history. Herbs can be distinguished from spices in that herbs are the leaves of nonwoody plants, used for their flavor or therapeutic properties. Spices, in contrast, are derived from other parts of plants, such as buds, stems, or bark, and are more strongly flavored, often because of the essential oils produced in these plant parts. More often than not, the leaf is the important plant part used for seasoning. The word "spice" also evokes a more exotic connotation, referring to plants obtained from distant places such as India or Ceylon. Spices are typically native to tropical areas.

History of Herbs

Herbs and spices have been used for thousands of years to add zest to meals, to help preserve food, and even to cover up the taste and smell of spoiled food. The Sumerians were known to have used laurel (*Laurus nobilis*), caraway (*Carum carvi*), and thyme (*Thymus vulgaris*) more than five thousand years ago. Other early records suggest that onion (*Allium cepa*) and garlic (*Allium sativum*) were also used. At least as early as 1000 B.C.E., the Egyptians used garlic and mint (*Mentha*) along with many other plants, for medicine, in religious ceremonies, or in embalming. The Greeks and Romans greatly expanded the uses of herbs to include their use as symbols, magical charms, cosmetics, dyes, perfumes, and air purifiers.

AP/WIDE WORLD PHOTOS

The benefits of herbs include their medicinal as well as culinary uses. In Hanoi, Vietnam, shoppers browse through tall paper sacks filled with herbs and dried roots at a traditional medicine shop.

Most of the historical information on herbs deals more directly with the medicinal properties of the plants. The first book available to the Europeans was *De materia medica* written by Pedanius Dioscorides, a Greek physician, in 100 C.E. This work on herbal medicine was so important that it was one of the first books printed after the printing press was developed. *De materia medica* continued to be the authoritative reference on the use of herbs in medicine for sixteen centuries. Another early work on medicinal herbs is *Pen-ts'ao* (*The Great Herbal*), a sixteenth century Chinese pharmacopoeia, attributed to Li Shih-chen, which lists more than eighteen hundred plants and medical preparations. The use of herbs as medicines began to de-

cline in the seventeenth century as new ways of treating illnesses developed, but herbal remedies continue even today in natural and homeopathic medicines.

Common Herbs

Among the herbs most commonly used in cooking are members of the *Lamiaceae*, or mint, family. Mint, rosemary, basil, oregano (*Origanum vulgare*), sage (*Salvia officinalis*), thyme (*Thymus vulgaris*), and marjoram (*Origanum majorana*) are the most frequently used members of this family. These herbs are important staples in most cuisines of the Mediterranean region. The key features that most mint family members share are square stems and simple leaves (having undivided blades) attached in groups of two leaves in an opposite position on the stem. The flowers are also distinct in that the tips of the petals are grouped into two clusters that bear a resemblance to lips.

The most commonly used herb in the United States today is parsley, *Petroselinum crispum*, a member of the *Apiaceae*, or carrot, family. Other important members of this family include dill (*Anethum graveolens*), cilantro (*Coriandrum sativum*, also known as coriander), fennel (*Foeniculum vulgare*), cumin (*Cuminum cyminum*), anise (*Pimpinella anisum*), celery (*Apium sativum*), and caraway. Members of this family are easy to recognize, particularly when in flower. The leaves are compound (having blades subdivided into leaflets), and the petiole (base of the leaf) is expanded such that the base wraps around the stem. The flowers are small and occur in clusters known as umbels. Umbels are clusters of flowers in which the flowers have stems of varying lengths so that all are located in one plane, forming a broad, flat inflorescence. Depending on the plant, the leaves or "seeds" or both might be used to impart characteristic musky flavors to a dish. Botanically, these are not simply seeds but seeds plus dried portions of the fruit. The fruit splits apart during maturation, forming two structures that appear to be "seeds." This type of fruit is called a schizocarp (*schizo* meaning "split," *carp* meaning "body") and is characteristic of members of the carrot family.

Chemistry of Herbs

Plants used as herbs contain compounds called aromatic oils, or *essential oils*. These are relatively small compounds of low molecular weight that are easily separated from the plant. The essential oil of a particular plant is usually mixture of compounds, rather than one single ingredient. Essential oils impart the characteristic taste and odor of the herbs. In nature, oils often serve as attractants to animals pollinating flowers or dispersing fruits. Many of the compounds may also act in defense against the invasion of fungi, bacteria, or predation by herbivores.

Additional Uses

A discussion of herbs would not be complete without mentioning some of the uses of these plants not related to food or medicine. Crushing the plant releases the aromatic volatiles, making them useful as perfumes to cover up odors on bodies or in spaces. Undoubtedly this practice began by simply crushing the plant to release the odor into the air or rubbing the crushed plant onto an object. Burning the plant is another way to release its odors. The Egyptians were quite skilled at the art of perfumery and passed these skills on to the Greeks and Romans. Additionally, herbs were commonly used in the nineteenth century as components of floral bouquets designed to deliver messages to recipients. The flowers had particular meanings; for example, roses meant love. The herbs were added for greenery to deliver additional messages. Rosemary was used to denote friendship, and basil meant hatred.

Joyce M. Hardin

See also: Biochemical coevolution in angiosperms; Medicinal plants; Spices.

Sources for Further Study

Levitin, Estelle, and Karen McMahon. *Plants and Society.* 2d ed. New York: McGraw-Hill, 1999. General information on herbs and other economically important plants.

Simpson, Beryl Brintnall, and Molly Conner Ogorzaly. *Economic Botany: Plants in Our World.* New York: McGraw-Hill, 1986. Useful information on herbs, spices, and perfumes.

Tucker, Arthur O., and Thomas DeBaggio. *The Big Book of Herbs: A Comprehensive Illustrated Reference to Herbs of Flavor and Fragrance.* Loveland, Colo.: Interweave Press, 2000. Presents useful information from cultivation to harvest; also contains detailed, specific information on more than one hundred of the most common herbs, including their aromatic chemistry.

Varney, Bill, and Sylvia Varney. *Herbs: Growing and Using the Plants of Romance.* Tucson, Ariz.: Ironwood Press, 1998. Practical information on growing and using herbs.

HETEROKONTS

Categories: Algae; economic botany and plant uses; microorganisms; pests and pest control; *Protista*; water-related life

Heterokonts are a group of closely related phyla with flagella in pairs, one long and one short. They include oomycetes, chrysophytes, diatoms, and brown algae.

The term "heterokont" refers either to the flagellar arrangement of biflagellate cells in which the two *flagella* differ in length (as in *anisokonts*), type of motion, or ornamentation, or to those organisms (and organisms evolutionarily derived from such lineages) in which biflagellate cells with heterokont flagella are produced at some point during their life cycle. The most common heterokont flagellar arrangement consists of a posteriorly directed whiplash flagellum and an anteriorly directed tinsel flagellum. The tubular tinsel flagellum characteristically bears two rows of stiff, glycoproteinaceous, tripartite hairs previously referred to as mastigonemes but now increasingly referred to as *stramenopili*. Such organisms are often referred to as stramenopiles.

Although alternative heterokont arrangements in which two nontinsel flagella differ are found in other unrelated groups (such as the dinoflagellates or endoparasitic slime molds (plasmodiophorids), these organisms are unrelated to the stramenopila-bearing (stramenopilous) heterokonts and are not generally referred to as heterokont genera.

Evolutionary Relationships

Stramenopilous heterokonts are a diverse group of protists containing tubulocristate mitochondria that are phylogenetically related by the presence of stramenopili (usually associated with the tinsel flagellum, otherwise on the cell surface). Their precise relationship to other eukaryotes is poorly defined, but they clearly represent one of the early independent lineages associated with the crown eukaryotic radiation (which included plant, animal, and fungal lineages). The inclusion of major autotrophic (many of the algal forms contain chlorophyll *a* and *c*) and heterotrophic groups has led to a conflicting nomenclature.

The initially proposed kingdom *Chromista* presumes heterotrophs were derived from ancestral autotrophs, while the more recently proposed kingdom *Stramenopila* implies a heterotrophic ancestor, with autotrophs as the derived forms. Relationships between the autotrophic and heterotrophic taxa, however, are still not clearly resolved. Heterotrophic heterokonts (heterokont fungi) include the fungal-like oomycetes, hyphochytriomycetes, thraustochytrids, and labyrinthulids. Autotrophic heterokonts (heterokont algae) include the chromophytic algal groups, represented by diatoms, brown algae, golden algae, and yellow-green algae.

Oomycetes

Often referred to as the oomycetous water molds, the oomycetes (*Peronosporomycetes*) form the largest and most ubiquitous group of heterotrophic stramenopiles. Free-living forms occupy diverse ecological niches ranging from freshwater and marine to terrestrial environments. Both terrestrial and marine forms include saprophytic and obligately, or facultatively, parasitic genera.

Hosts include a range of nematodes, arthropods, molluscans, algae, and plants. Several of the plant pathogens have impacted the cultural and economic history of humans. These include the causative agents for a variety of root rots, downy mildews, white rusts, and late blights. Downy mildews of grapes (*Plasmopara*) and tobacco (*Peronospora*) were responsible for the near-decimation of the French wine industry and the Cuban tobacco industry in the late 1870's and the 1980's, respectively. Similarly, *Phytophthora infestans*, the causative agent of potato late blight, was responsible for the Irish Potato Famine of the mid-1840's and, during World War I, for the starvation of German civilians in 1915-1916.

Diatoms

Diatoms are the most diverse and abundant of the photosynthetic heterokonts. They are among the most important aquatic photosynthesizers and are probably the most numerous aquatic eukaryotes. Found as single cells or chains of cells in marine, freshwater, and terrestrial environments, they often dominate the phytoplankton of nutrient-rich waters. The cells are enclosed in a highly ornamented silica box (*frustule*). These frustules are of either the centric type, with apparent radial symmetry, or of the pennate type, with apparent bilateral symmetry. The chloroplasts of diatoms commonly contain the carotenoid fucoxanthin as an accessory photosynthetic pigment. A single tinsel flagellum, but no whiplash flagellum, is evident in sperm cells of those species in which sexual reproduction is known. A few species have been associated with fish kills, while some species of *Pseudo-nitzschia* produce the toxin domoic acid, responsible for amnesiac shellfish poisoning in humans. In the newer taxonomic systems they are included in the phylum *Ochrophyta* rather than the phylum *Bacillariophyta*.

Chrysophytes

Chrysophytes are often referred to as the golden algae because of the dominance of the pigment fucoxanthin. The phylum *Chrysophyta* had previously included a loosely related assemblage of algal forms, such as haptophytes, synurophytes, and diatoms. The term "chrysophycean," referring to the class *Chrysophyceae* of the *Ochrophyta*, now more appropriately includes those chrysophytes phylogenetically related by ultrastructural, pigment, and molecular analyses. These chrysophyceans are unified by the use of chlorophylls a, c_1, and c_2, the accessory pigments fucoxanthin and violaxanthin, and a silica-walled resting stage (stomatocyst, stomatopore). They are present as unicellular or colonial forms. Representative genera include *Ochromonas*, *Dinobryon*, and *Chrysocapsa*.

Brown Algae

The brown algae (phaeophyceans) are primarily marine algae and range in linear length from microscopic filaments to several meters (up to 60 meters and 300 kilograms in the giant kelp *Macrocystis pyrifera*). Many species exhibit specialized organs and tissues such as leaflike *blades*, rootlike *holdfasts*, and cells resembling sieve elements of higher land plants. Like other ochrophytes, they have two forms of chlorophyll c as well as fucoxanthin and violaxanthin. Species of *Laminaria*, *Fucus*, and *Macrocystis* are primary sources of alginates used in a variety of industrial applications, from the textile industry (prints) to food processing (as a thickening agent).

M. E. S. Hudspeth

See also: Algae; Brown algae; Chrysophytes; Diatoms; Oomycetes; Phytoplankton; *Protista*.

Sources for Further Study

Alexopolous, Constantine J., Charles W. Mims, and Meredith Blackwell. *Introductory Mycology*. 4th ed. New York: John Wiley and Sons, 1996. Basic mycology text includes sections on the fungal-like heterotrophic heterokonts. Includes illustrations, bibliographical references.

Dick, Michael W. *Straminipilous Fungi*. Boston: Kluwer Academic, 2001. Extensive discussion of all aspects of heterotrophic heterokont biology, including ecology, physiology, and systematics. Includes illustrations, bibliographical references, and index.

Graham, Linda E., and Lee W. Wilcox. *Algae*. Upper Saddle River, N.J.: Prentice-Hall, 2000. Basic algal text contains chapters on the heterokont algae. Includes illustrations, bibliographical references.

Margulis, Lynn, John O. Corliss, Michael Melkonian, and David J. Chapman, eds. *Handbook of Protoctista*. Boston: Jones and Bartlett, 1990. Monographs include representative autotrophic and heterotrophic heterokont groups. Includes illustrations, bibliographical references.

HIGH-YIELD CROPS

Categories: Agriculture; economic botany and plant uses; food

High-yield agricultural crops are those that have been bred, genetically modified, or fertilized to increase their production yields.

The health and well-being of the world's growing population are largely dependent on the ability of the agricultural industry to raise high-yielding food and fiber crops. No one knows for certain when the first crops were cultivated. At some time in the past, people discovered that seeds from certain wild grasses could be collected and later planted where they could be controlled during the growing process and eventually harvested for food. Agriculture was firmly established in Asia, India, Mesopotamia, Egypt, Mexico, Central America, and South America at least six thousand years ago.

Development of Modern Agriculture

Farming practices have undergone many changes, but until the nineteenth century most farms and ranches were family-owned, and people primarily practiced subsistence agriculture. Just as in almost all other industries, the arrival of the Industrial Revolution dramatically changed the agriculture industry. Inventions such the cotton gin in 1793, the mechanical reaper in 1833, and the steel plow in 1837 led the way to mechanization of most farms and ranches. The Industrial Revolution produced significant societal changes, as people involved in agricultural production left the farms to work in city factories. Fewer and fewer people were required to produce more and more agricultural crops for an increasing number of consumers.

As the population continued to grow, it became necessary to select and produce higher-yielding crops. The *Green Revolution* of the twentieth century helped make this possible. Agricultural scientists developed new, higher-yielding varieties of numerous crops, particularly the seed grains. Tremendous increases in the global food supply resulted from the use of these higher-yielding crops, along with improved farming methods.

Improved Yields

There are two major ways to improve yield in seed grains such as wheat: produce more seed per seed head or produce larger seeds. Numerous agricultural practices are required to produce higher yields, but one of the most important is the selection and breeding of genetically superior cultivars. When a plant with a potentially desirable gene mutation appears to improve a yield characteristic, seeds are collected and studied to see whether they consistently produce plants with higher yields.

This selection process remains one of the major means of improving yield in agricultural crops. Advances in the understanding of genetics have made it possible to breed some of the desirable characteristics that have resulted from mutation into plants that lacked the characteristic. In addition, the advent of *recombinant DNA technology* makes it possible to transfer genetic characteristics from one plant to any other plant.

The high-yielding varieties led to an increased reliance on *monoculture*. This practice of growing only one crop over large areas has made efficient, mechanized farming possible, with relatively few workers. It has also decreased the genetic variability of many agricultural plants, increased the need for commercial fertilizers and pesticides, and produced an increased susceptibility among crops to damage from a host of biotic and abiotic factors. For example, most pests prey upon specific plant species. If many fields are planted in one crop, the entire farm—and possibly the whole community—is susceptible should a pest strike. The corn blight that destroyed more than 15 percent of the North American crop in 1970 would have been less severe if a single cultivar of corn had not been so heavily planted in the United States.

The major high-yielding crops, in terms of land devoted to their culture and the total amount of produce, are wheat, corn, soybeans, rice, potatoes, and cotton. Each of these crops originated from a low-yielding native plant that was gradually converted into one of the highest-yielding plants in the world.

High-Yield Grains

Throughout the world, large portions of agricultural land are devoted to the production of wheat (*Triticum sativum*). Wheat is the staple of forty-three countries and 35 percent of the people of the world. It also provides 20 percent of the total food calories for the world's population.

The cultivation of wheat is older than the written history of humankind. Its place of origin is unknown, but many authorities believe wheat may

The World Food Prize was awarded in 2000 for the development of a corn containing nearly twice as much protein as normal corn and yielding 10 percent more grain.

have grown wild in the Tigris and Euphrates Valleys and spread from there to the rest of the Old World. Wheat is mentioned in the first book of the Bible, was grown by Stone Age Europeans, and was reportedly produced in China as far back as 2700 B.C.E. Wheat is widely adapted throughout the world and can grow in many climates; about the only places where wheat does not grow are those with climates that are continually hot and moist.

The total production of rice (*Oryza sativa*) is similar to that of wheat, and it is the principal food crop for nearly half of the world's population. Rice has been under cultivation for so long that its country of origin is unknown; however, botanists believe that the plant originated in Southeast Asia. Rice was being cultivated in India as early as 3000 B.C.E. and spread from there throughout Asia.

Although rice is currently produced on all continents, more than 90 percent of the total world crop is grown in Asia. Rice was introduced into the United States in 1694, and the total U.S. crop is produced in just six states. The per capita consumption of rice is less than 10 pounds per person in the United States; however, in some Southeast Asian countries, the per-capita consumption can be as high as 325 pounds per person.

Corn (*Zea mays*) ranks third behind wheat and rice in world production of cereal grain. Corn may be the Americas' greatest contribution to agriculture. The earliest traces of the human use of corn date back to about 5200 B.C.E. It was probably first cultivated in the high plateau region of central or southern Mexico and was the basic food plant of all pre-Columbian advanced cultures and civilizations, including the Inca of South America and the Maya of Central America.

Corn is still produced primarily in the Western Hemisphere, although some is produced in Europe. It makes up more than 50 percent of the acreage devoted to all seed grains in the United States. Of the total grain crop produced, approximately 85 percent is fed to animals.

High-Yield Potatoes, Soybeans, and Cotton

The potato (*Solanum tuberosum*) is the New World's second gift to world agriculture. The white potato is capable of nourishing large populations, especially in cooler regions where many other crops will not grow. It contains large stores of energy, high-quality protein, and valuable minerals and vitamins.

Potatoes are indigenous to South America and probably originated in the central Andean region. The potato has become one of the world's great food crops because it combines, in one crop, the desirable characteristics of high yield, low cost, and nutrition as well as palatability. Potatoes are grown in practically all countries of the world. As the potato is better adapted to cool, humid climates, total production and average yields are much higher in northern Europe than in the United States.

Soybeans (*Glycine max*) have been an important food crop in Asian countries, particularly China and Japan, since long before the time of recorded history. Soybeans will grow in nearly all soil types except extremely deep sand and are adapted mainly to temperate regions with fairly humid, warm growing seasons. Hence, soybeans are now grown in much of the New World.

Cotton (*Gossypium*), with a total annual production of more than 13 million tons, is by far the most important fiber crop in the world. Humans heavily rely on it for clothing and other textiles. Cotton enters the daily life of more people than any other product except salt.

Cotton fiber has been known and highly valued by people throughout the world for more than three thousand years. Like most crop plants that have been in cultivation a long time, cotton has an obscure origin. A vigorous cotton industry was present in India as early as 1500 B.C.E. From India, the cultivation of cotton spread to Egypt and then to Spain and Italy. In the New World, a different species of cotton was being grown in the West Indies and South America long before the Europeans arrived. Although cotton is native to tropical regions, it has been adapted to humid, subtropical climates where there are warm days, relatively warm nights, and a frost-free season of at least two hundred days.

Genetic Engineering

The importance of plant nutrition was discovered during the nineteenth century, and the use of fertilizers has dramatically increased crop yields. Advances in the understanding of genetics in the early part of the twentieth century made it possible for people to breed desirable characteristics into plants, and people began to breed plants to increase yields. At about the same time, the development of large-scale chemical synthesis and processing made it possible to produce a variety of chemical agents to control plant pests, but many of these chemicals were found to be hazardous to the environment.

Developing plants with high natural resistance would decrease the reliance on chemical pesticides, so plant breeders started to breed not only for increased yields but also for increased resistance. Plant breeders began to examine older, abandoned cultivars, discarded breeders' stock, wild plant relatives, and native or foreign cultivars and occasionally induced mutations for increased yields or resistance to pests. When a particularly attractive trait was found, genetic crosses were made until progeny that showed improved yields and increased resistance were obtained. These developments in plant resistance have been partially responsible for the tremendous advances in the production of high-yield crops during the last half of the twentieth century.

To cope with the need for crop varieties that will both provide higher yields and exhibit greater resistance to pests and disease (thus avoiding the use of too many chemicals), new biotechnologies, including genetic engineering, are being refined. Total crop production can also be raised by using biotechnology to engineer plants that will flourish under what may have previously been considered marginal growing conditions, such as salty soil. As the population grows, the agricultural industry will feel the pressure to produce more food, with the added element of pressure to make crop production more friendly to the environment. Biotechnology has the potential to play a major role in the development of a long-term, sustainable, environmentally friendly agricultural system.

The future development of higher-yielding crops that can be harvested mechanically and the production of new types of equipment to facilitate the harvesting process will also be important improvements in the production of high-yielding crops.

D. R. Gossett

See also: Agricultural crops: experimental; Agriculture: modern problems; Agriculture: world food supplies; Biopesticides; Biotechnology; Corn; Fertilizers; Genetically modified foods; Green Revolution; Horticulture; Hybridization; Monoculture; Plants with potential; Resistance to plant diseases; Rice; Wheat.

Sources for Further Study

Brown, L. V. *Applied Principles of Horticultural Science*. Boston: Butterworth-Heinemann, 1996. Presents detailed practical exercises for horticulture students with no previous knowledge of science, supported by brief summaries of basic principles. Includes sections on plant science, soil science, and pests and disease.

Campbell, Neil A., and Jane B. Reece. *Biology*. 6th ed. San Francisco: Benjamin Cummings, 2002. Introductory textbook written for the biology major provides an excellent discussion of the biology associated with plant reproduction. Offers superb graphics, lists of suggested readings, and a glossary.

Chrispeels, M. J., and D. E. Sadava. *Plants, Genes, and Agriculture*. Boston: Jones and Bartlett, 1994. A treatise on the use of biotechnology in crop production. Contains sections related to the use of biotechnology to transfer resistance to susceptible plants. Although an advanced text, the book provides much of interest to the general reader. The well-illustrated text offers a supplemental reading list.

Janick, Jules. *Horticulture Science*. 4th ed. New York: W. H. Freeman, 1999. This text, intended for the beginning horticulture students, contains sections on horticultural biology, environment, technology, and industry. Covers the fundamentals associated with the production of high-yield crops. Well illustrated; includes references.

Lurquin, Paul F. *The Green Phoenix: A History of Genetically Modified Plants*. New York: Columbia University Press, 2001. History of the field gives equal weight to the science behind developing improved crop strains and the multinational corporations marketing the results.

Metcalfe, D. S., and D. M. Elkins. *Crop Production: Principles and Practices*. 4th ed. New York: Macmillan, 1980. This text for the introductory agriculture student is one of the most valuable sources available on the practical aspects of high-yield crop production. Well illustrated; includes references and glossary.

HISTORY OF PLANT SCIENCE

Categories: Disciplines; history of plant science

The study of plants has, since the fifth century B.C.E., expanded from the broad-based discipline of natural philosophy to include the specialized studies of genetics, ecology, and microbiology.

Plant Science in Antiquity

Natural philosophy, the "science" of its day, arose in Greece during the 500's B.C.E. and included speculations about plants, such as whether domesticated plants descended from wild plants. Full-fledged science began at the Lyceum in Athens, founded by Aristotle in 335 B.C.E. Upon Artistotle's departure from Athens in 323, his colleague Theophrastus became head of the Lyceum. He was author of two botanical treatises which cover general natural history. *Historia plantarum* ("Enquiry into Plants" in *Enquiry into Plants and Minor Works on Odours and Weather Signs*, 1916) emphasizes description of plants, their parts, and their locations, and *De causis plantarum* (*De Causis Plantarum*, 1976-1990) emphasizes *physiology*. The works include accounts of crops, fruit trees, and medicinal plants, but the stronger emphasis is on abstract knowledge. Greek physicians and pharmacists were more practical-minded, as is seen in Pedanius Dioscorides' *De materia medica*, compiled during the first century C.E. Although it concerns medicinal plants, practically all plants had medicinal uses, and Dioscorides was the first to give a species-by-species description of the different plants known in much of the Mediterranean region.

The Romans were heavily influenced by Greek civilization, yet Romans were more practical and less abstract thinkers than the Greeks; there were no important Roman scientists. Instead of botanical treatises, Romans wrote agricultural manuals, of which the longest and most thorough was Lucius Junius Moderatus Columella's in the early first century C.E. He used Greek sources, especially the work of Theophrastus of Eresos, and also Roman sources and his own experiences. Columella's near contemporary, Gaius Plinius Secundus (also known as Pliny the Elder), was one of the Roman compilers of encyclopedias; he also used Greek and Roman sources, but his goal was to educate and entertain. However, that did not preclude dispensing practical knowledge. His lengthy accounts of plants emphasized useful species and included curious folklore.

Middle Ages and Renaissance

Before the western Roman Empire declined in the later 400's C.E., original contributions to natural history were already rare. The eastern Empire persisted for another millennium, but it was politically and economically static, if not stagnant, and its main cultural contribution was to transmit Greek learning to the Arabs during the 700's and to the Italians during the 1400's. Arabic-language science was superior to Roman science and sometimes equal to ancient Greek science. It was studied and enlarged over a much broader geographical area than ancient science had been. Generally, Arabic-language authors did not use Roman sources, the exception being on the Iberian Peninsula, where Ibn al-ʿAwwam compiled his agricultural treatise during the second half of the 1100's. Dioscorides' manual was translated into Arabic and became the foundation of Arabic pharmacopoeias. Arabic-language science surpassed Greek science in the extent of its knowledge but rarely in theoretical understanding.

In the year 1000 C.E. both the Byzantine and Islamic civilizations were more prosperous and sophisticated than Latinized Western Europe, and yet by 1400 Western Europe had surpassed both adjacent civilizations in cultural achievements. There is no simple answer as to why Western Europe made a greater investment in higher education than any other civilization in the world, but it did so, and that investment paid off in science. While Western Europe's culture was rather rudimentary, scholars took the opportunity to translate important works of an-cient or medieval authors from Arabic or Greek into Latin and then built upon them. The most impressive botanical example was Albertus Magnus's *De vegetabilibus et plantis*, written during the 1200's without access to Theophrastus's works but in the Lyceum tradition. Albertus used Greek and Arabic sources available in Latin translations and also Latin authors and his own experiences. However, medicine was much more important than natural history in the universities, and *De materia medica* was where most botanical studies focused.

During the 1450's Johann Gutenberg developed a printing press that used movable type. It was an important step for botany and medicine because science books were prominent among early printed works. Scholars of the Italian Renaissance were enthusiastic about ancient Greek and Roman civilizations and studied living plants to decide which ancient names and descriptions fit the species they found. This concern then spread northward during the 1530's, as seen in Euricius Cordus's *Botanologicon*, Otto Brunfels's illustrated herbal, and Jean Ruel's development of terminology to describe parts of plants. Leonhard Fuchs combined Ruel's concern for terminology with Brunfels's concern for illustrations in his elaborate herbal (1542). In Italy, Pier Andrea Mattioli tied the old text of Dioscorides to the new trend in illustration (1544), while in France and the Low Countries three scholars—Rembert Dodoens, Charles de L'Écluse, and Matthias L'Obel—sought new species to name and describe for their herbals. Konrad Gesner and Valerius Cordus did equally outstanding research, yet the influence of each was impaired by death before their works were fully published.

Scientific Revolution

The precise investigations contrived by Nicolaus Copernicus, Galileo Galilei, and Isaac Newton made ancient astronomy and physics obsolete, and Andreas Vesalius and William Harvey did the same for ancient anatomy and physiology. Innovations in botany were more evolutionary than revolutionary. Luca Ghini developed the botanic garden and herbarium as teaching and research aids, and his student Andrea Cesalpino wrote the first true textbook of botany (1583). Joachim Jung developed morphological terminology beyond that of Ruel and Cesalpino but without benefit of a microscope.

When the microscope became available to Robert Hooke, he described plant cells and mold fungi

(1665). Following his lead during the 1670's were Nehemiah Grew and Marcello Malpighi, who described the anatomy of plant tissues. The Dutch microscopist Antoni van Leeuwenhoek examined small plants and discovered bacteria but not bacteria's role in disease. Cesalpino chose fruit as a key to plant classification, but Fabio Colonna decided that flowers provided a better guide to relationships. Colonna's judgment was strengthened by the discovery of the sexuality of plants, which Grew suspected and Rudolph Jakob Camerer demonstrated. Johann Hedwig later demonstrated sexuality in bryophytes and algae but failed to understand sexuality in ferns and fungi.

The age of exploration and the scientific revolution coalesced when explorers returned to Europe with previously unknown plants. Brothers Gaspard and Jean Bauhin facilitated knowledge of the increase in known species by using binomial nomenclature in their descriptions of about six thousand species. Engelbert Kämpfer botanized in Iran and Japan, published a travel book, and brought home a herbarium that eventually reached the British Museum. John Ray traveled only in Western Europe, including Britain, but his *Historia generalis plantarum* (1686-1704; *A Catalogue of Mr. Ray's English Herbal*, 1713) provided a synthesis of knowledge from many explorers' observations and specimens. Joseph Pitton de Tournefort's contributions to botany were enhanced by his explorations in the Levant, yet his main influence came from careful studies on plant genera.

In the eighteenth century the Swedish naturalist Carl Linnaeus formalized the system of binomial nomenclature and his sex-based classification system in response to all the newly discovered plants. The ease with which species could then be identified stimulated the search for still more species, which he named and classified. Some of his students, including Pehr Kalm and Carl Peter Thunberg, traveled far abroad and returned with observations and specimens. Michel Adanson explored Senegal within a French context and also devised a natural classification. John Bartram and his son William responded to European interest by discovering American plants to describe and send abroad. Joseph Banks, sailing with Captain James Cook, initiated an important partnership between British botanists and the navy, which blossomed during the 1800's. Seeds and live plants brought back to Europe commonly were planted in botanic gardens

located at universities and in capital cities, where they were accessible to botanists. Directorships of these gardens provided employment for some botanists.

Experimentation came late to botany. Edme Mariotte set a good example (1679) with his experiments in plant physiology, but his results were rather inconclusive. More successful was Stephen Hales, whose inspiration came from animal physiology; he devised methods to measure the movements of sap in vines and saplings and correlate movements with sunlight. Although his experiments were well published (1727), he had no followers, and his experiments during the later 1700's addressed different issues. John Turberville Needham and Georges-Louis Leclerc de Buffon thought they had demonstrated spontaneous generation of life, but Lazzaro Spallanzani was skeptical and conducted superior experiments that showed they were mistaken. Spallanzani's work inspired Jean Senebier, who translated Spallanzani's works into French and became himself an outstanding experimentalist. Another experimental tradition arose within a chemical context, which utilized Hales's apparatus. Chemists explored ways to generate gases and to identify them. This led Joseph Priestley, Horace Bénédict de Saussure, and Jan Ingenhousz to conduct experiments on live plants in glass enclosures during the 1770's. Camerer (1694) and Josef Gottlieb Kölreuter in the later 1700's were among the early experimenters on sexual reproduction in flowers.

Taxonomic Botany

Taxonomic botany, as formally organized by Linnaeus, became the main scientific specialty during the 1800's because many new species were still being discovered in all parts of the world. Outstanding collectors included Alexander von Humboldt, Robert Brown, Joseph Dalton Hooker, George Englemann, and Edward Lee Greene. The accumulation of herbaria specimens and plantings in botanic gardens enabled botanists to produce encyclopedic accounts of species, such as those by Augustin Pyramus de Candolle and his son Alphonse Louis de Candolle and by George Bentham and Hooker.

Phytogeography

Phytogeography developed using those same plant collections and taxonomic encyclopedias. Linnaeus had begun writing on this subject, and

Karl Ludwig Willdenow, who trained Humboldt, carried it further. However, Humboldt's extensive explorations, collections, environmental measurements, and scientific publications became the real foundation of phytogeography (1808). In 1820, Augustin de Candolle introduced the concept of competition among species as a factor in the distribution of species, and his son Alphonse de Candolle wrote an important synthesis of phytogeography, *Géographie botanique raisonée* (1855, 2 volumes). August Heinrich Rudolph Grisebach achieved another world synthesis in 1872. Hewett Cottrell Watson founded British phytogeography (1835) and devoted his career to a study of the distribution of all British species and their variability. Asa Gray, an American who specialized in taxonomic botany, became, through his association with Joseph Hooker and Charles Darwin, the United States' leading phytogeographer.

Evolution and Heredity

Ideas on the evolution of species made little headway until the French Revolution (1789); then two naturalists wrote books discussing it in plants and animals. Although the British became enemies of the Revolution, Erasmus Darwin favored it. Both he and Jean-Baptiste Lamarck published speculative hypotheses that viewed struggle as a source of new traits, to be followed by the inheritance of the new traits. Johann Wolfgang von Goethe speculated in a more idealistic way about species varying from a basic type.

Paleobotany was less popular than vertebrate paleontology, but Alexandre and Adolphe-Théodore Brongniart, father and son, made important discoveries. Franz Unger followed in their footsteps. It became obvious to them that fossil species differed from living species, causing Unger to speculate on changes in species. Charles Darwin, who read his grandfather's books, outdid Erasmus Darwin with a carefully developed theory of evolution by natural selection. Alfred Russel Wallace had read many of the same scientific works as Charles Darwin, and he had also read *Darwin's Journal of Researches* (1839) before publishing his own ideas on evolution that resembled Darwin's theory of natural selection. That theory remained controversial for decades after its publication in Darwin's *On the Origin of Species by Means of Natural Selection* (1859), but it also stimulated much study of both fossil and living species to discover the details of evolution. Charles

Edwin Bessey took up the challenge of explaining the history of plant evolution.

Heredity was a weak link in Darwin's theory of evolution, awaiting the advent of *genetics*. Karl Gärtner conducted numerous empirical studies but without achieving a theoretical breakthrough. When Gregor Mendel, an inconspicuous monk, developed experiments on peas which clarified patterns of inheritance, his thinking was beyond his audience's comprehension, including Karl Wilhelm von Nägeli, to whom he turned for encouragement. Mendel's 1866 article was only appreciated in 1900, when three botanists—Karl Franz Joseph Erich Correns, Erich Tschermak, and Hugo de Vries—independently rediscovered it and Mendel's laws. Meanwhile, others followed cell division more and more closely, demonstrating that chromosomes carry hereditary material (genes) and that those chromosomes divide in a regular way during both mitosis and meiosis.

Cytology, Fertilization, and Alternation of Generations

Studies on *cytology*, *fertilization*, and *alternation of generations* advanced considerably during the 1800's because advances in the quality of microscopes and slide-preparation techniques allowed botanists to achieve a more precise understanding than had been possible earlier. Charles-François Brisseau de Mirbel initiated French cytology with studies on plant anatomy, seeds, and embryos (1800-1832). Brown used a microscope to discover the cell nucleus. Matthias Jakob Schleiden, who used the improved microscopes, is credited with establishing the cell theory in plants—after his colleague. Theodor Schwann had done so for animals, though Schleiden misunderstood cell division. Nevertheless, his botany textbook (1942-1943) inspired others to investigate cellular processes. Hugo von Mohl advanced microscopy and developed the protoplasm concept. Wilhelm Friedrich Benedict Hofmeister and Nathanael Pringsheim studied fertilization and alternation of generations in diverse groups of plants. Later, Walther Flemming and Eduard Adolf Strasburger used improved techniques to study chromosomes during mitosis and meiosis.

Microbiology and Mycology

Microbiology and *mycology* also benefited from advances in microscopy and cytology. Ferdinand

Julius Cohn, Anton de Bary, and Louis Pasteur all made their main contributions during the 1850's, 1860's, and 1870's. Cohn studied unicellular algae and bacteria. De Bary founded mycology with studies on sexual reproduction in fungi and on the two-stage life cycle of wheat rust. Pasteur vindicated Schwann's claim that alcohol fermentation is caused by yeast and discovered anaerobic metabolism. Pasteur also investigated the causes of various diseases, several of which were bacterial. Robert Koch developed bacteriological techniques that enabled him to demonstrate clearly the bacterial cause of several diseases.

Physiology and Agronomy

Physiology and *agronomy* received less attention than several other botanical specialties during the first half of the nineteenth century, though Henri Dutrochet showed that plant respiration and animal respiration are essentially the same. In the second half of the century, Julius von Sachs and John Bennet Lawes made these specialties more conspicuous. Sachs was a brilliant experimentalist, teacher, and author of textbooks, making him the founder of modern plant physiology. Lawes used private resources to found modern agricultural research in Britain at a time when the U.S. Congress was establishing land grant colleges and state agricultural and forestry research stations. By the end of the century, American scientists were doing as much or more agricultural research as the rest of the world combined. Vasily Vasilievich Dokuchaev developed soil science (agronomy) as an aid to Russian agriculture.

Twentieth Century

All of the specialties from the 1800's continued throughout the 1900's. In addition, *genetics, ecology,* and *molecular biology* became important specialties. Plant sciences advanced at an unprecedented rate and in more countries than ever before.

Evolution became the organizing theory for all of biology. However, *evolutionary biology* retained a close relationship with taxonomy and phytogeography. Bessey, and later John Hutchinson, advanced the understanding of the evolution of vascular plants as a whole, while Marie Stopes contributed paleobotanical evidence. Two Russians took the lead in their subspecialties: Nikolai Ivanovich Vavilov used genetics and cytological evidence to clarify the history and phytogeography of domesti-

cated plants, and Aleksandr Ivanovich Oparin used physiology and biochemistry to investigate the origin of life.

Genetics did not develop within a strong evolutionary context during its first four decades, though de Vries had an evolutionary motive to study heredity and Wilhelm Ludwig Johannsen's studies on breeding homozygous versus heterozygous strains of peas had evolutionary implications. Genetics was advanced by both botanists and zoologists. The three re-discoverers of Mendel's laws and article were botanists. Thomas Hunt Morgan studied gene linkage on chromosomes using fruit flies, and later Barbara McClintock continued these studies using maize. Albert Francis Blakeslee initially studied sexual fusion in fungi but is remembered more for discovering that colchicine produces polyploidy in plant chromosomes. A new specialty, molecular biology, arose out of James Watson's and Francis Crick's struggle to understand the structure and function of the gene, as represented by deoxyribonucleic acid (DNA).

Four English botanists advanced plant anatomy, particularly at the level of cellular biology. Ethel Sargent studied intracellular structures relating to cell division and vascular bundles as clues to evolution. Vernon Herbert Blackman studied plant cytology and alternation of generations in rust fungi. Agnes Robertson Arber wrote monographs on monocotyledons as a whole and on particular groups of them. Irene Manton used an electron microscope to study chromosomes and cell organelles. An Italian animal histologist, Camillo Golgi, discovered the "Golgi body" within an owl's brain cell (1898), but explanation of its function came much later. Christian De Duve was a Belgian cytologist and biochemist who used a centrifuge and electron microscope in his research to discover new organelles: lysosomes and peroxisomes.

Physiology flourished during the 1900's, beginning with Frederick Frost Blackman's studies on respiration, Jagadis Chandra Bose's on biophysics, and Mikhail Semenovich Tsvet's on cytophysiology. In the 1930's Paul Jackson Kramer began investigating water usage, and Hans Adolf Krebs began investigating cyclic metabolic pathways. After World War II, François Jacob studied the functioning of DNA and RNA, the genetic control of enzymes, and helped develop the concept of the operon in cellular physiology. Melvin Calvin took advantage of availability of radioactive tracers, particularly carbon 14, to clarify biochemical steps in photosynthesis.

Ecology arose simultaneously with genetics but at a slower pace. Johannes Warming, Gottlieb Haberlandt, and Andreas Franz Wilhelm Schimper were botanists who laid a foundation for plant ecology shortly before 1900, and the self-consciously ecological researches by Christen Raunkiaer, Felix Eugen Fritsch, Frederic Edward Clements, and Henry Chandler Cowles built upon their foundation. Fritsch studied periodicity in phytoplankton and helped found the Freshwater Biological Association. Andrew Ellicott Douglass was interested in sunspot cycles, and to document their occurrence he developed *dendrochronology*, which also was used to clarify vegetation cycles. Clements was an enthusiastic theoretician; Cowles was more cautious, and Henry Allan Gleason and Arthur George Tansley attacked Clements's concepts of climatic climax and plant communities as overinterpretations of the evidence. Gleason preferred the concept of association, which John Thomas Curtis modified into the continuum. Tansley, who helped found the British Ecological Society and the *Journal of Ecology*, preferred his own ecosystem concept, which became extremely important after World War II. Early examples of the use of ecological concepts for environmental protection are Paul Bigelow Sears's *Deserts on the March* (1935) and Rachel Carson's *Silent Spring* (1962).

Microbiology expanded beyond what can be indicated here. Five virologists illustrate the involvement of plant scientists in microbiology. Dmitri Iosifovich Ivanovsky investigated tobacco mosaic disease in the 1890's and early 1900's and found evidence that the causative pathogen passed through a porcelain filter. His microscopic observations were excellent, yet he concluded that the pathogen was a bacterium, whereas it was later determined to be a virus. Martinus Willem Beijerinck investigated the puzzle and found the pathogen is not a toxin and could not be cultivated in vitro; he concluded it must be a molecular pathogen. Three other filterable pathogens were also identified in the same period. Later, Louis Otto Kunkel studied viral diseases in potatoes and sugarcane in order to inhibit their transmission. Frederick Charles Bawden and N. W. Pirie discovered that plant viruses are nucleoproteins.

Agronomy benefited enormously from the advance of many sciences, as indicated by the work of four plant scientists. George Washington Carver drew upon chemistry to find new uses for peanuts and sweet potatoes. Rowland Harry Biffen used the new genetics to improve crops, including rust-resistant wheat. Vavilov used genetic variability in wild plant populations to modify closely related domesticates to achieve varieties best suited to the Soviet Union's diverse environments. Norman E. Borlaug, known as the father of the Green Revolution, bred varieties of rice, corn, and wheat that greatly increased yields in tropical countries.

Frank N. Egerton

See also: Agricultural revolution; Biotechnology; Botany; Cell theory; Dendrochronology; DNA: historical overview; Ecology: concept; Ecology: history; Environmental biotechnology; Evolution: historical perspective; Genetics: Mendelian; Genetics: post-Mendelian; Green Revolution; Paleobotany; Paleoecology; Plant science.

Sources for Further Study
Ainsworth, Geoffrey C. *Introduction to the History of Mycology.* New York: Cambridge University Press, 1976.

_____. *Introduction to the History of Plant Pathology.* New York: Cambridge University Press, 1981. Well-illustrated topical surveys, with contemporary portraits and plant illustrations, notes, and bibliographies.

Greene, Edward Lee. *Landmarks of Botanical History.* 2 vols. Edited by Frank N. Egerton. Stanford, Calif.: Stanford University Press, 1983. A detailed survey of major works from antiquity to about 1700; editor added three appendices to fill gaps in coverage. Includes contemporary portraits and plant illustrations, extensive bibliography.

Hessenbruch, Arne, ed. *Reader's Guide to the History of Science.* London: Fitzroy Dearborn, 2000. Includes discussions and bibliographies on a several developments in the history of botany. Index.

Morton, A. G. *History of Botanical Science: An Account of the Development of Botany from Ancient Times to the Present Day*. New York: Academic Press, 1981. Concise and well balanced. A few contemporary portraits and illustrations, notes.

Stafleu, Frans A., and Richard S. Cowan. *Taxonomic Literature: A Selective Guide to Botanical Publications and Collections with Dates, Commentaries and Types*. 2d ed. 8 vols. Utrecht, Netherlands: Bohn, Scheltema and Holkema, 1976-1992. A vast encyclopedia that is essential for the history of taxonomic botany. Coverage is from about 1500 to recent times.

Waterson, A. P., and Lise Wilkinson. *An Introduction to the History of Virology*. New York: Cambridge University Press, 1978. A topical survey with some contemporary portraits and illustrations, notes, and bibliography.

HORMONES

Categories: Cellular biology; physiology

Plant hormones are the major group of chemical messengers by which most plant activities are controlled. The five different groups of hormones regulate virtually every aspect of plant growth and development.

The majority of higher plants begin life as seeds. When seeds germinate, the embryonic tissues begin to grow and undergo differentiation until, ultimately, the various parts of the mature plant are formed. Every aspect of this growth and development is regulated by a group of chemical messengers called hormones. These plant hormones, or *phytohormones*, function as *plant regulators*. A plant regulator is an organic compound, other than a nutrient, which in small amounts promotes, inhibits, or otherwise modifies a basic plant process. Hormones are produced in one area of the plant and transported to another area, where their effects are exerted. There are five major classes of identifiable plant hormones, and others will surely be identified in the future: *auxins, cytokinins, gibberellins, ethylene,* and *abscisic acid.*

Auxins

Although there are numerous plant responses to the auxins, one of the primary functions of this group of hormones is to cause increases in cell length by loosening cell walls and increasing the synthesis of cell-wall material and protein. In order for plants to grow, cells produced at the stem or root *meristems* must undergo this process of elongation. No cell elongation can take place in the absence of auxin.

The cell elongation promoted by auxins results in regular growth, and it is also responsible for various *tropisms*. For example, *phototropism* causes plants

Plant Hormones and Their Functions	
Hormone	*Responsible For*
Abscisic acid	Leaf abscission
Auxins	Cell elongation, tropisms, root growth
Brassinolides	Cell division, cell elongation
Cytokinins	Cell division, increased metabolism, chlorophyll synthesis
Ethylene	Fruit ripening, fruit abscission, dehiscence
Gibberellins	Stem elongation, breaking dormancy
Jasmonates	Seed germination, root growth, accumulation of storage and defense proteins
Salicylic acid	Pathogen defense activator
Systemin	Stimulates defense genes after tissue wounding

to grow toward a source of light. *Gravitropism* is a response to gravity and causes the roots to grow downward.

Besides cell elongation, auxins will initiate root growth at the base of the stem. Auxins inhibit growth of the lateral buds; as long as auxins are being transported down the stem from the apical bud, the lateral buds will not develop. This phenomenon, known as *apical dominance*, accounts for the fact that plants will not bush out until the apical buds are removed. While there are a number of natural and synthetic compounds that exhibit auxin activity, the major, naturally occurring auxin is a compound called indole acetic acid (IAA).

Cytokinins

Cytokinins are referred to as the cell division hormones, and while cell division will take place only in the presence of one of these hormones, the cytokinins stimulate a number of other plant responses as well. These hormones have been shown to retard senescence in detached leaves and to create *metabolic sinks*, areas within a tissue where increased metabolism takes place. As a result, there is an increase in transport of metabolites to the area. For example, amino acids will be transported to a site where the presence of cytokinins has increased protein synthesis.

In many plants, these hormones stimulate the production of larger, greener leaves by causing leaf cells to expand and by promoting both chloroplast development and chlorophyll synthesis. A number of naturally occurring substances that exhibit cytokinin activity have been identified. Of these, *ribosyl zeatin* is one of the most abundant in plants.

Gibberellins

Many dwarf plants exhibit a decrease in height because they contain low levels of the stem elongation hormones called gibberellins. These hormones, of which more than thirty have been identified, increase the amount of water taken up by the cells of stems. As the individual cells swell from the increased water content, the stem grows longer. In addition to stem elongation, the gibberellins elicit a number of other plant responses.

The seeds of numerous plant species exhibit *dormancy*, which can be broken by gibberellins. The length of the daylight period (*photoperiod*) is crucial to the flowering response in many plants. Fall- and winter-flowering plants require short days, while the plants that flower in the spring and summer must be exposed to long days. Some plants must also be subjected to a prolonged period of cold before flowering can occur. The gibberellins can substitute for the long day or the cold requirement in many plants. Additionally, the gibberellins can produce thicker stem growth in certain woody plants and increase the number of fruits that develop in some species.

Ethylene

One of the most important functions of ethylene is fruit ripening. Some fruits produce almost no ethylene until a few days before ripening and then release large amounts. Such fruits are said to be *climacteric*. *Nonclimacteric* fruits continuously produce more moderate amounts of the hormone throughout the ripening period. In all types of fruit, ethylene must be present before ripening can occur.

Ethylene also causes ripened fruit to *abscise* (separate from the parent plant) and is even involved in *dehiscence* (removal of the husk) of some types of fruit, such as pecans and walnuts. In addition to its role in fruiting, ethylene can initiate root development, cause leaves to droop, inhibit plant motion, and increase metabolic activity in some plants. Because ethylene inhibits auxin transport, it will also release apical dominance.

Abscisic Acid

The name "abscisic acid" was chosen because one of the major activities of this compound is to promote *leaf abscission*. During the autumn of the year, the concentration of this hormone increases in many plants. This high concentration causes the leaves to senesce, turn yellow, and abscise. Abscisic acid also inhibits growth and promotes both bud and seed dormancy.

Hormone Mechanisms

In the vast majority of hormone responses, *enzyme activation* or *gene induction* (the turning on of genes that produce new enzymes or other proteins) can be detected. This suggests that the hormones are initially acting as "first messengers." These first messengers react with the cellular membranes, and as a result of these reactions, the membranes activate "second messengers" within the cell. The second messengers then activate a group of substances referred to as *inducers*.

One group of inducers is the protein kinases, a

The plant hormone ethylene has been applied in commercial operations to husk walnuts.

tively. The cytokinins have been used to enhance sex ratios among the plants in the cucumber family, producing more female plants and thus more fruit. For years, ethylene has been used to enhance flowering and, in turn, increase yields in pineapples. Ethylene has also been used to promote root formation in stem cuttings and has been applied in commercial operations to husk walnuts.

Because gibberellins increase growth in the cellular layer that produces thicker stems in pines, seedlings can be sprayed with the hormone and be made ready for transport in two years rather than the normal three or four. The action of gibberellins on the production of amylase (an enzyme that breaks down starch to glucose) has proved to be useful in the brewing industry. The malting process can be accelerated because barley seeds treated with a gibberellin exhibit an increased rate of starch digestion.

One of the most economically important uses of the gibberellins has been in the grape industry. Yields have been increased by as much as one-third in grape vineyards sprayed with gibberellins. Gibberellins have also been used to increase length and water content in celery and sugar content in Hawaiian sugarcane.

Normally, hormone action is associated with the promotion of, enhancement of, or increase in some plant activity. This is the case, however, only when the concentration of the hormone is extremely low. At high concentrations, hormones actually have detrimental effects and can cause plant death. The knowledge of this has led to the development of a number of hormone-type *herbicides*. The most effective, by far, are the auxin-type herbicides such as 2,4-dichlorophenoxy acetic acid and picloram. These herbicides have provided excellent control of broadleaf weeds in grasses and food grains, because grasses are unaffected by them, while broadleaf weeds die as a result of overgrowth. There are probably a number of plant hormones that have not yet been identified. As these are discovered, and as those currently known are better understood, it will become possible to control even more aspects of plant growth and development.

D. R. Gossett

class of enzymes that add phosphate to other proteins. Protein kinases activate various enzymes or induce DNA (deoxyribonucleic acid) regulatory proteins to control genes responsible for cell elongation, cell division, flowering, fruit ripening, or one of the many other responses to the hormones. Each class of hormones activates a different set of protein kinases to produce a different set of responses. Most of the information concerning protein kinases is based on studies of animal systems, but it is highly likely that a similar type of mechanism is also present in plants.

Hormone Uses

Horticultural and agricultural applications of hormone technology are widespread. A dilute solution of auxin is used to promote the root development necessary to propagate stem cuttings vegeta-

See also: Dormancy; Flowering regulation; Germination and seedling development; Growth and growth control; Leaf abscission; Pheromones; Roots; Stems; Tropisms.

Sources for Further Study

Campbell, Neil A., and Jane B. Reece. *Biology*. 6th ed. San Francisco: Benjamin Cummings, 2002. An introductory, college-level textbook for science students. The chapter "Control Systems in Plants," provides a clear, concise description of the hormone groups. The well-written text combined with superb graphics furnish the reader with a clear understanding of hormone action. Includes list of suggested readings at the end of the chapter and a glossary.

Davies, Peter J., ed. *Plant Hormones: Physiology, Biochemistry, and Molecular Biology*. 2d ed. Boston: Kluwer, 1995. Describes how hormones are synthesized and metabolized, how they act at the organismal and molecular levels, how scientists measure them, and some of the roles they play.

Moore, Thomas C. *Biochemistry and Physiology of Plant Hormones*. 2d ed. New York: Springer-Verlag, 1989. Richly illustrated textbook. This updated edition includes a new chapter on brassinosteroids.

Raven, Peter H., Ray F. Evert, and Susan E. Eichhorn. *Biology of Plants*. 6th ed. New York: W. H. Freeman/Worth, 1999. An introductory, college-level textbook. The chapter "Regulating Growth and Development" provides an in-depth discussion of each group of hormones. Well illustrated, with a good pictorial record of various plant responses to the different hormones. Includes glossary.

Salisbury, Frank B., and Cleon W. Ross. *Plant Physiology*. 4th ed. Belmont, Calif.: Wadsworth, 1992. An intermediate, college-level textbook. The chapters "Hormones and Growth Regulators" and "Cytokinins, Ethylene, Abscisic Acid, and Other Compounds" give an in-depth view of the physiological role of plant hormones. An excellent explanation of hormones and hormone action is provided in text and graphics. Contains a detailed bibliography for each chapter.

HORNWORTS

Categories: Nonvascular plants; paleobotany; *Plantae*; taxonomic groups; water-related life

Hornworts are small, short, nonflowering, nonvascular plants which live both on land and in water. They represent an early land plant group and belong to three hundred species in the order Bryophyta. *Related to mosses and liverworts, hornworts are sometimes called horned liverworts.*

Evolutionary biologists have found a few fossils that could be the first known hornwort specimens. Fossilized spores dating from the late Cretaceous period, about 100 million years ago, are considered to be from hornworts. Other fossils dated earlier in that period resemble hornworts but might actually have been liverworts. Spores from the Miocene epoch, approximately twenty-six million years ago, are the most common hornwort fossils.

Botanists theorize that hornworts first appeared with earliest land plants prior to the Devonian period, 395 million years ago, and that fossil evidence is limited because the plants were so fragile. Many taxonomists consider hornworts to be more primitive than mosses and not as closely related to those plants as previously classified. A separate phylum, *Anthocerophyta*, was designated for hornworts, to separate them from mosses and liverworts.

A common and widely distributed plant, hornworts are blue-green members of six genera in two families, *Notothylales* and *Anthocerotales*, in the class *Anthocerotopsida*, derived from Greek words meaning "horn flower." Sometimes hornworts are classified in the subclass *Anthocerotidae* of the class *Hepaticae* as horned liverworts.

Found globally, usually in moist, shady, and

sometimes rocky environments, most hornworts belong to the genus *Anthoceros*. Species in the genera *Dendroceros* and *Megaceros* live primarily in tropical regions. Hornworts are found in forests, fields, ponds, streams, and riverbanks in small clumps or large patches and sometimes grow on tree trunks. They are nonvascular and do not have vessels to transport nourishment and moisture. In 1989 scientists reported they had detected a chemical compound with a structural and antimicrobial function in a hornworts species which was also found in the alga *Coleochaete*. This discovery suggested that green algae and hornworts were more closely related than previously thought.

Life Cycle

Hornworts' reproductive cycle occurs in an alternation of generations, in which a gametophyte creates a plant body (thallus) in the sexual generation, and the sporophyte produces a spore-containing capsule in the asexual generation. In the sexual generation, hornworts' gametophytes are flat and as small as 1 to 2 centimeters (0.4 to 0.8 inch) in diameter.

The thallus has hornworts' sexual organs, the male antheridium and female archegonium, on its top surface (whether the male and female organs occur on the same plant varies by species) and rhizoids on the bottom, which secure plants to soils. The hornworts' gametophytes resemble a clump of small leaves. Occasionally, the cyanobacteria *Nostoc* can be found in holes in the thallus, where it fixes nitrogen to exchange for carbohydrates.

Classification of Hornworts

Subdivision *Anthocerotae*
Class *Anthocerotopsida*
Order *Anthocerotales*
Family *Anthocerotaceae*
Genus *Anthoceros*
Genus *Phaeoceros*
Family *Dendrocerotaceae*
Genus *Megaceros*
Family *Notothyladaceae*
Genus *Notothylas*

Source: Data are from U.S. Department of Agriculture, National Plant Data Center, *The PLANTS Database*, Version 3.1, http://plants.usda.gov. National Plant Data Center, Baton Rouge, LA 70874-4490 USA.

Hornworts differ from liverworts in that hornworts lack cellular oil bodies and have mucilage instead of air chambers.

During the asexual generation, the sporophyte relies on the gametophyte for food and moisture, remaining connected to it during its life. Water is essential for the transport of sperm from the antheridium to the archegonium, where fertilized eggs become sporangia. From a basal sheath on the thallus, the sporophyte creates a slender, hornlike cylinder, which can be as high as 12 centimeters (4.75 inches) and gives the hornwort its common name.

Filled with spores, the sporangium elongates and gradually splits open from top to base, scattering spores. Internally, horn tissue continues to divide to produce thousands of spores. Other cells, called *pseudoelaters*, help spread spores, which germinate to begin the cycle again. Hornwort spores can survive as long as one decade in soil. Hornwort reproduction can also be achieved vegetatively if the thallus is fragmented.

Physically, hornworts resemble liverworts. They subtly differ from liverworts by having a place on the sporophyte's base, a near-basal meristem, where cells continue to divide and grow during the sporophyte's life. Hornworts also have a tissue column on the capsule, called a columella.

Hornwort gametophytes can be differentiated from those of liverworts and ferns by examination under a microscope to detect one large chloroplast in each hornwort cell. This chloroplast sometimes surrounds the nucleus and contributes to the color and translucency of hornworts. Magnification also reveals the mucilage between cells, instead of air, where cyanobacteria thrive. Hornwort sporophytes also deviate from those found in mosses and liverworts because they exist even after the gametophyte dies.

Taxonomy

In the late 1990's some botanists, deviating from the Linnean system, pursued new classification methodology in which molecular data was used to evaluate plants' taxonomical descriptions. This evidence was used to investigate and hypothesize about the origins and diversification of land plants. Such studies led some scientists to theorize that although hornworts, liverworts, and mosses physically resemble and grow near one another, liverworts and mosses are more closely related to each other than they are related to hornworts and should be assigned to different divisions.

Plant evolutionary biologists seek to determine terrestrial plants' primary lineages and phylogenetic relationships to one another. Botanists are especially interested in how the three bryophyte groups are genetically connected and their relationships to land plant lineages. Scientists disagree on whether hornworts or liverworts comprise land plants' sister group and whether mosses and liverworts are descended from the same lineage. Researchers want to determine which plant is the most primitive and whether one group is an ancestor of the others. Based on phylogenetic studies, botanists have developed hypotheses regarding hornworts. They suggest that hornworts are bryophytes derived from the earliest land plant lineages and that hornworts have developed specializations which differentiate them from other bryophytes and terrestrial plants.

In the early 1980's B. D. Mishler and S. P. Churchill presented phylogenetic trees which placed the three bryophyte lineages in the sequence of liverworts, hornworts, and mosses at the base of land plants, indicating that liverworts are a sister group to all land plants, and mosses are the vascular plants' sister group. Although many botanists accepted this interpretation, others developed dissenting theories. David J. Garbary and his colleagues examined male gametogenesis in the sperm of bryophytes, which share similar development characteristics. They determined that bryophytes collectively are a sister group to vascular plants and that mosses and liverworts are more closely related to each other than to hornworts.

Additional molecular sequence analysis of morphological characteristics reinforces the concept that hornworts are the basal lineage for terrestrial plants and that mosses and liverworts are the monophyletic sister group to vascular plants. Other botanists are convinced by molecular examination of chloroplast genetic sequences that hornworts are vascular plants' sister group. Scientists continue to test theories to prove that hornworts represent the oldest surviving land plant lineage.

Uses

The name hornwort is also used to describe an aquatic flowering plant, *Ceratophyllum demersum*, often called the coontail, from the family *Ceratophyllaceae*. Usually considered a weed or herb, this plant has many rigid, toothed, horn-shaped leaves, produced in groups of five. Sometimes looking like a raccoon's tail, it has no roots and lives beneath or floats upon the water surface, occasionally in colonies. This plant can grow as long as 4.6 meters (15 feet) and has branches. For pollination, both tiny male and female flowers are located on plants, which produce fruits with seeds. Most reproduction occurs vegetatively by plant fragments.

These hornworts are popular aquarium plants. Cuttings and shoots grow from placement of rhizoids in gravel-covered tank bottoms and quickly reach water surfaces in thick clumps, which need to be thinned. These hornworts provide cover for fish fry and also oxygenate bodies of water in which they are submerged.

Elizabeth D. Schafer

See also: *Archaea*; Bryophytes; Chloroplasts and other plastids; Cladistics; Evolution: historical perspective; Evolution of plants; Forests; Liverworts; Mosses; Oil bodies; Paleobotany; Systematics and taxonomy; Systematics: overview.

Sources for Further Study

Asthana, A. K., and S. C. Srivastava. *Indian Hornworts: A Taxonomic Study*. Berlin: J. Cramer, 1991. An edition of the series Bryophytorum Bibliotheca, which describes and classifies hornworts indigenous to the Indian subcontinent. Includes illustrations, map, and bibliography.

Crum, Howard A. *Liverworts and Hornworts of Southern Michigan*. Ann Arbor: University of Michigan Herbarium, 1991. A study focusing on the identification and classification of bryophytes growing in Michigan. Includes illustrations, map, bibliography, and index.

Kenrick, Paul, and Peter R. Crane. *The Origin and Early Diversification of Land Plants: A Cladistic Study*. Washington, D.C.: Smithsonian Institution Press, 1997. The role of hornworts in plant evolution was examined for this paleobotany text, published as part of a Smithsonian series in comparative evolutionary biology. Includes illustrations, bibliography, and indexes.

Lucas, Jessica R., and Karen S. Renzaglia. "Anatomy, Ultrastructure, and Physiology of Hornwort Stomata: An Evaluation of Homology." *The American Journal of Botany* 87 (June, 2000): S36-S37. Discusses how plant stomata function for water transport and gas exchange and explains that stomata are found in three hornwort genera, comparing these stomata with those of mosses and lycophytes.

Wakeford, Tom. *Liaisons of Life: From Hornworts to Hippos, How the Unassuming Microbe has Driven Evolution.* New York: John Wiley, 2001. Studies how microscopic organisms and symbiosis have contributed to evolutionary development. Includes illustrations, bibliography, and index.

Walstad, Diana L. *Ecology of the Planted Aquarium: A Practical Manual and Scientific Treatise for the Home Aquarist.* Chapel Hill, N.C.: Echinodorus, 1999. Credits hornworts with creating a healthy environment for aquarium fish, which are less likely to develop gill problems and suffer fewer toxic reactions when hornworts are present to purify tank water. Includes illustrations, bibliographical references, and index.

HORSETAILS

Categories: *Plantae*; seedless vascular plants; taxonomic groups

The plants known as horsetails or scouring rushes belong to the genus Equisetum, *the only remaining genus in the phylum* Sphenophyta, *a group of seedless vascular plants.*

Members of the phylum *Sphenophyta*, the horsetails, reached their maximum diversity during the Late Devonian and Carboniferous periods. One fossil group of the *Sphenophyta*, the calamites, grew from 12 to 18 meters (24 to 60 feet) in height, with trunks as much as 45 centimeters (more than 3 feet) in diameter. Today these ancient plants survive in the single genus *Equisetum*. *Equisetum* is found throughout the world and, depending on the classification scheme, comprises between fifteen and twenty-five species. Species that have branching forms are commonly called horsetails; *scouring rushes* are the unbranched species. *Equisetum* prefers moist or wet habitats, although some species do grow in drier areas. They can often be found growing along streams and the edges of woodlands.

Appearance

The stems of *Equisetum* are distinctly ribbed, with obvious nodes and internodes; there are numerous stomata in the grooves beneath the ribs. While it is the stems that carry out photosynthesis, *Equisetum* does have true leaves. The scalelike leaves are found at the nodes and form a collarlike structure just above the node. They are fused into a sheath around the stem.

Within the stems, especially of the scouring rushes, one can find significant deposits of silica,

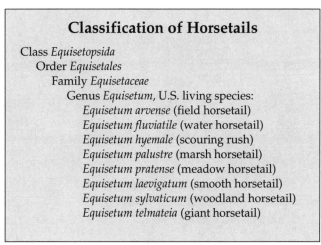

Classification of Horsetails

Class *Equisetopsida*
  Order *Equisetales*
    Family *Equisetaceae*
      Genus *Equisetum*, U.S. living species:
        Equisetum arvense (field horsetail)
        Equisetum fluviatile (water horsetail)
        Equisetum hyemale (scouring rush)
        Equisetum palustre (marsh horsetail)
        Equisetum pratense (meadow horsetail)
        Equisetum laevigatum (smooth horsetail)
        Equisetum sylvaticum (woodland horsetail)
        Equisetum telmateia (giant horsetail)

Source: Data are from U.S. Department of Agriculture, National Plant Data Center, *The PLANTS Database*, Version 3.1, http://plants.usda.gov. National Plant Data Center, Baton Rouge, LA 70874-4490 USA.

which gives the stems a rough, gritty texture. The name "scouring rush" indicates that, in a pinch, campers, like their pioneer forebears, can scrub dishes clean with stems of *Equisetum*. American Indians used the abrasive stems to smooth and polish bows and arrows. Stems of some species have often been used in the manner of sandpaper to polish wood.

A cross-section of a stem of *Equisetum* will show that the pith in the stem breaks down as the plant matures, leaving a hollow central canal in the stem. Two cylinders of smaller canals are located outside the pith. The inner cylinders, called carinal canals, are associated with strands of xylem tissue, which conducts water, and strands of phloem tissue,

which conducts the products of photosynthesis to places of storage. The outer cylinder contains vallecular canals, which contain air.

Reproduction

Asexual reproduction in *Equisetum* can take place through fragmentation of the stems, although sexual reproduction is the more common process. In the spring, some species produce, from the rhizomes, special cream- to buff-colored, nonphotosynthetic stems. Small, conelike strobili develop at the tips of these special stems or, in other species, at the tips of regular photosynthetic stems.

The strobili are usually about 2 to 4 centimeters (1 to 2 inches) long and look somewhat like little cones on top of the stems, with each conelike tip covered with hexagon-shaped plates. Each hexagon marks the top of a sporangiophore, which has five to ten elongate sporangia connected to the rim. When the sporangiophores separate slightly at maturity, the spores are released.

When the spore mother cells in the sporangia undergo meiosis, distinctive green spores are produced. The outer spore wall differentiates from the inside of the spore, forming coiled bands called *elaters*. Elaters uncoil when they dry out, functioning like wings to carry the spore along in the wind. If the spores are blown into a moist habitat, the elaters coil up, allowing the spore to drop and land in habitat suitable for germination.

After germination, spores produce lobed, cushionlike, photosynthetic green gametophytes that are seldom more than 8 millimeters in diameter but may range in size from a few millimeters up to 3.5 centimeters (about 1.5 inches) in diameter. Rhizoids anchor the gametophytes to the surface of the earth. The optimum habitats for gametophyte germination are recently flooded, nutrient-rich mudflats.

At first, about half of the gametophytes are male, with antheridia, and the other half are female, with archegonia. After a month or two, however, the female gametophytes of most species become bisexual, producing only antheridia. When water contacts mature antheridia, sudden changes in water pressure cause the sperm cells pro-

The extant representatives of the ancient plants known as horsetails survive in the single genus Equisetum.

PHOTODISC

duced within the antheridia to be explosively ejected. Sperm have several flagella, which aid them in swimming to the archegonia to fertilize the eggs. Several eggs on a female or bisexual gametophyte may be fertilized, and the development of more than one sporophyte is common.

Toxic Properties

Aerial stems develop from horizontal rhizomes, which are highly branched and perennial. Because the rhizomes can grow quite rapidly, *Equisetum* is often invasive. If livestock pastures are invaded by *Equisetum*, it can cause problems for farmers and ranchers, as the aerial stems of *Equisetum* are poisonous. *Equisetum arvense*, a species found in the United States and southern Canada, is especially toxic in dry hay. Horses, sheep, and cows are all susceptible to *Equisetum* poisoning. Symptoms include trembling, rapid pulse, weakness, excitability, diarrhea, staggering, and cold in the extremities. The major toxic substance is called *thiaminase*, which interferes with thiamine metabolism.

Medicinal Uses

Some members of the genus *Equisetum* have been found useful in folk and herbal medicine. Both *Equisetum arvense* and *Equisetum robustum* are known to exert diuretic effects. Poultices of crushed sterile stems can be applied to wounds to stop bleeding. In Ukraine, decoctions of *Equisetum heleocharis* have been used for this purpose. Boiling the stems yields a liquid extract that can be used as a mouthwash. There is a possibility that horsetail has some antibiotic properties. Roots were often given to teething babies, and *Equisetum* tea was often used to treat tubercular lung lesions.

Carol S. Radford

See also: Ferns; Fossil plants; Paleobotany; Seedless vascular plants.

Sources for Further Study

Abbe, Elfriede Martha. *The Fern Herbal: Including the Ferns, the Horsetails, and the Club Mosses*. Ithaca, N.Y.: Cornell University Press, 1985. Primarily interested in the therapeutic and medicinal uses of herbs, including the horsetails. Illustrations.

Fenton, Carroll Lane, Mildred Adams Fenton, Patricia Vickers Rich, and Thomas Hewitt Rich. *The Fossil Book: A Record of Prehistoric Life*. New York: Doubleday, 1989. While most of this book is devoted to fossil animals, it does contain material on plants of fossil importance.

Hauke, Richard Louis. *A Taxonomic Monograph of the Genus Equisetum, Subgenus Hippochaete*. Weinheim, Germany: J. Cramer, 1963. Monograph on the subgenus *Hipposhaete*, with maps, illustrations.

Raven, Peter H., Ray F. Evert, and Susan E. Eichhorn. *Biology of Plants*. 6th ed. New York: W. H. Freeman/Worth, 1999. This standard college textbook discusses the systematics and morphology of the different divisions of the plant kingdom.

Stern, Kingsley R. *Introductory Plant Biology*. 8th ed. Boston: McGraw-Hill, 2000. Introductory botany textbook includes discussions of the different divisions of the plant kingdom.

HORTICULTURE

Categories: Agriculture; disciplines; economic botany and plant uses; gardening

Horticulture is the branch of agriculture that is connected with the production of plants that are directly used by people for food, medicine, and aesthetic purposes.

The ability to produce crops, particularly those crops associated with food and fiber, is the multidisciplinary science of intensively cultivating plants to be used by humans for food, medicinal purposes, or aesthetic satisfaction. Crop production is largely determined by a variety of environmental conditions, including soil, water, light, temperature, and atmosphere. Therefore, horticulture science is primarily concerned with the study of how to manipulate the plants or these environmental factors to achieve maximum yield.

Because there is tremendous diversity in horticultural plants, the field is subdivided into *pomology*, the growth and production of fruit crops; *olericulture*, the growth and production of vegetable crops; *landscape horticulture*, the growth and production of trees and shrubs; and *floriculture*, the growth and production of flower and foliage plants. Each of these subdivisions is based on a fundamental knowledge of plant-soil interactions, soil science, plant physiology, and plant morphology.

Propagation

Horticulture science is concerned with all aspects of crop production, from the collection and germination of seed to the final marketing of the products. Plant propagation, protection, and harvesting are three areas of particular interest to horticulturists.

Generally, *propagation from seed* is the most common and least expensive way of propagating plants. In order to prevent cross-pollination from undesirable varieties, plants to be used for seed production are grown in genetic isolation from other, similar plants. At maturity, the seed is collected and is usually stored at low temperatures and less than 50 to 65 percent relative humidity to maintain full viability. The seed is often tested for viability prior to planting to determine the percentage of seed that should germinate. At the appropriate time, the seed is usually treated with a fungicide to ensure an adequate crop stand and planted under proper temperature, water, and light conditions. For most crops, the seed is germinated in small containers, and the seedlings are then transplanted to the field or greenhouse.

For many horticultural crops it is not feasible to produce plants from seed. For some, the growth from seed may require too much time to be economically practical. In other cases, the parent plants may produce little or no viable seed, and in still others, there may be a desire to avoid hybridization in order to maintain a pure strain. For some plants, almost any part of the root, stem, or leaf can be *vegetatively propagated*, but chemical treatment of the detached portion to ensure regeneration of the missing tissue is often required.

For other plants, a variety of specific vegetative plant tissues, including the roots, bulbs, corms, rhizomes, tubers, and runners, must be used for propagation. Individual runners are used for propagation purposes, but a number of cuttings can be propagated from one rhizome. Tubers are propagated by slicing the organ into several pieces, each of which must contain an "eye," or bud. Corms and bulbs are propagated by planting the entire structure.

A relatively new process of generating plants from cell cultures grown in the laboratory, called *tissue culture*, is a method often used to propagate pure lines of crops with a very high economic value. *Grafting*, a specialized form of vegetative propagation, is particularly useful in tree farming. The shoot from one plant with a desirable fruit quality can be grafted onto the root stock of another, more vigorous plant with a less desirable fruit quality.

Pest Control

Because plants are besieged by a panoply of biological agents that use plant tissues as a food source, protecting plants from pests is a major concern in the horticulture industry. Microbial organisms, nematodes, insects, and weeds are the major plant pests. Weeds are defined as unwanted plants and are considered to be pests because they compete with crop plants for water, sunlight, and nutrients. If left unchecked, weeds will drastically reduce crop yields because they tend to produce a large amount of seed and grow rapidly. Weed control is generally accomplished either by physically removing the weed or by use of a variety of herbicides that have been developed to chemically control weeds. Herbicides are selected on the basis of their ability to control weeds and, at the same time, cause little or no damage to the desired plant.

Plant protection from microbes, nematodes, and insects generally involves either preventing or restricting pest invasion of the plant, developing plant varieties that will resist or at least tolerate the invasion, or a combination of both methods. The application of chemicals, use of biological agents, isolation of an infected crop by quarantine, and cultural practices that routinely remove infected plants or plant tissues are examples of control methods. A large number of different bactericides, fungicides, nematocides, and insecticides has been developed in recent years. Because many of these chemicals are harmful to other animals, including humans, the use of pesticides, and insecticides in particular, requires extreme caution. There is an increasing interest in the use of biological control methods because many of the chemical pesticides pose a threat to the environment. The development and use of pest-resistant crop varieties and the introduction of natural enemies that will not only reduce the pest population but also live harmoniously in the existing environment are two of the more promising biological measures being employed.

Harvest

A crop must be harvested once it has grown to maturity. Harvesting is one of the most expensive aspects of crop production because it is usually very labor-intensive. For almost all crops, there is a narrow window between the time the plants are ready to harvest and the time when the plants are too ripe to be of economic value. Hence, the harvesting process requires considerable planning to ensure that the appropriate equipment and an adequate labor supply are available when the crop is ready to be harvested. Predicting the harvest date is of paramount importance in the planning process. The length of the harvest window, the length of the growing season that is necessary for a given plant to mature under normal environmental conditions at a given geographic location, and the influence of unexpected weather changes on the growing season all have to be considered in the planning process. Because nature is unpredictable, even the best planning schedules sometimes have to be readjusted in midseason.

Some crops are picked from the plant by hand and then mechanically conveyed from the field,

A horticulturist examines a climbing aster vine in a native plant garden at the University of Florida's research center. Native plants are gaining popularity with consumers and landscapers as concerns about biological invasions receive more publicity.

while other crops are harvested entirely by hand. New mechanical harvesting equipment is continually being developed by agricultural engineers, and crops that lend themselves to mechanical harvesting are growing in importance as the manual labor force continues to shrink. After harvest, most crops are generally stored for varying lengths of time, from a few days to several months. Because postharvest storage can affect both the quality and appearance of the product, considerable care is given as to how the crop is stored. Sometimes storage improves the quality and appearance, while in other cases, it causes them to deteriorate. The ideal storage conditions are those that maintain the product as close to harvest condition as possible.

Genetic Engineering

In order for horticulture to remain a viable resource in the future, advances in horticulture technology will have to continue to keep pace with the needs of an increasing population. However, horticulturists will also have to be mindful of the fragile nature of the environment. New technologies must be developed with the environment in mind, and much of this new technology will center on advances in genetic engineering. New crop varieties that will both provide higher yields and reduce the dependency on chemical pesticides by exhibiting greater resistance to a variety of pests will have to be developed. The future development of higher-yielding crops that can be harvested mechanically and the production of new types of equipment to facilitate the harvesting process will also be important improvements in the horticulture industry.

D. R. Gossett

See also: Agriculture: experimental; Biopesticides; Biotechnology; Genetically modified foods; High-yield crops; Hydroponics; Plant biotechnology; Plants with potential.

Sources for Further Study

Everett, Thomas H. *The New York Botanical Garden Illustrated Encyclopedia of Horticulture*. 10 vols. New York: Garland, 1980-1982. One of the most comprehensive standard resources, fully illustrated.

Fennema, O. R. *Principles of Food Science*. New York: Dekker, 1975. An authoritative presentation of numerous topics in food science, including the harvesting, preservation, and marketing of a variety of horticultural food crops.

Hartmann, H. T., and D. E. Kester. *Plant Propagation: Principles and Practices*. 6th ed. Upper Saddle River, N.J.: Prentice Hall, 1997. One of the most valuable sources available on the practical aspects of plant propagation.

Janick, Jules. *Horticultural Science*. 4th ed. New York: W. H. Freeman, 1986. Contains sections on horticulture biology, environment, technology, and industry. The focus is on the science on which the art of horticulture is based.

HUMAN POPULATION GROWTH

Categories: Animal-plant interactions; environmental issues

Since the Industrial Revolution of the nineteenth century, human populations have experienced a period of explosive growth. Overpopulation now poses a real threat to plant lives, ecosystems, and the long-term sustainability of the earth's current ecological balance.

Just eleven thousand years ago, there were only about five million humans who lived on the planet Earth. The initial population growth was slow, largely because of the way humans lived—by hunting. Such a mobile lifestyle limited the size of families for practical reasons. When simple means of birth control, often abstention from sex, failed, a woman would elect abortion or, more commonly, infanticide to limit her family size. Furthermore, a high mortality rate among the very young, the old, the ill, and the disabled acted as a natural barrier to rapid population growth.

Thus, it took more than one million years for the world's population to reach one billion. The second billion was added in about one hundred years, the third billion in fifty years, the fourth in fifteen years, and the fifth billion in twelve years. When humans became sedentary, some limits on the family size were lifted. With the development of agriculture, children became an asset to their families by helping with farming and other chores.

By the beginning of the common era (1 C.E.), human population had grown to about 130 million, distributed all over the earth. By 1650, the world population had reached 500 million. The process of industrialization had begun, bringing about profound changes in the lives of humans and their interactions with the natural world. With improved living standards, lower death rates, and prolonged life expectancies, human population grew exponentially. By 1999, there were about 6 billion people, compared with 2.5 billion in 1950. By 2002, the world population was well on its way to 7 billion, with an annual growth rate of nearly 100 million.

Plants, Agriculture, and Human Population

Starting about eleven thousand years ago (five million people), humans began to cultivate such plants as barley, lentils, wheat, and peas in the Middle East—an area that extends from Lebanon and Syria in the northwest eastward through Iraq to Iran. In cultivating and caring for these crops, early farmers changed the characteristics of these plants, making them higher yielding, more nutritious, and easier to harvest. Agriculture spread and first reached Europe by approximately six thousand years ago.

Agriculture might also have originated independently in Africa in one or more centers. Many crops were domesticated there, including yams, okra, coffee, and cotton. In Asia, agriculture based on staples such as rice and soybeans and many other crops such as citrus, mangos, taro, and bananas was developed. Agriculture was developed independently in the New World. It began as early as nine thousand years ago in Mexico and Peru.

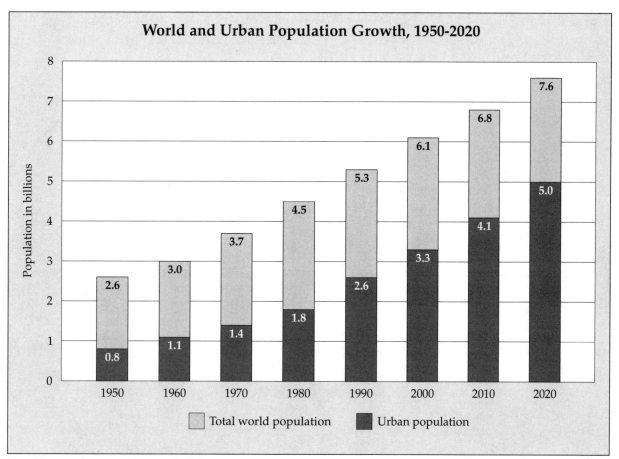

World and Urban Population Growth, 1950-2020

Note: The world's population passed 6 billion in the year 2000.

Sources: Data are from U.S. Bureau of the Census International Data Base and John Clarke, "Population and the Environment: Complex Interrelationships," in *Population and the Environment* (Oxford, England: Oxford University Press, 1995), edited by Bryan Cartledge.

Christopher Columbus and his followers found many new crops to bring back to the Old World, including corn, kidney beans, lima beans, tomatoes, tobacco, chili peppers, potatoes, sweet potatoes, pumpkins and squashes, avocados, cacao, and the major cultivated species of cotton.

For the last five to six centuries, important staple crops have been cultivated throughout the world. Wheat, rice, and corn, which provide 60 percent of the calories people consume, are cultivated wherever they will grow. Other crops, including spices and herbs, have also been brought under cultivation.

The growing population has changed the landscape, distribution, and diversity of plants dramatically. *Clear-cutting* and *deforestation* have driven many species (both plant and animal) to extinction. Relatively little has been done to develop agricultural practices suitable for tropical regions. As a result, the tropics are being devastated ecologically, with an estimated 20 percent of the world's species likely to be lost by the mid-twenty-first century.

A Threat to Sustainability

Without effective measures of control, the human population could exceed the earth's carrying capacity. Humans are, at present, estimated to consume about 40 percent of the total net products generated via photosynthesis by plants. Human activities have reduced the productivity of earth's forests and grasslands by 12 percent. Each year, millions of acres of once-productive land are turned into desert through *overgrazing* and deforestation, especially in developing countries. Due to overfertilization and aggressive practices in agriculture, loss of topsoil occurs at an annual rate of 24 billion metric tons. Collectively, these practices caused the destruction of 40 million acres of rain forest each year during the 1960's and 1970's and the extinction of enormous numbers of species.

Through technological innovation and aggressive practices in agriculture, a 2.6-fold increase in world grain production has been achieved since 1950. However, this increase in food output is not nearly enough to feed the population. Based upon

A busy street in Barcelona, Spain, photographed in 2002, as the world's human population passed 6.1 billion, with an annual growth rate of nearly 100 million.

an estimate by the World Bank and the Food and Agriculture Organization of the United Nations, one out of every five people is living in absolute poverty, unable to obtain food, shelter, or clothing dependably. About one out of every ten people receives less than 80 percent of the daily caloric intake recommended by the United Nations. In countries such as Bangladesh and Haiti and in regions such as East Africa, humans are dying in increasing numbers because of the lack of food. This lack of food may stem from drought, soil depletion, or soil loss; more often, famine results from inequitable distribution of resources among populations. Situations exacerbated by a growing population also

pose threats to the environment, aggravating the problems of acid rains, toxic and hazardous wastes, water shortages, topsoil erosion, ozone layer punctuation, greenhouse effects and groundwater contamination.

Ming Y. Zheng

See also: Acid precipitation; Agricultural revolution; Agriculture: modern problems; Agriculture: traditional; Agriculture: world food supplies; Erosion and erosion control; Green Revolution; Plant domestication and breeding; Rain forests and the atmosphere; Sustainable agriculture; Sustainable forestry.

Sources for Further Study
Brown, L. *State of the World 2000*. New York: W. W. Norton, 2000. Provides a comprehensive framework for the global debate about the future of the world in the new century. Presents invaluable analysis of negative environmental trends and a guide to emerging solutions.
Heiser, C. B., Jr. *Seed to Civilization: The Story of Food*. 2d ed. San Francisco: W. H. Freeman, 1981. A concise account of the plants that feed people.
National Research Council. *Lost Crops of the Incas: Little-Known Plants of the Andes with Promise for Worldwide Cultivation*. Washington, D.C.: National Academy Press, 1990. Presents an excellent example of ways in which economically cultivatable plants may be used for food.
Weiner, J. *The Next One Hundred Years: Shaping the Future of Our Living Earth*. New York: Bantam Books, 1990. Well-written account of the ecological problems that confront the human population as it continues to expand.

HYBRID ZONES

Categories: Ecosystems; genetics; reproduction and life cycles

Hybrids are offspring of parents from different species. Hybrid zones are areas where such different species overlap and crossbreed. Hybridization and hybrid zones have played a major role in the formation of new species in a number of plant groups and represent major factors in the evolutionary process.

A *hybrid* individual is produced from successful matings (cross-pollination) between individuals from different species or between individuals from different populations that differ markedly in one or more heritable traits. The mating process by which hybrid offspring are produced is *hybridization*. A distinction needs to be made between natural and artificial hybrids. An artificial hybrid typically involves direct human intervention in an effort to obtain plants with agricultural or horticultural properties superior to those of either parent.

Natural hybrids do not involve human intervention; they occur naturally.

In many cases hybridization between individuals that belong to different species is prevented by barriers or impediments to cross-pollination, known as *reproductive isolating mechanisms*. These mechanisms can be either prezygotic, preventing the formation of hybrid zygotes, or postzygotic, preventing or greatly reducing gamete exchange after a hybrid zygote has been formed. Hybridization typically occurs between species in which reproductive barriers (such as impediments to cross-pollination between members of different species) are not fully formed or are incomplete.

A hybrid zone is a geographic location in which two or more populations of individuals that differ in one or more heritable traits (either of the same or of different species) overlap, cross-breed, and produce viable and sometimes fertile offspring. The formation of hybrid zones involves *sympatric species*, that is, species whose geographic ranges overlap. Typically, *allopatric species* (those whose geographic ranges do not overlap) do not form hybrid zones unless some event, such as wind dispersal of seeds, brings individuals of the two species together.

Hybrid zones can be either continuous zones or a mosaic of scattered groups across a geographic range. They can also differ markedly in size. For example, the common herb *Gaillardia pulchella* (*Asteraceae* family) forms narrow hybrid zones in Texas, where transition progeny formed with neighboring *Gaillardia* species occurs over a few meters. In contrast, individuals found in hybrid zones involving the Bishop pine, *Pinus muricata*, in California can be several kilometers wide

In many cases hybrid zones are the result of human disturbance of the natural landscape. Such disturbance can lead to unique and novel habitat conditions in which the hybrid species might have a selective advantage over the parental species. One example occurs in Washington and Idaho, where certain hybrid zones are incubators for the speciation of *Tragopogon mirus* and *T. miscellus*.

Speciation Dynamics of Hybrid Zones

If natural selection eliminates hybrid offspring in the hybrid zone, then the reproductive barriers present in the parental species will be reinforced. If, however, the environment within the hybrid zone allows for persistence and reproduction of hybrid taxa, then these hybrids can persist through time, with several possible results. One result is the eventual establishment of reproductive barriers between the hybrid offspring and the parental species, with the formation of a new species from the hybrid lineage.

In some hybrid zones *allopolyploidy* can lead to the formation of a sterile hybrid. This occurs when two chromosome sets from different parents are present within one hybrid individual. The sterility is due to irregularities at meiosis, as there is only

one chromosome of each type leading to irregular segregation at meiosis. However, if the chromosome set is doubled through *autopolyploidy*, meiotic regularity is restored because each chromosome then has a homolog, which allows for successful chromosome segregation and gamete formation. Because of the difference in chromosome number between the polyploid hybrid derivative and the parental species, a reproductive barrier is established, and a new species will be established. This is the case with two species of goat's beard, *Tragopogon mirus* and *T. miscellus*, from southeastern Washington and adjacent Idaho. The progenitors of the polyploid *T. mirus* are the diploid species *T. dubius* and *T. porrifolius*, and of the polyploid *T. miscellus* are *T. dubius* and the diploid *T. pratensis*.

Another mechanism responsible for the formation of species within a hybrid zone is *recombination speciation*. In this process the parental genomes present within the semisterile hybrid offspring undergo rearrangement and recombination events over several generations, eventually producing mixed genomes in hybrid individuals. Over time, fertility is restored in the hybrid individuals, which are then reproductively isolated from the parental species. A prime example of recombination speciation occurs within a group of sunflowers found in the western United States. *Helianthus anomalus* arose as a consequence of hybridization between two sympatric parental species, *H. annuus* and *H. petiolaris*. Because of genome incompatibilities, the immediate hybrid offspring were reproductively isolated from either parental species and were semisterile. The genome arrangements that occurred over time within hybrids and their offspring resulted in increased fertility within the hybrid individual and breeding incompatibility between the hybrids and either parental species.

Introgression and Hybrid Swarms

Introgression occurs when the hybrid offspring engage in *backcrossing* with either one or both of the parental species. A *hybrid swarm* is usually a complex mixture of parental forms, F1 hybrids, and backcross individuals. The Louisiana irises provide a striking example of an introgressive swarm. Two parental species, *Iris fulva* and *I. hexagona*, have produced numerous hybrid populations in southern Louisiana. The hybrid individuals found in the hybrid zones are not true F1 hybrids but are the progeny resulting from numerous backcrosses to the parental species. A mixture of phenotypes is present in the hybrid swarm, with differing levels of similarity to the parental species among hybrid offspring.

Pat Calie

Sources for Further Study
Arnold, M. L. *Natural Hybridization and Evolution.* New York: Oxford University Press, 1997. An extensive analysis of the role of plant hybridization in the evolutionary process. Includes figures, references, and index.
Briggs, D., and S. M. Walters. *Plant Variation and Evolution.* New York: Cambridge University Press, 1997. Review of plant reproductive biology, with reference to hybridization in natural populations. Includes figures, tables, references, and index.
Grant, V. *The Origin of Adaptations.* New York: Columbia University Press, 1963. Discussion of plant isolating mechanisms and hybridization in natural populations. Includes references, index.
Ridley, M. *Evolution.* Boston: Blackwell Scientific Publications, 1993. Discussion of hybrid zones with their role in speciation events. Includes figures, tables, references, and index.
Stebbins, G. L., Jr. *Variation and Evolution in Plants.* New York: Columbia University Press, 1950. The historical standard for discussion of issues of plant evolution, with an extensive review of the role of hybridization in plant evolution. Includes references and index.

See also: Animal-plant interactions; Flower structure; Gene flow; Hybridization; Pollination; Polyploidy and aneuploidy; Population genetics; Reproductive isolating mechanisms; Species and speciation.

HYBRIDIZATION

Categories: Agriculture; genetics

Hybridization is the process of crossing two genetically different individuals to result in a third individual with a different, often preferred, set of traits. Plants of the same species cross easily and produce fertile progeny. Wide crosses are difficult to make and generally produce sterile progeny because of chromosome-pairing difficulties during meiosis.

Hybridization is the process of crossing two genetically different individuals to create new genotypes. For example, a cross between a parent 1, with the genetic makeup (genotype) *BB*, and parent 2, with *bb*, produces progeny with the genetic makeup *Bb*, which is a hybrid (the first filial generation or F_1). Hybridization was the basis of Gregor Mendel's historic experiments with garden peas. Inheritance studies require crossing plants with contrasting or complementary traits.

Hybridization of plants occurs in nature through various mechanisms. Some plants (such as the oil palm) are insect-pollinated, and others (such as maize, or corn) are wind-pollinated. Such plants are referred to as *cross-pollinated plants*. Natural hybridization has played a significant role in producing new genetic combinations and is the norm in cross-pollinated plants. It is a common way of generating genetic variability.

In plants with perfect flowers (*autogamous*, having flowers with both stamens and pistils), cross-pollination rarely occurs. Such plants (such as wheat and rice) are called *self-pollinated plants.* Flowers bearing only pistils or stamens are said to be imperfect flowers. Plants that have separate pistillate and staminate flowers on the same plant (such as maize) are called *monoecious*. Plants that have male and female flowers on separate plants (such as asparagus) are called *dioecious*.

Through artificial means (controlled pollination), hybridization of both cross-pollinated and self-pollinated plants can be accomplished. Artifi-

AP/WIDE WORLD PHOTOS

An apple grower checks on his recently planted root-stalks; a root-stalk is the result of a hybrid grafting technique combining a budding branch with a tree root to produce a fast-growing dwarf tree. Artificial hybridization is an important aspect of improving both cross-pollinated and self-pollinated plants.

Kölreuter was the first to report on hybrid vigor in interspecific crosses of various species of *Nicotiana*. He concluded that cross-fertilization was generally beneficial and self-fertilization was not. In 1799 T. A. Knight conjectured that because of widespread existence of cross-pollination in nature, it must be the norm. Charles Darwin reported the results of his experiments with maize. He indicated that in twenty-four crosses, there was an increase in plant height, which was attributed to hybridization, and that decrease in plant height was associated with self-pollination (or selfing). He also noted that crossing of inbred plants could reverse the deleterious effects of selfing or *inbreeding*. In 1862 Darwin wrote, "Nature tells us, in the most emphatic manner, that she abhors perpetual self-fertilization." In the late 1800's William J. Beal evaluated hybrids between maize varieties. He observed that some hybrids yielded 50 percent more than the mean of their parents. S. W. Johnson provided an explanation for hybrid vigor in 1891. G. W. McClure reported in 1892 that hybrids between maize varieties were superior to the mean of the two parents.

Exploitation of Heterosis

The phenomenon of heterosis has been exploited in crop plants, such as maize, sorghum, sunflower, onion, and tomato. Maize (corn) was the first crop in the United States in which hybrids were produced from *inbred lines*. It was George Shull who, following the rediscovery of Mendel's laws of inheritance in 1900, conducted the first experiments on inbreeding and crossing, or hybridizing, of inbred lines. Shull suggested that inbreeding within a maize variety resulted in pure (*homozygous*) lines and that hybrid vigor resulted from crossing of pure lines because *heterozygosity* was created at many allelic sites. Hybrid maize was introduced in the United States in the late 1920's and early 1930's, after which U.S. maize production increased dramatically from the use of hybrids.

Heterosis now drives a multibillion-dollar business in agriculture. Yield improvement made in various crops in which heterosis was detected has been tremendous. In 1932 in the United States, 44.8 million hectares (111 million acres) were required to produce 51 million metric tons of maize grain, with

cial hybridization is an important aspect of improving both cross-pollinated and self-pollinated plants. The breeder must know the time of development of reproductive structures of the species, treatments to promote and synchronize flowering, and pollinating techniques.

Applications to Agriculture

The concept of *hybrid vigor*, or *heterosis*, resulted from hybridization. Heterosis (or heterozygosis) occurs when the hybrid outperforms its parents for a certain trait. Around 1761 Joseph Gottlieb a mean yield of 1.66 metric tons per hectare. In 1994 it took only 32 million hectares (79 million acres) to produce 280 million metric tons of grain, with a mean yield of 8.69 metric tons per hectare. In the United States in 1996, twenty-one vegetable crops occupied 1,576,494 hectares (3.9 million acres), with a mean of 63 percent of the crop in hybrids. Heterosis saved an estimated 220,337 hectares (544,459 acres) of agricultural land per year, feeding 18 percent more people without an increase in land use. From 1986 to 1995, the best rice hybrids showed a 17 percent yield advantage over the best inbred-rice varieties at the International Rice Research Institute.

Despite the impact that heterosis has had on crop production, its molecular genetic basis is still not clear. It is hoped that with the progress being made in the genetic sequencing of various plant species, a better understanding of heterosis will emerge. Plant breeding entails hybridization within a species as well as hybridization between species or even genera, called *wide crosses*. The latter are important for generating genetic variability or for in-corporating a desirable gene not available within a species. There are barriers, however, for accomplishing *interspecific* and *intergeneric* crosses. Plants of the same species cross easily and produce fertile progeny. Wide crosses are difficult to make and generally produce sterile progeny because of chromosome-pairing difficulties during meiosis.

Triticale is the only human-made cereal crop, which is a cross between the genus *Triticum* (wheat) and the genus *Secale* (rye). The first fertile triticale was produced in 1891. Some of the interspecific and intergeneric barriers should be overcome via the newer techniques of gene transfer. It is expected that genes from wild relatives of cultivated plants will continue to be sought to correct defects in otherwise high-yielding varieties.

Manjit S. Kang

See also: Agriculture: traditional; Alternative grains; Corn; Genetics: Mendelian; Grains; Green Revolution; Hybrid zones; Pollination; Species and speciation; Wheat.

Sources for Further Study

Basra, A. S., ed. *Heterosis and Hybrid Seed Production in Agronomic Crops*. Binghamton, N.Y.: Food Products Press, 1999. This book discusses current research in some of the most important crops of the world.
Coors, J. G., and S. Pandey. *Genetics and Exploitation of Heterosis in Crops*. Madison, Wis.: American Society of Agronomy and Crop Science Society of America, and Soil Science Society of America, 1999. Provides an account of the various issues related to hybrid vigor, or heterosis.
Fehr, W. R., and H. H. Hadley. *Hybridization of Crop Plants*. Madison, Wis.: American Society of Agronomy and Crop Science Society of America, 1980. Brings together the experience of plant breeders and scientists in a form that can be used by the layperson.

HYDROLOGIC CYCLE

Categories: Biogeochemical cycles; ecology

The hydrologic cycle is a continuous system through which water circulates through vegetation, in the atmosphere, in the ground, on land, and in surface water such as rivers and oceans.

The sun and the force of gravity provide the energy to drive the cycle that provides clean, pure water at the earth's surface. The total amount of water on earth is an estimated 1.36 billion cubic kilometers. Of this water, 97.2 percent is found in the earth's oceans. The ice caps and glaciers contain 2.15 percent of the earth's water. The remainder, 0.65 percent, is divided among rivers (0.0001 percent), freshwater and saline lakes (0.017 percent), groundwater (0.61 percent), soil moisture (0.005

percent), the atmosphere (0.001 percent), and the biosphere and groundwater below 4,000 meters (0.0169 percent). While the percentages of water appear to be small for these water reservoirs, the total volume of water contained in each is immense.

Evaporation

Evaporation is the process whereby a liquid or solid is changed to a gas. Heat causes water molecules to become increasingly energized and to move more rapidly, weakening the chemical force that binds them together. Eventually, as the temperature increases, water molecules move from the ocean's surface into the overlying air. The rate of evaporation is influenced by radiation, temperature, humidity, and wind velocity.

Each year about 320,000 cubic kilometers of water evaporate from these oceans. It is estimated that an additional 60,000 cubic kilometers of water evaporate from rivers, streams, and lakes or are *transpired* by plants each year. A total of about 380,000 cubic kilometers of water is *evapotranspired* from the earth's surface every year.

Condensation and Precipitation

Wind may transport the moisture-laden air long distances. The amount of water vapor the air can hold depends upon the temperature: The higher the temperature, the more vapor the air can hold. As air is lifted and cooled at higher altitudes, the vapor in it condenses to form droplets of water. *Condensation* is aided by small dust and other particles in the atmosphere. As droplets collide and coalesce, raindrops begin to form, and *precipitation* begins.

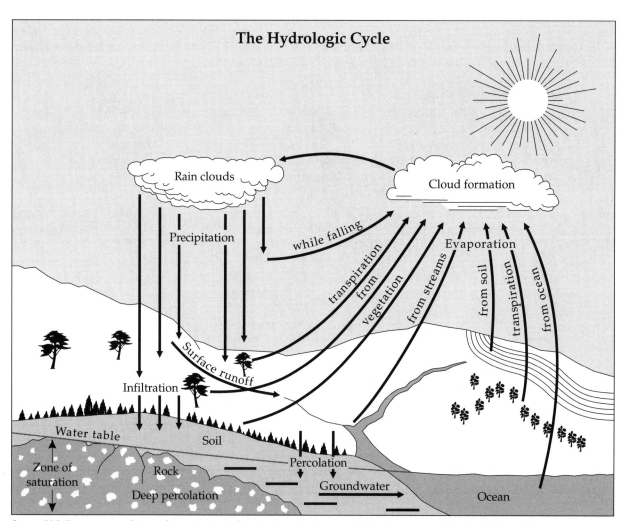

Source: U.S. Department of Agriculture, *Yearbook of Agriculture* (Washington, D.C.: Government Printing Office, 1955).

A wall of fog blows into the bay in Grand Portage, Minnesota. Wind may transport the moisture-laden air long distances. The amount of water vapor the air can hold depends upon the temperature: The higher the temperature, the more vapor the air can hold.

Most precipitation events are the result of three causal factors: frontal precipitation, or the lifting of an air mass over a moving weather front; convectional precipitation related to the uneven heating of the earth's surface, causing warm air masses to rise and cool; and orographic precipitation, resulting from a moving air mass being forced to move upward over a mountain range, cooling the air as it rises.

Each year, about 284,000 cubic kilometers of precipitation fall on the world's oceans. This water has completed its cycle and is ready to begin a new cycle. Approximately 96,000 cubic kilometers of precipitation fall upon the land surface each year. This precipitation follows a number of different pathways in the hydrologic cycle. It is estimated that 60,000 cubic kilometers evaporate from the surface of lakes or streams or transpire directly back into the atmosphere. The remainder, about 36,000 cubic kilometers, is intercepted by human structures or vegetation, infiltrates the soil or bedrock, or becomes surface runoff.

Interception

In cities, the amount of water intercepted by human structures may approach 100 percent. However, much urban water is collected in storm sewers or drains that lead to a surface drainage system or is spread over the land surface to infiltrate the subsoil. *Interception* loss from vegetation depends upon interception capacity (the ability of the vegetation to collect and retain falling precipitation), wind speed (the higher the wind speed, the greater the rate of evaporation), and rainfall duration (the interception loss will decrease with the duration of rainfall, as the vegetative canopy will become saturated with water after a period of time). Broad-leaf forests may intercept 15 to 25 percent of annual precipitation, and a bluegrass lawn may intercept 15 to 20 percent of precipitation during a growing season.

Transpiration

Plants are continuously extracting soil moisture and passing it into the atmosphere through a process called *transpiration*. Moisture is drawn into the plant rootlet through osmotic pressure. The water moves through the plant to the leaves, where it is passed into the atmosphere through the leaf openings, or stomata. The plant uses less than 1 percent of the soil moisture in its metabolism; thus, transpiration is responsible for most water vapor loss from the land in the hydrologic cycle. For example, an oak tree may transpire 151,200 liters per year.

Overland Flow and Infiltration

When the amount of rainfall is greater than the earth's ability to absorb it, excess water begins to run off, a process termed *overland flow*. Overland flow begins only if the precipitation rate exceeds the infiltration capacity of the soil. *Infiltration* occurs when water sinks into the soil surface or into fractures of rocks; the amount varies according to the characteristics of the soil or rock and the nature of the vegetative cover. Sandy soils have higher infiltration rates than clay rock soils. Nonporous rock has an infiltration rate of zero, and all precipitation that reaches it becomes runoff. The presence of vegetation impedes surface runoff and increases the potential for infiltration to occur.

Water infiltrating the soil or bedrock encounters two forces: capillary force and gravitational force. A capillary force is the tendency of the water in the subsurface to adhere to the surface of soil or sediment particles. Capillary forces are responsible for the soil moisture a few inches below the land surface.

The water that continues to move downward under the force of gravity through the pores, cracks, and fissures of rocks or sediments will eventually enter a zone of water saturation. This source of underground water is called an aquifer—a rock or soil layer that is porous and permeable enough to hold and transport water. The top of this aquifer, or saturated zone, is the *water table*. This water is moving slowly toward a point where it is discharged to a lake, spring, or stream. Groundwater that augments the flow of a stream is called *base flow*. Base flow enables streams to continue to flow during droughts and winter months. Groundwater may flow directly into the oceans along coastlines.

When the infiltration capacity of the earth's surface is exceeded, overland flow begins. Broad, thin sheets of water a few millimeters thick are called *sheet flow*. After flowing a few meters, the sheets break up into threads of current that flow in tiny channels called rills. The rills coalesce into gullies and, finally, into streams and rivers. Some evaporation losses occur from the stream surface, but much of the water is returned to the oceans, thus completing the hydrologic cycle.

Residence Time

Residence time refers to how long a molecule of water will remain in various components of the hydrologic cycle. The average length of time that a water molecule stays in the atmosphere is about one week. Two weeks is the average residence time for a water molecule in a river, and ten years in a lake. It would take four thousand years for all the water molecules in the oceans to be recycled. Groundwater may require anywhere from a few weeks to thousands of years to move through the cycle. This time period suggests that every water molecule has been recycled millions of times.

Methods of Study

Several techniques are used to gather data on water in the hydrologic cycle. These data help scientists determine the water budget for different geographic areas. Together, these data enable scientists to estimate the total water budget of the earth's hydrologic cycle.

Scientists have developed a vast array of mathematical equations and instruments to collect data on the hydrologic cycle. Variations in temperature, precipitation, evapotranspiration, solar radiation, vegetative cover, soil and bedrock type, and other factors must be evaluated to understand the local or global hydrologic cycle.

Precipitation is an extremely variable phenomenon. The United States has some thirteen thousand precipitation stations equipped with rain gauges, placed strategically to compensate for wind and splash losses. Precipitation falling on a given area is determined using a rain-gauge network of uniform density to determine the arithmetic mean for rainfall in the area. The amount of water in a *snowpack* is estimated by snow surveys. The depth and water content of the snowpack are measured and the extent of the snow cover mapped using satellite photography.

The amount of precipitation lost by interception can be measured and evaluated. Most often, interception is determined by measuring the amount above the vegetative canopy and at the earth's surface. The difference is what is lost to interception.

The volume of water flowing by a given point at a given time in an open stream channel is called discharge. Discharge is determined by measuring the velocity of water in the stream channel, using a current meter. The cross-sectional area of the stream channel is determined at a specific point and multiplied by the stream velocity. Automated stream-gauging stations are located on most streams to supply data for various hydrologic investigations.

The U.S. National Weather Service maintains about five hundred stations using metal pans, mimicking reservoirs, to measure free-water evaporation. Water depths of 17 to 20 centimeters are maintained in the pans. Errors may result from splashing by raindrops or birds. Because the pans will heat and cool more rapidly than will a natural reservoir, a pan coefficient is employed to compensate for this phenomenon. The wind velocity is also determined. A lake evaporation nomograph determines daily lake evaporation. The mean daily temperature, wind velocity, solar radiation, and mean daily dew point are all used in the calculation.

The amount of evapotranspiration can be measured using a lysimeter, a large container holding soil and living plants. The lysimeter is set outside, and the initial soil moisture is determined. All precipitation or irrigation is measured accurately. Changes in the soil moisture storage determine the amount of evapotranspiration.

Samuel F. Huffman

See also: Calvin cycle; Carbon cycle; Leaf anatomy; Nitrogen cycle; Phosphorus cycle; Plant tissues; Water and solute movement in plants.

Sources for Further Study

Berner, Elizabeth K., and Robert A. Berner. *Global Environment: Water, Air, and Geochemical Cycles*. Upper Saddle River, N.J.: Prentice Hall, 1996. Cycles of the major chemical components of water, air, rocks, and life are discussed as they occur naturally and as they are disturbed by humans.

Moore, J. W. *Balancing the Needs of Water Use*. New York: Springer-Verlag, 1989. Reviews components of the water use cycle and, to a lesser extent, the hydrologic cycle. Requirements of various water users are examined, such as natural fish and wildlife populations, water storage, agriculture, forestry, and municipalities.

Viessman, Warren Jr., and Gary L. Lewis. *Introduction to Hydrology*. 4th ed. Glenview, Ill.: HarperCollins, 1995. Covers development, management, and protection of water resources. Accessible to students but also useful for water scientists and engineers.

HYDROPONICS

Categories: Agriculture; disciplines; economic botany and plant uses; gardening; water-related life

Literally "water culture," hydroponics originally referred to the growth of plants in a liquid medium. It now applies to all systems used to grow plants in nutrient solutions with or without the addition of synthetic soil for mechanical support.

Hydroponics has become an important method of crop production with the increase in the number of commercial greenhouse operations. Greenhouses are utilized in the production of a wide array of bedding plants, flowers, trees, and shrubs for commercial as well as home and garden use. Cash receipts from greenhouse and nursery crops total more than $4 billion annually. In some arid regions, the majority of vegetable crops are produced in greenhouses.

A farmer culls dead leaves from tomato plants, which are raised in a greenhouse in a soilless environment. Like all hydroponically grown plants, these tomatoes are grown with nutrient solutions.

Types of Systems

The four most commonly used hydroponic systems are *sand-culture systems*, *aggregate systems*, *nutrient film techniques*, and *floating systems*. While these systems are similar in their use of nutrient solutions, they vary in both the presence and type of supporting medium and in the frequency of nutrient application. In sand culture, coarse sand is used in containers or spread over a greenhouse floor or bed, on top of a recirculating drain system. A drip irrigation system is used to apply nutrient solution periodically, and a drainage system is used to collect the excess solution as it drains through the sand.

In an aggregate open system, plants are transplanted into plastic troughs filled with an inert supporting material, and nutrient solution is supplied via drip irrigation. The aggregate system and the sand culture are open systems because the nutrient solution is not recycled.

In the nutrient film technique, there is no supporting material. Seedlings are transplanted into troughs through which the nutrient solution is channeled, and the plants are in direct contact with the nutrient solution. In this closed system, the nutrient solution is channeled past the plant, collected, and reused. The floating hydroponic system involves the floating of plants over a pool of nutrient solution.

While the nutrient film technique and floating hydroponic systems are primarily used in research applications, the sand culture and aggregate systems are commonly used in commercial plant production. These two systems require the use of a nutrient solution and synthetic soil for mechanical support. A variety of nutrient solutions have been formulated since the first was developed in 1950.

Mechanical Support Materials

A large variety of both *organic* and *inorganic* materials have been used to formulate the synthetic soils used for mechanical support in hydroponic systems. Commonly used organic materials include sphagnum moss, peat, manure, wood, and other plant residues. Sphagnum moss, the shredded, dehydrated remains of several species of moss in the genus *Sphagnum*, is harvested for the purpose of producing synthetic soil. "Peat" is a term normally used to describe partially decomposed remains of wetlands vegetation that has been preserved under water. Peat moss is the only type of peat suitable for synthetic soil mixes. Peat moss is harvested from peat bogs, dried, compressed into bales, and sold. Animal manures are almost never used in commercial synthetic soil mixtures because they require costly handling and sterilization procedures.

Wood residues such as tree bark, wood chips, shavings, and sawdust are generally produced as by-products of the timber industry. A variety of other plant residues, including corn cobs, sugarcane stems, straw, and peanut and rice hulls have been substituted for peat in synthetic soil mixtures where there is a supply of these materials.

Commonly used inorganic materials include vermiculite, sand, pumice, perlite, cinders, and calcined clay. Vermiculite is a very lightweight material produced by heating mica to temperatures above 1,090 degrees Celsius (nearly 2,000 degrees Fahrenheit). Sand is a preferred material for formulating synthetic soils because it is inert and inexpensive but very heavy compared to other commonly used materials. Pumice, a natural, glasslike material produced by volcanic action, provides a good inert supporting material when ground into small particles. Perlite, a porous material that will hold three to four times its weight in water, is produced by heating lava at temperatures above 760 degrees Celsius (1,400 degrees Fahrenheit). Cinders are derived from coal residues that have been thoroughly rinsed to remove harmful sulfates. Calcined clay is derived from the mineral montmorillonite baked at temperatures above 100 degrees Celsius.

D. R. Gossett

See also: Fertilizers; Horticulture; Nutrients; Nutrition in agriculture.

Sources for Further Study

Janick, Jules, ed. *Horticulture Science*. Englewood Cliffs, N.J.: Prentice Hall, 1989. Contains a section on hydroponics and synthetic soils and their use in horticulture.

Jensen, M. H., and W. L. Collins. "Hydroponic Vegetable Production." *Horticultural Reviews* 7 (1985). An excellent review with an interesting discussion of hydroponic technology.

Jones, J. Benton. *Hydroponics: A Practical Guide for the Soilless Grower*. Boca Raton, Fla.: St. Lucie Press, 1997. Explains the basics of plant growth and development, different methods of preparing and using hydroponic nutrient solutions, and hydroponic options for various environmental conditions. Gives the reader instructions for simple experiments and a number of helpful charts, tables, and illustrations.

Schwarz, M. *Soilless Culture Management*. New York: Springer-Verlag, 1995. A guide for students, agriculture instructors, and soilless-culture farmers. Provides information on optimal plant nutrition, deficiencies and toxicities of nutrients, plant growth media, optimal root environment, environmental control, carbon dioxide requirements, saline conditions, and use of sewage in soilless culture.

INFLORESCENCES

Categories: Anatomy; angiosperms; gardening

The term "inflorescence" refers to the arrangement of flowers on a floral axis. Most schemes that define inflorescence types separate solitary flowers from flower clusters and stipulate that an inflorescence is a cluster of two or more flowers.

It is not always easy to distinguish between *solitary flowers* and an inflorescence. An examination of the evolutionary development of the flower and the inflorescence provides some insight into the problem. It generally is accepted that the flower arose as a modified stem tip that bore male and female reproductive structures at its apex. These reproductive structures became the pistils and stamen of the flower. Leaves that immediately subtended the reproductive structures became the sterile parts of the flower (petals and sepals) and are typically more leaflike as distance from the apex increases.

If leaves subtending the flower are much smaller or distinctly different from regular leaves, they are referred to as bracts. If a second, considerably smaller set is present, its component parts are termed *bracteoles*. The determination of whether subtending leaflike structures are leaves or bracts may establish a flower as solitary or as part of an inflorescence.

Sometimes woody branches that support flowers are modified. They may grow much more slowly than branches that support only vegetative structures. The latter pattern is observed in many fruit tree species, such as apples and pears, where the fruit is supported by short, modified branches called *spurs*. Clusters of flowers issuing from such spurs may resemble inflorescence types, although the flowers are solitary.

Flowers may be *complete*, possessing all four sets of floral appendages, or they may be reduced to as little as one set of reproductive structures (stamen or pistils). If clusters of many reduced flowers are borne on very short stems, the resulting aggregation may superficially resemble a single flower. This type of inflorescence is associated with daisies and asters (*Asteraceae*) and is often mistaken for a single flower by those unfamiliar with flower and inflorescence structure. In spite of these confusing elements, most common inflorescence patterns can easily be recognized.

Parts of an Inflorescence

The following terms, some of which already have been introduced, are features or structures that are used to classify inflorescences. An *axillary bud* occurs in the angles between a stem and a leaf petiole. A *bract* is a small or modified leaf immediately beneath a flower or inflorescence. A bracteole is a bract that is much smaller in size.

An *involucre* is a series of bracts or bracteoles subtending a flower or inflorescence. A *pedicel* is a stalk supporting a single flower of an inflorescence. A *peduncle* is a stalk of a solitary flower or of an inflorescence. A *rachis* is the main branch or axis within a complex inflorescence.

Inflorescence Types

Parameters used to classify basic inflorescence types include

(1) number and position of flowers

(2) sequence of flower development, and

(3) the nature of inflorescence branching.

Because the inflorescence type of a given species may result from evolutionary reduction, classification schemes are typically artificial and do not reflect evolutionary significance. The form of an inflorescence, however, is determined largely by two patterns of development.

If the growing tip of the stem (*apical meristem*) continues to grow and produce new flowers as it elongates, the inflorescence is said to be *indeterminate*. A raceme (defined below) is a typical indeterminate inflorescence. If the apical meristem quickly matures into a flower, it can no longer grow in length, and the inflorescence exhibits a limited growth pattern. This type of inflorescence is said to be *determinate* and is best represented by a cyme (defined below). The following descriptions of inflorescence types represent most of the basic types. Any vascular plant taxonomy text will provide a more comprehensive list.

Indeterminate Inflorescences

A *catkin* (also known as an *ament*) is a spikelike inflorescence. Dissection may reveal the presence of minute, and possibly branched, pedicels. The flowers are typically unisexual and are hidden by bracts. This inflorescence is typical of trees such as oaks, hickories, and birches.

A *corymb* is a flat- or rounded-top inflorescence. The pedicels of flowers are attached along the length of the peduncle. Corymbs may be simple or compound. Examples include hydrangea and hawthorn.

A *head*, or *capitulum*, is a tight cluster of sessile flowers (flowers with no pedicel) borne on a flattened or short stem tip (receptacle). Heads are a diagnostic feature of the sunflower family (*Asteraceae*),

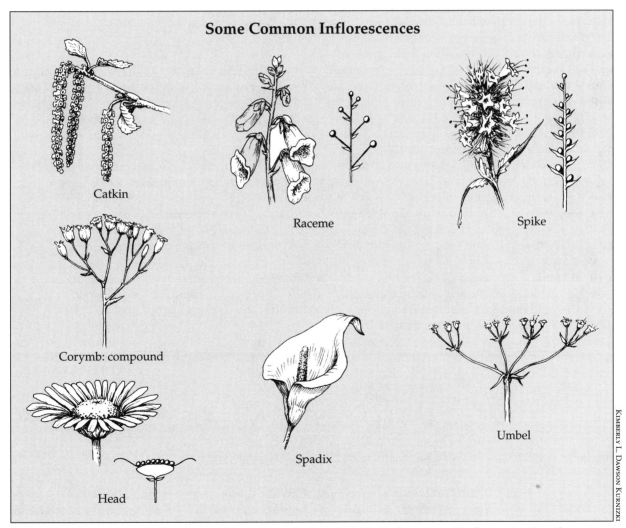

Some Common Inflorescences

Catkin

Raceme

Spike

Corymb: compound

Spadix

Umbel

Head

KIMBERLY L. DAWSON KURNIZKI

The variety of arrangements flowers may take on a floral axis, called the "inflorescence," is far greater than those examples displayed here, also including cyme, panicle, cluster, cyathium, mono-, di-, and pleiochasium, thyrse, and verticillaster inflorescences and variations on these. Solitary flowers are technically not inflorescences because they are not clusters of flowers.

examples of which are daisies, chrysanthemums, and sunflowers.

A *panicle* has a branched floral axis (rachis), which may re-branch prior to bearing flower pedicels (described sometimes as a compound raceme).

A *raceme* has pedicellate flowers borne on an elongate rachis. It is often confused with a spike when the pedicels are small and inconspicuous. Examples include foxglove and lupine.

A *spike* has sessile flowers borne on a single rachis. Examples are ladies' tresses (a type of orchid) and plantain. A *spikelet* is a small spike. The flowers are inconspicuous and often hidden by a series of modified bracts. This is the basic inflorescence unit of grasses and sedges.

A *spadix* is a spike with flowers embedded in a fleshy rachis. Typically, the spadix is subtended and surrounded by a large modified bract termed a spathe. The spadix is characteristic of the arum family (*Araceae*), examples of which are jack-in-the-pulpit and elephant ear.

Determinate Inflorescences

An *umbel* can be determinate or indeterminate, with a flat or rounded top. The pedicels of flowers are attached to a common point on the peduncle. Umbels may be simple or compound. Compound umbels are the typical inflorescence of most members of the carrot family (*Apiaceae*). Examples include onion, carrot, and dill.

A *cyme* is a branching inflorescence with individual flowers at the end of each branch. A simple cyme is determinate, with a grouping of three flowers on a peduncle. The central flower matures first. Examples include champions and some of the anemones. A compound cyme is composed of two or more cymes together. Examples include chickweed and phacelia.

John F. Logue

See also: Angiosperm cells and tissues; Angiosperm evolution; Angiosperms; Flower structure; Flower types; Flowering regulation; Garden plants: flowering; Pollination; Shoots; Stems.

Sources for Further Study

Heywood, V. H. *Flowering Plants of the World*. Rev. ed. New York: Oxford University Press, 1993. Reference work on families of flowering plants. Contains a comprehensive glossary, which includes inflorescence types. An additional section provides colored drawings and diagrams of inflorescences. Family descriptions contain reference to inflorescence.

Moore, Randy, et al. *Botany*. 2d ed. New York: WCB/McGraw-Hill, 1998. This basic textbook presents inflorescences as a separate topic, with ample photographs, diagrams, and colored drawings.

Walter, Dirk R., and David J. Keil. *Vascular Plant Taxonomy*. 4th ed. Dubuque, Iowa: Kendall/Hunt, 1996. Basic taxonomy text that includes descriptive sections on the flower, perianth, and inflorescences. It also provides a key to inflorescence types. Includes diagrams, bibliographic references, glossary, and index.

INTEGRATED PEST MANAGEMENT

Categories: Agriculture; environmental issues; pests and pest control

Integrated pest management (IPM) is the practice of integrating insect, animal, or plant management tactics, such as chemical control, cultural control, biological control, and plant resistance, to maintain pest populations below damaging levels in the most economical and environmentally responsible manner.

In the past, pest management strategies in agriculture focused primarily on eliminating all of a particular pest organism from a given field or area. These strategies depended on the use of chemical pesticides to kill all of the pest organisms. Prior to the twentieth century, farmers used naturally occurring compounds such as kerosine or pyrethrum for this purpose. During the second half of the twentieth century, synthetic pesticides began playing a prominent role in controlling crop pests.

Chemical Effects

After 1939 the use of pesticides such as dichloro-diphenyl-trichloroethane (DDT) was so successful in terms of controlling pest populations that farmers began to substitute a heavy dependence on pesticides for sound pest management strategies. Soon pests in high-value crops became resistant to one pesticide after another. In addition, outbreaks of secondary pests occurred because either they developed resistance to the pesticides or the pesticides killed their natural enemies. This supplied the impetus for chemical companies to develop new pesticides, to which the pests also eventually developed resistance.

Rationale for IPM

Certain pests have developed resistance to all federally registered materials designed to control them. In addition, many pesticides are toxic to humans, wildlife, and other nontarget organisms and therefore contribute to environmental pollution. For these reasons, and because it is very expensive for chemical companies to put a new pesticide on the market, many producers began looking at alternative strategies such as IPM for managing pests. The driving forces behind the development of IPM programs are concern about the contamination of groundwater and other nontarget sites, adverse effects on nontarget organisms, and development of pesticide resistance. Pesticides will probably continue to play a vital role in pest management, even in IPM, but it is believed that their role will be greatly diminished over time.

An agricultural ecosystem consists of the crop environment and its surrounding habitat. The interactions among soil, water, weather, plants, and animals in this ecosystem are rarely constant enough to provide the ecological stability of nonagricultural ecosystems. Nevertheless, it is possible to use IPM to manage most pests in an economically efficient and environmentally friendly manner. IPM programs have been successfully implemented in the cropping of cotton and potatoes, and they are being developed for other crops.

Developing IPM Programs

There are generally three stages of development associated with IPM programs, and the speed at which a program progresses through these stages is dependent on the existing knowledge of the agricultural ecosystem and the level of sophistication desired. The first phase is referred to as the *pesticide management phase*. The implementation of this phase requires that the farmer know the relationship between pest densities and the resulting damage to crops so that the pesticide is not applied excessively. In other words, farmers do not have to kill all of the pests all of the time. They must use pesticides only when the economic damage caused by a number of pest organisms present on a given crop exceeds the cost of using a pesticide. This practice alone can reduce the number of chemical applications by as much as half.

The second phase is called the *cultural management phase*. Implementation of this phase requires knowledge of the pest's biology and its relationship to the cropping system. Cultural management includes such practices as delaying planting times, rotating crops, altering harvest dates, and planting resistance cultivars. It is necessary to understand pest responses to other species as well as abiotic factors, such as temperature and humidity, in the environment. If farmers know the factors that control population growth of a particular pest, they may be able to reduce the impact of that pest on a crop. For example, if a particular pest requires short days to complete development, farmers might be able to harvest the crop before the pest has a chance to develop.

The third phase is the *biological control phase*, which involves the use of biological organisms rather than chemicals to control pests. This is the most difficult phase to implement because farmers must understand not only the pest's biology but also the biology of the pest's natural enemies and the degree of effectiveness with which these agents control the pest.

In general, it is not possible to rely completely on biological control methods. A major requirement in using biological agents is to have sufficient numbers of the control agent present at the same time that the pest population is at its peak. It is sometimes possible to change the planting dates so that the populations of the pests and the biological control agents are synchronized. Also, there is often more than one pest species present at the same time within the same crop, and it is extremely difficult to control simultaneously two pests with biological agents.

D. R. Gossett

Sources for Further Study

Pedigo, Larry P. *Entomology and Pest Management*. 4th ed. Upper Saddle River, N.J.: Prentice Hall, 2002. Discusses insects and the strategies used to control them.

Romoser, William S., and John G. Stoffolano. *The Science of Entomology*. 4th ed. Boston: McGraw-Hill, 1998. General text on insects. Contains an excellent overview of the general aspects of IPM.

Western Regional IPM Project. *Integrated Pest Management for Potatoes*. Berkeley: Regents of the University of California, Division of Agriculture and Natural Resources. IPM Manual Group, 1986. Part of a series of books that contain useful information on IPM use for a variety of crops, including potatoes, citrus, cotton, rice, and tomatoes.

See also: Agriculture: modern problems; Biopesticides; Pesticides; Sustainable agriculture.

INVASIVE PLANTS

Categories: Environmental issues; poisonous, toxic, and invasive plants; water-related life

Nonnative (also termed "exotic") fungi and plants that can outcompete native species are called invasive plants. Invasive plants cause irreversible changes to ecosystems, threaten plant and animal species, and cost billions of dollars to control.

Between the damage they cause and the cost of control efforts, invasive plants cost the United States more than $140 billion every year. Nearly half of the threatened and endangered plant species listed for the United States in 1999, 400 of 958, are in peril because of competition from invasive species. Thus, invasive plants are capable of causing irreparable changes in ecosystems.

Most invasive species invading the United States originated in Asia or Europe. A key factor in this problem is that seeds or spores of these plants are accidentally transported into new habitats by humans, but the plants' natural enemies and competitors are left behind. Without natural biological controls, the alien species can thrive and outcompete the native flora, driving the native plants toward extinction and creating a near monoculture of the invader.

Invasive plants are weedy species that grow rapidly, produce large numbers of long-lived seeds, and frequently have perennial roots, or rhizomes, that enhance asexual propagation. Invasive plants have a variety of effects on invaded ecosystems. Many invasive species deplete soil moisture and nutrient levels, either by growing more vigorously than native plants early in the growing season or by being more tolerant of reduced levels of water and nutrients than are natives. Some invasive species produce toxic chemicals (*allelopathy*) that are released into the soil and inhibit the growth of competitors. By outcompeting native plants, the invader decreases species diversity as it replaces many native species. As a result, animal species dependent on native flora are also affected. Fungi and seed plants are among the most disruptive invasive plants in the United States today.

Control Methods

Invasive species are carried to new habitats, either in or on machinery or organisms, and are usually transported by humans, so prevention is the most cost-effective method of control. Once an invasive species has entered an area, plant quarantine is an effective first line of defense. For example, living plants and animals brought into the United States must pass inspection by the U.S. Department of Agriculture Animal and Plant Health Inspection Service (APHIS) to ensure that they are not carrying

Most plant species invading the United States originated in Asia or Europe, such as this Scotch broom. In Oregon and Washington, complete Douglas fir plantation failures have been attributed to Scotch broom.

potentially invasive species. Particular care is taken to ensure that imports from known areas of infestation are clean of seeds, spores, or propagules.

The next most effective strategy is detection and control of small infestations. When there is a known threat of invasion, the affected area should be surveyed periodically and individual plants removed by hand or, in extreme cases, by "spot-spraying" herbicide. Eradication is possible when the infestation is small.

Once an invasive species becomes established, the only means of management are expensive chemical or biological controls which, at best, will only minimize damage. A variety of chemicals may be used to kill invasive plants. Most chemicals, however, affect a broad spectrum of plants, including native species. Biological controls, including natural enemies from the invasive plant's native ecosystem, can be more specific but may also be ca-

pable of displacing native species and becoming "invaders."

Fungi

Many of the most serious plant pathogens are invasive species introduced into the Americas since the beginning of European settlement. Two classic examples are Dutch elm disease, caused by the fungi *Ophiostoma ulmi* and *Ophiostoma novo-ulmi* and chestnut blight, caused by the fungus *Cryphonectria parasitica*. At the beginning of the twentieth century, the most common street tree growing in the cities of the eastern United States was the American elm. About 1910, the European bark beetle was introduced into the United States. It was not until the 1930's that Dutch elm disease was observed in Ohio and a few eastern states. The fungal spores are carried by the beetles, which burrow under the elm bark. The native elm has little resistance to this fungus, whose spores rapidly germinate and form extensive mycelia within the phloem of the host tree, killing it within a few years. After its initial contact, the fungus spread throughout the cities and forests of the East and gradually westward, so that by 1990 nearly all of the native American elm trees in the United States had been killed.

American chestnut was also one of the early dominant trees of the eastern U.S. forest. In addition to providing edible fruit, the chestnut became a commercially important timber tree. Chestnut blight fungus was first reported in 1904 on chestnut trees in the New York Zoological Garden and quickly began to spread. This infestation led directly to passage of the Plant Quarantine Act of 1912, the forerunner of APHIS. By 1950 most native chestnut trees were reduced to minor understory shrubs. Biological control using virus strains first isolated in Italy show promise for controlling the blight.

Terrestrial Green Plants

Virtually all of the plants commonly called "weeds" are foreign invaders that are difficult, if not impossible, to control. Some of the most severe include Canada thistle (*Circium arvense*), leafy spurge (*Euphorbia esula*), and purple loosestrife (*Lythrum salicaria*). Canada thistle is the most widespread and difficult species of thistle to control. It was introduced to Canada from Europe in the 1600's and in 1795 was listed as a noxious weed in Vermont. It is now found in most of the United States as well as in Canada. Single herbicide applications do not provide long-term control, and there are no effective biological controls that do not also attack native species.

Leafy spurge was first reported in Newbury, Massachusetts, in 1827, where it arrived in ship ballast. By 1900 it had reached the West Coast, and it now thrives in more than half the states and in Canada. Thirteen species of insects are approved for biological control, and several herbicides can be used to control infestations effectively. Sheep and goats will browse on spurge.

Purple loosestrife was introduced into the United States as an ornamental plant in the early 1800's and became established in New England by 1830. Its early spread into the Great Lakes region was by barge and other canal traffic. Rapid expansion of the pest, particularly in the West, occurred after 1940, primarily due to the plant's "escape" from ornamental cultivation into irrigation projects. It is now found in all the lower forty-eight states except Florida. At present, there are no effective controls.

Aquatic Green Plants

Invasive plants are not limited to the terrestrial habitat or to vascular plants. One dramatic example is the alga *Caulerpa taxifolia*, the so-called killer alga. This attractive tropical alga was found to be easy to grow in saltwater aquaria and useful as a secondary food source for herbivorous tropical fish. It began to be used this way at the Oceanographic Museum of Monaco in 1982. Two years later, a meter-square patch was found growing in the Mediterranean Sea, visible from a win-

---

Backyard Solutions

Public understanding is fundamental to successful control of invasive plant species. Gardeners and landscapers have a personal role to play. Many reputable garden suppliers offer problem plants. Additionally, government agencies will recommend these plants for specific purposes, presuming they will be maintained and controlled in the landscape.

When considering new plants for gardens or landscapes, check references for warnings about high seed production, rank growth, or other invasive characteristics. The Web site http://plants.usda.gov/ is a database of standardized information about plants including identification, distribution, and growth information; a list of plants considered invasive can be searched by common or scientific name.

If the plant being considered has been known to pose a problem, avoid it or be vigilant in keeping it under control, particularly if cultivating an area near natural and unmaintained habitats. Horticulturalists should remember that their industry can be a source of invasive escapes. Additional ways gardeners can help are by eradicating invasive weeds on their property and volunteering for local groups that monitor or eradicate invasive plants.

According to the National Tropical Botanical Garden, botanic gardens, museums, herbaria, and protected areas should take responsibility for publicizing threats from invasive species. Garden clubs, the horticultural and forestry industries, and botanic gardens should have policies and activities to prevent inadvertant introduction of invasive plants.

dow of the museum. By 1990 the alga had reached France, and by 1995 it could be found from Spain to Croatia. *Caulerpa* produces a number of toxins that inhibit foraging by native fish, and it is a prolific vegetative reproducer. Fragments of the alga, stuck on an anchor for example, can start a new infestation wherever the anchor is next dropped. This species has been discovered in Southern California, and a related species has become dominant in Sydney Harbor, Australia. Other aquatic invasive plants in the United States include the mosquito fern (*Azolla*), the Eurasian water milfoil (*Myriophyllum*), and the water hyacinth (*Eichhornia crassipes*).

Marshall D. Sundberg

See also: Algae; Allelopathy; Animal-plant interactions; Aquatic plants; Biopesticides; Biological invasions; Community-ecosystem interactions; Competition; Dinoflagellates; Endangered species; Eutrophication; Fungi; Herbicides; Pesticides.

Sources for Further Study

Meinesz, Alexandre. *Killer Algae*. Chicago: University of Chicago Press, 1999. A noted scientist describes the spread of an exotic seaweed in the Mediterranean Sea.

Randall, John M., and Janet Marinelli, eds. *Invasive Plants: Weeds of the Global Garden*. Brooklyn, N.Y.: Brooklyn Botanic Garden, 1996. Simply written, useful book for identification of and removal techniques for invasive species. Discusses weed origins, environmental impacts, and U.S. distribution. Includes illustrations of about fifty species.

Sheley, Roger L., and Janet K. Petroff. *Biology and Management of Noxious Rangeland Weeds*. Corvalis: Oregon State University Press, 1999. General information about invasive plants with individual chapters on specific species.

IRRIGATION

Category: Agriculture

Irrigation techniques supply additional water to arid and semiarid horticulture or farming regions where few, if any, crops could otherwise be grown.

Approximately 350 million acres (142 million hectares) of land worldwide are irrigated. In the United States more than 10 percent of the crops, encompassing approximately 50 million acres (20 million hectares), receive water through irrigation techniques; 80 percent of these are west of the Mississippi River. In countries such as India, Israel, North Korea, and South Korea, more than one-half of food production requires irrigation. From 1950 to 1980, the acreage of irrigated cropland doubled worldwide. Increases since then have been more modest.

Imperial Valley and Israel

An often-cited example of irrigation success is that of the Imperial Valley of Southern California. The valley, more than 5,000 square miles (12,900 square kilometers) in size, was originally considered to be a desert wasteland. The low annual rainfall resulted in a typical desert ecosystem with cacti and other arid-adapted plants and animals. In 1940, however, engineers completed the construction of the All-American Canal, which carries water 80 miles (130 kilometers) from the Colorado River to the valley. The project converted the Imperial Valley into a fertile, highly productive area where farmers grow fruits and vegetables all year.

Successful agriculture in Israel also requires irrigation. As a result of settlement of the area throughout the twentieth century, large amounts of food must be produced. To fulfill this need, a system of canals and pipelines carries water from the northern portion of the Jordan Valley, where the rainfall is heaviest, to the arid south.

Methods and Technology

Irrigation is an expensive operation that requires advanced technology and large investments of capital. In many cases, irrigation systems convey water from sources hundreds of miles distant. Such vast engineering feats are largely financed by taxpayers. Typically, water from a river is diverted into a main canal and from there into lateral canals that supply individual farms. From the lateral canals, various systems are used to supply water to the crop plants in the field.

Utah farmer Ed Brewer checks the irrigation water flow in an onion field. "The first time is always the hardest," he says of getting the water in a field flowing optimally the first time it is irrigated every growing season.

Flood irrigation supplies water to fields at the surface level. Using the *sheet irrigation* method, land is prepared so that water flows in a shallow sheet from the higher part of the field to the lower part. This method is especially suitable for hay and pasture crops. Row crops are better supplied by *furrow irrigation*, in which water is diverted into furrows that run between the rows. Both types of flood irrigation cause soil *erosion* and loss of nutrients. However, erosion can be reduced in furrow irrigation by contouring the furrows.

Sprinkler irrigation, though costly to install and operate, is often used in areas where fields are ungraded or steeply sloped. Sprinklers are supplied with water by stationary underground pipes or a center pivot system in which water is sprinkled by a raised horizontal pipe that pivots slowly around a pivot point. A disadvantage of sprinkler irrigation is loss of water by *evaporation*. In *drip irrigation*, water is delivered by perforated pipes at or near the soil surface. Because it is delivered directly to the plants, much less water is wasted by evaporation compared to other methods.

Much of the water used in irrigation never reaches the plants. It is estimated that most practices deliver only about 25 percent of the water to the root systems of crop plants. The remaining water is lost to evaporation, supplies weeds, seeps into the ground, or runs off into nearby waterways.

Soil Salinization

As fresh water evaporates from irrigated fields over time, a residue of salt is left behind. The process, called *salinization*, results in a gradual decline in soil productivity and can eventually render fields unsuitable for agricultural use. Correcting saline soils is not a simple process. In principle, large amounts of water can be used to leach salt away from the soil. In practice, however, the amount of water required is seldom available, and if it is used, it may waterlog the soil. Also, the leached salt usually pollutes groundwater or streams. One way to deal with salinization is to use genetically selected crops adapted to salinized soils.

Water Resources

As the number of acres of farmland requiring irrigation increases, so does the demand for water. When water is taken from surface streams and rivers, the normal flow is often severely reduced, changing the ecosystems downstream and reducing their biodiversity. Less water becomes available for other farmers downstream, and that situation can lead to disputes over water rights.

In other cases water is pumped from deep wells or aquifers. Drilling wells and pumping water from such sources can be expensive and may lead to additional problems, such as the sinking of land over aquifers. Such land subsidence is a major problem in several parts of the southern and western United States. Subsidence in urban areas can cause huge amounts of damage as water and sewer pipes, highways, and buildings are affected. In coastal areas, depletion of aquifers can cause the intrusion of salt water into wells, rendering them unusable. The federal government spends millions of dollars to repair damage to irrigation facilities each year.

Like many modifications to natural ecosystems, the use of water for irrigation achieves some remarkable but temporary advantages that are complicated by long-term environmental problems. Assessments of total financial costs and environmental impacts are continuously weighed against gains in production.

Thomas E. Hemmerly

See also: Agriculture: modern problems; Erosion and erosion control; Hydrologic cycle; Soil salinization.

Sources for Further Study
Ortho's All About Sprinklers and Drip Systems. Des Moines, Iowa: Meredith Books, 1998. Outlines the irrigation systems available to homeowners.

Postel, Sandra. *Pillar of Sand: Can the Irrigation Miracle Last?* New York: W. W. Norton, 1999. Examines challenges to modern irrigation society, such as mounting water scarcity, soil, salinization, and water disputes. Shows how innovative irrigation strategies can alleviate both hunger and environmental stress.

Reisner, Marc P. *Cadillac Desert: The American West and Its Disappearing Water.* Rev. ed. New York: Penguin Books, 1993. History of the struggle to control water in the American West; describes rivers diverted and dammed, political corruption, and economic and ecological disaster.

Roy, Arundhati. *The Cost of Living*. New York: Modern Library, 1999. Attacks dams built for irrigation purposes worldwide and the policies that created them. Examples named include displacement of people in India and regional threats to plant and animal life.

KREBS CYCLE

Categories: Cellular biology; photosynthesis and respiration; physiology

The Krebs cycle, also called the citric or tricarboxylic acid cycle, is a series of chemical reactions that completes the aerobic breakdown of glucose and facilitates the transfer of energy to the electron transport system.

Every living organism must process chemical energy to survive. The series of metabolic pathways known as *cellular respiration*, which obtains most of the energy needed for cellular metabolism, consumes both organic fuel and oxygen. Respiratory processes ultimately produce the *adenosine triphosphate* (ATP) that drives metabolic processes. The Krebs cycle, named for biochemist Hans A. Krebs, is a basic chemical process that is found in the mitochondria of all eukaryotic cells.

The Krebs cycle is the crucial second part in the breakdown of glucose to water and carbon dioxide—a part of *cellular respiration*. Cellular respiration begins in the cytoplasm (the fluid within the cell) with the breakdown of glucose to form pyruvic acid, in a process known as *glycolysis*. Pyruvic acid is then transported across the mitochondrial membranes into the matrix, where it loses a molecule of carbon dioxide and is converted into acetyl coenzyme A (acetyl CoA). The Krebs cycle completes the breakdown of glucose by joining the acetyl portion of acetyl CoA to an organic acid which then, through a series of steps, releases the equivalent of what was left of the glucose as carbon dioxide.

Together, these steps supply energized electrons, which are necessary for the final step, *oxidative phosphorylation*, where the bulk of the cell's ATP is produced. In addition to its central role in *catabolism*, or the breakdown of organic molecules, the Krebs cycle plays a central role in *anabolism*, or the synthesis of organic molecules. Many of the intermediate molecules in the Krebs cycle can be used in other biochemical pathways to produce amino acids, carbohydrates, and lipids.

Oxidation and Electron Transfer

The fuel for running the Krebs cycle is the two-carbon fragments known as *acetyl groups*. The overall chemistry of the Krebs cycle involves the *oxidation* of the acetyl group's two carbon atoms to two molecules of carbon dioxide. As oxidation occurs in the Krebs cycle, electrons are released, in the form of hydrogen atoms, and picked up by electron carriers. The release of these electrons, which have a high energy content, is the primary goal of the Krebs cycle. They are used later as the energy source for oxidative phosphorylation.

The electron acceptors are two coenzymes similar to coenzyme A; *nicotinamide adenine dinucleotide* (NAD$^+$) and *flavin adenine dinucleotide* (FAD). Both NAD$^+$ and FAD have a ring containing nitrogen that shares four electrons with an adjacent carbon atom. Such arrangements are especially suitable for accepting electrons and protons and then releasing them later. In their oxidized states, NAD$^+$ and FAD are each capable of accepting two electrons, donated initially as two hydrogen atoms. When NAD$^+$ reacts with two hydrogen atoms, it keeps one and strips away the electron from the other, releasing what is left as a free proton (H$^+$). In the process NAD$^+$ is made into its reduced form, NADH. When FAD reacts with two hydrogen atoms, it accepts them both, along with their two electrons, to become FADH$_2$. Thus, the coenzymes NAD$^+$ and FAD, as NADH and FADH$_2$, serve as electron transfer agents, connecting the Krebs cycle with the *electron transport system* embedded in the inner mitochondrial membrane.

Principal Steps

After years of research, a detailed picture of the chemistry of the Krebs cycle is available. For each of the eight principal steps, the structure of the reactants and products, as well as the enzymes that catalyze the reactions, has been determined. During one turn of the Krebs cycle, the equivalent of one acetyl group is converted into two carbon dioxide molecules.

The first step of the Krebs cycle occurs when acetyl CoA reacts with oxaloacetate, the ionic form of oxaloacetic acid, to form citrate. (All of the acids in the Krebs cycle occur in their ionic forms.) This first product is tricarboxylic acid; hence, one of the other names of this cycle, the tricarboxylic acid cycle. In addition to citrate, the first step in the Krebs cycle releases a molecule of coenzyme A, which is ready to react with another acetyl group from a pyruvate molecule. While the overall result of the Krebs cycle is degradation, this initial step is one of building up, or synthesis.

In the second step, citrate is made into isocitrate by a complex rearrangement involving the loss of a molecule of water and then the addition of a water molecule. The net effect is to move a hydroxyl or alcohol group from one carbon to an adjacent one. The starting citrate and the product, isocitrate, have the same molecular formula but have different molecular structures. Such molecules are called *isomers*.

In the third step, isocitrate is oxidized. It passes two hydrogen ions to NAD+, thus reducing it to NADH and releasing a free proton. Isocitrate also loses a molecule of carbon dioxide and becomes alpha-ketoglutarate. With the loss of this carbon dioxide molecule, the equivalent of only one of the two original acetyl carbon atoms remains.

The fourth step involves the loss of another carbon, in the form of carbon dioxide, equivalent to another of the original two acetyle carbon atoms. Alpha-ketogluterate bonds with a molecule of coenzyme A to form succinyl CoA. The remaining steps involve the remaking of oxaloacetate so the cycle can occur again with another acetyl group. In the process, a few more high-energy electrons are passed off to electron carriers.

The fifth step involves splitting succinyl CoA to produce free coenzyme A and succinate. The splitting of this bond releases enough energy to drive a substrate-level phosphorylation reaction which takes place in two steps. First, a molecule of guanosine diphosphate (GDP) reacts with inorganic phosphate to form guanosine triphosphate (GTP).

Then GTP transfers its phosphate to adenosine diphosphate (ADP), to produce ATP. These two nucleotides are very similar in having high-energy phosphate bonds, but they differ in their nitrogenous bases; GTP has guanine, and ATP has adenine.

In step six, succinate is oxidized to a fumarate. In the process two hydrogen atoms, with their high-energy electrons, are passed to FAD to form FADH$_2$.

In step seven, fumarate is transformed to malate by the addition of a molecule of water.

In the last step, malate is made into oxaloacetate, which is ready to start the process all over again. A further consequence of this reaction is the release of another two hydrogen atoms, with high-energy electrons, that are picked up by NAD+, with the consequent production of the usual free proton.

Advantages

Detailed studies of these chemical reactions reveal that the carbon atoms that are actually oxidized and released as carbon dioxide come from the oxaloacetate portion of the citrate ion rather than from the acetyl group. The acetyl group is now one-half of the new oxaloacetate and, after another turn of the Krebs cycle, these acetyl carbons will be released as carbon dioxide.

The mechanics of the Krebs cycle have additional advantages. The cycle's chief function of obtaining energy for the cell's needs is accomplished in small, discrete increments rather than in one large burst. This stepwise process allows finer control of the entire reaction sequence, with a large number of points at which control can be exercised. Finally, as the acetyl group passes through a cycle, a variety of molecules are produced, which can provide raw materials for the synthesis of essential biological molecules.

Bryan Ness

See also: ATP and other energetic molecules; Glycolysis and fermentation; Oxidative phosphorylation; Respiration.

Sources for Further Study

Gilbert, Hiram F. *Basic Concepts in Biochemistry: A Student's Survival Guide.* 2d ed. New York: McGraw-Hill, 2000. Part of McGraw-Hill's Basic Sciences series; includes bibliography and index.

Igelsrud, Donald E. "How Living Things Obtain Energy: A Simpler Explanation." *The American Biology Teacher* 51, no. 2 (February, 1989): 89-93. This well-presented discussion is important for two reasons: It is written from a biologist's point of view, and it makes the chemistry much easier for the nonspecialist to grasp. While it is directed at biology teachers and includes valuable classroom suggestions, it can be appreciated by anyone interested in biochemical oxidation and energy transfer.

Krebs, Hans A. "The History of the Tricarboxylic Cycle." *Perspectives in Biology and Medicine* 14 (Autumn, 1979): 154-170. While this journal is probably not easily available in public libraries, it is worthwhile to seek it out at a college. Krebs not only describes the origin of his ideas and their development but also provides rare insight concerning his success. A revealing analysis of how one scientist found a solution which others, equally brilliant, had overlooked.

Metzler, David. *Biochemistry*. 2d ed. New York: Harcourt, 2002. Provides a detailed and technical view of all aspects of biochemistry. Readable by high-school students, with some background, and by scientists.

LEAF ABSCISSION

Categories: Cellular biology; physiology; reproduction and life cycles

In the process of leaf abscission, plants periodically shed their leaves. Leaf abscission involves a number of biochemical and physical changes that are largely controlled by plant hormones.

Plants are primarily categorized as *annuals, biennials,* or *perennials,* based on their growth patterns. Annuals are those plants that undergo a complete life cycle from seed to seed in one growing season. Biennials require two growing seasons to complete a life cycle; during the first year, only vegetative growth takes place. The aboveground portion dies through the winter, and in the next growing season the roots send up a reproductive shoot that produces the seeds. Perennials have the capacity to live through many successive growing seasons. In this group, those plants referred to as *deciduous* species shed all their leaves at the same time. The *evergreen* species shed leaves throughout the year, yet never shed the entire complement at any one time.

Preparation and Precursors

Prior to any natural abscission that may take place, leaves (or any other plant organ subject to abscission, such as flowers or fruit) undergo *senescence.* Senescence can be defined as the deterioration that occurs in conjunction with aging, and it results in the death of an organ or organism. Senescence can occur throughout the entire plant, as it does in annuals, or in only the aboveground portion, as it does in perennial herbs. In woody perennials, however, only the leaves senesce, while the bulk of the stem and roots remain alive.

From the time that leaves begin to grow, biochemical activities such as photosynthesis increase. This increase will continue until the leaves expand to maximum size. Soon after they reach maximum size, senescence begins, and photosynthetic rates begin to decrease. As the photosynthetic ability of the leaves declines, there is an accompanying decrease in other metabolic activities. Respiration rates begin to subside dramatically, and leaf protein levels drop sharply because of increased *proteolytic* activity (enzymatic breakdown of proteins). Protein synthesis diminishes, and there is an increase in the enzymatic degradation of ribonucleic acid (RNA). There is also an increase in the hydrolytic breakdown of carbohydrates. Finally, destruction of the green pigment, *chlorophyll*, is accompanied by increased visibility of the yellow or orange pigments called *carotenoids*, which were previously masked by chlorophyll. Most of the protein, carbohydrates, RNA, and chlorophyll degradation products are rapidly transported out of the senescing leaf. The final result is the production of yellowish, dead leaves.

The senescence process is a natural progression of the normal plant life cycle; however, environmental conditions can influence the process. Lack of water will speed the senescence process in most species. Higher-than-normal temperatures also cause an increase in senescence-related reactions. Darkness dramatically hastens senescence: Most leaves will senesce two or three times faster in darkness than if growing under normal light conditions.

Numerous studies strongly suggest that senescence is under hormonal control. Both *ethylene* and *abscisic acid* enhance senescence, but ethylene is the more effective of the two. The *gibberellins, cytokinins,* and *auxins* (other types of hormones) have all been shown to delay the process in various plant species. The exact role of each of these hormones in senescence has not yet been determined, but it is apparent that the process involves the interaction of several of these growth-regulating substances.

Onset of Abscission

Following senescence, abscission of the leaves inevitably takes place. This process usually involves the formation of an *abscission layer* at the base of the leaf petiole. During the early life of a leaf, auxin is produced in relatively high concentrations and is steadily transported out of the leaf through the petiole. As long as the auxin level remains high in the leaf and a sufficient amount of the hormone is transported across the petiole, both senescence and abscission are delayed.

In addition, gibberellins and cytokinins are produced elsewhere in the plant and then sent to the leaves to help retard the destructive processes. As the leaf matures, however, the level of the senescence-retarding hormones, especially auxin, decreases.

With the decrease in auxin levels, the catabolic (breakdown) reactions begin to outnumber the ana-bolic (synthetic) processes. In conjunction with the increase in catabolic reactions, there is a rise in the levels of abscisic acid and, especially, ethylene. Ethylene is particularly important in producing the abscission layer.

In most species, the abscission layer is formed from one or perhaps several layers of cells across the base of the petiole. In the earlier stages of abscission, there is a noticeable rise in the respiration rates of the cells of the abscission layer closest to the stem (the proximal cells). As the respiration rates increase to supply additional energy, ethylene stimulates one or more of these cell layers nearest the stem to increase in size. Along with the increase in size of those cells, the cells in the abscission layer farthest from the stem (the distal cells) increase the production of enzymes that break down polysaccharides in the cell walls. With the secretion of these enzymes into the cell walls, digestion of the cell-wall materials begins. The pressure created by the expansion of the cells in the proximal region of the abscission layer (causing them to grow against the weakened senescing cells of the distal region) results in the two layers breaking apart. Thus, the leaf detaches and falls from the plant.

Abscission, like senescence, is a natural order of progression during the life cycle of most plants. Although the process is closely correlated with regular seasonal changes, variations in environmental conditions can enhance abscission. Deficiencies in certain nutrients, such as nitrogen, or lack of water can stimulate abscission. These conditions also hasten senescence, and senescence always precedes abscission. Hence, adverse environmental conditions perhaps trigger only the onset of senescence, and abscission occurs as a secondary result of the aging process.

DIGITAL STOCK

Throughout much of the world trees and other plants are subjected to freezing temperatures. Leaves age and fall prior to winter. Without the abscission process to cause the plant to prepare for winter by shedding its leaves, the senescent tissues would shade the new spring growth that appears the following growing season on what are shown here as bare branches.

Function of Abscission

Throughout much of the world, plants are subjected to freezing temperatures. The leaves of most plants are unable to withstand the cold weather and face certain death during the winter. If the leaves did not prepare for the onset of cold weather by undergoing senescence, the first freeze would kill the leaves before materials within them could be salvaged. Without the abscission process to remove the dead leaves, the senescent tissues would shade the new spring growth that appears the following growing season. Hence, senescence and abscission provide a means by which perennials can recycle a major portion of leaf materials as the plants prepare for both the cold weather and the following growing season.

Competition for nutrients from other parts of the plant may initiate the senescence process. The pull of nutrients to another part of the plant such as roots, flowers, or fruit would reduce the amount of these materials bound for the leaves. The reduced supply of nutrients could very well decrease synthetic rates, and the overall result would be a decline in major leaf macromolecules such as proteins, chlorophyll, and nucleic acids.

Competition alone, however, cannot account for the senescence and abscission phenomena, because even in plants that do not produce fruit, the leaves experience aging and the loss of leaves. In addition, numerous studies have shown that leaf senescence will still occur when flowers are removed from the plant soon after being formed. Although competition for nutrients may not be the sole cause of the phenomenon, the mobilization of substances such as amino acids and carbohydrates from the leaves to other metabolic sinks, such as the fruit, is definitely linked to the initiation of senescence. Several of the plant hormones or other factors that stimulate mobilization also hasten senescence. Hence, it is possible that the competition for nutrients triggers the production of some unknown senescence hormone by the fruit or some other competing plant part. This theoretical substance would be transported to the leaves, where it would initiate mobilization of leaf contents. This mobilization might enhance senescence, which, in turn, might trigger the metabolic reactions that lead to abscission of the leaves.

D. R. Gossett

See also: Angiosperm life cycle; Hormones; Leaf anatomy; Plant life spans.

Sources for Further Study

Addicott, Fredrick T. *Abscission*. Berkeley: University of California Press, 1982. An excellent discussion of the biochemical and biological aspects of the abscission process. A fairly technical book, it gives the reader a comprehensive overview of the phenomenon. Includes extensive bibliography.

Campbell, Neil A., and Jane B. Reece. *Biology*. 6th ed. San Francisco: Benjamin Cummings, 2002. An introductory college-level textbook for science students. The chapter on control systems in plants provides a clear, concise description of senescence and abscission in plants. The well-written text, combined with superb graphics, furnishes the reader with a clear understanding of the process. List of suggested readings at the end of the chapter. Includes glossary.

Davies, Peter J., ed. *Plant Hormones: Physiology, Biochemistry, and Molecular Biology*. 2d ed. Boston: Kluwer, 1995. Revised edition of *Plant Hormones and Their Role in Plant Growth and Development*. With bibliographical references and index.

Fosket, Donald E. *Plant Growth and Development: A Molecular Approach*. San Diego: Academic Press, 1994. Advanced undergraduate textbook approaches the topic of plant development from the perspective of plant molecular biology and genetics. Chapters end with study and review questions as well as suggestions for further reading. No special knowledge of plant biology is needed.

Lyndon, R. F. *Plant Development: The Cellular Basis*. Boston: Unwin Hyman, 1990. From the Topics in Plant Physiology series; discusses plant cells, tissues, molecular genetics. Includes bibliography, index.

Salisbury, Frank B., and Cleon Ross. *Plant Physiology*. 4th ed. Belmont, Calif.: Wadsworth, 1992. An intermediate college-level textbook for science students. The chapter "Hormones and Growth Regulators: Cytokinins, Ethylene, Abscisic Acid, and Other Compounds," gives an in-depth view of the physiological role of senescence and abscission. An excellent explanation of the phenomenon is provided in text and graphics. Contains a detailed bibliography at the end of the chapter.

LEAF ANATOMY

Categories: Anatomy; photosynthesis and respiration

The leaf has evolved as the chief part of the plant for gathering light energy from the sun and conducting photosynthesis to transform that light energy into biochemical energy. Hence, its structure is adapted to that function.

Leaves are formed by a plant to manufacture food. *Photosynthesis*—a complicated chemical reaction in which carbon dioxide from the air and water from the soil, in the presence of light, produce sugar—is carried out in the *chloroplasts* found packed within the leaf cells. Because energy is derived from light by chlorophyll, either the leaf must be thin enough for light to penetrate all the cell layers or, in the case of plants with succulent leaves, chloroplasts must be most concentrated near the surface of the leaves.

Orientation to Light

No matter how many leaves a plant has, each is arranged in respect to light. Although some plants that are adapted to hot, dry conditions may orient their leaves to minimize exposure to the sun, most arrange their leaves to maximize exposure to the sun. Some are exposed to direct rays of the sun; others may face only a portion of the sky. A plant may be forced into various growing patterns to allow the leaves to be exposed to light. Some species germinate their seeds high above the ground, in the cracks of bark on tree trunks and branches. Plants that climb wind their way around larger plants until they gain a place in the sun. Therefore, leaf shape may be determined by where a plant species is best adapted to grow.

The leaves of plants that grow mostly in shade, called *shade plants*, have a larger surface area, tend to be thinner, and have a higher concentration of chlorophyll. The leaves of plants growing mostly in the sun, called *sun plants*, have a smaller surface area, tend to be thicker, and have a lower concentration of chlorophyll. Even leaves on the same plant can vary in structure depending on whether they spend most of their time in the sun or the shade, showing similar traits as seen in sun and shade plants. Taxonomists use leaf pattern, leaf arrange-

ment, and leaf shape to help identify and classify plants.

Leaf Margins

As leaves vary in size and shape, so do their edges, or *margins*. Many plants, such as holly and thistles, have prickly margins; prickles are the outgrowths of the leaf's vein endings. The prickly boundary acts as a defense against grazing animals. Some margins have razor-sharp, sawlike teeth that cut anything that brushes across them. The margins of many tropical leaves terminate in finely pointed tips, called *drip tips*. When rainwater accumulates on the leaf's surface, the tip helps the water to drain off the leaf so that accumulated water will not weigh down the leaf to the point of breakage. Many leaves, however, have no distinctive margins; when the margin is even all around, it is called an "entire" margin.

Leaf Parts

Normally, a mature leaf has three parts: the *petiole*, a stemlike portion that grows from a node on the stem and supports the leaf; the *blade*; and, at the base of the petiole, a *sheath*, which attaches the petiole to the stem. If the blade has no petiole, it is attached directly to the stem and is referred to as being *sessile*. Petioles are able to move to some extent, so that leaves can be arranged to receive sufficient light.

Vascular tissue that makes up *leaf veins* is composed of *xylem*, which brings water up from the roots, and *phloem*, which is responsible for transporting the products of photosynthesis. Xylem and phloem extend from the branch or stem into the leaf by a *leaf trace*. This strand is continuous through the petiole into the leaf veins that intersect the entire leaf. The veins are the point of contact between root and chloroplasts, ensuring that water can be continually furnished during the photosynthetic process. The veins may also act as a structural support within the leaves. If a blade has a *midrib*, it appears that the petiole extends onward to the tip of the leaf. Often, secondary veins branch off from this central vein, forming a reticulate pattern; other times, many of the same-sized veins branch out from the base of the blade in a fan-shaped pattern.

The upper surface of a leaf is covered by a continuous, transparent sheet of cells called the *cuticle*. The cuticle may perform three functions: to help prevent excess water loss, to protect against physical damage or damaging organisms, and to aid in reflecting intense sunlight. Cuticle cells are generally thin-walled, except, perhaps, on the margin, where thicker cells reinforce the leaf and aid in preventing tearing of the leaf by wind currents.

The cell layer immediately beneath the cuticle layers is the *epidermis*. Cells between the upper and lower epidermis form *mesophyll* tissues (from the Greek *mesos*, "middle," and *phyll*, "leaf"). Mesophyll in dicot leaves forms two observable layers; the upper layer is the *palisade mesophyll*, composed of cells that are columnar and closely packed; they are also rich in chloroplasts. Below this layer is the *spongy mesophyll*, so named because of numerous air spaces surrounding the small, oval cells. The air spaces are important in circulating carbon dioxide and oxygen that enters and leaves from *stomata*, small openings, generally confined to the underside of the leaf, where gas exchange is regulated.

Transpiration Organs

The stomata are bounded and controlled by two kidney-shaped cells called *guard cells*. Water evaporates from the leaf cells and goes into the air through these ventilation sites by a process referred to as *transpiration*. Normally, during light hours, the stomata are open (and losing water). At nighttime, the cells close, and water is retained. To open, potassium ions (K$^+$) are pumped into the guard cells, and water follows by osmosis, which causes an increase in internal pressure, called *turgor pressure*. As pressure increases, the water pushes against and stretches the guard-cell walls, bowing the cells outward. The filling and stretching of the guard cells opens the stoma.

Stomata close when *water stress* (lack of water in the plant) occurs, which can result from insufficient water in the soil or excessive transpiration rates. The most likely physiological mechanism for stomatal closing involves the hormone *abscisic acid* (ABA). The effects of water stress seem directly to trigger the release of ABA. The exact mechanism is unclear, but in some way ABA causes K$^+$ ions to move out of the guard cells and, again, water pas-

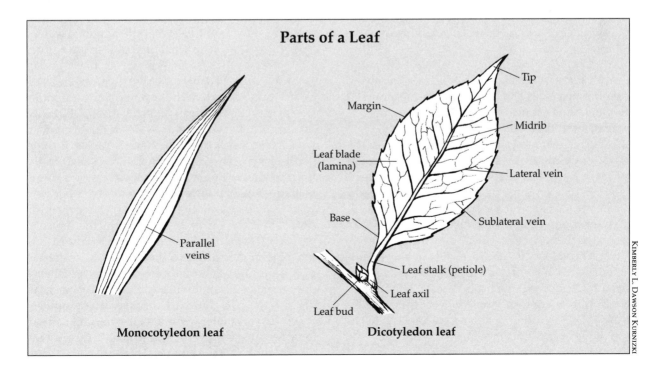

Parts of a Leaf

Tip

Margin

Midrib

Leaf blade (lamina)

Lateral vein

Base

Sublateral vein

Leaf stalk (petiole)

Leaf axil

Leaf bud

Parallel veins

Monocotyledon leaf

Dicotyledon leaf

KIMBERLY L. DAWSON KURNIZKI

Cross-Section of Leaf Tissue

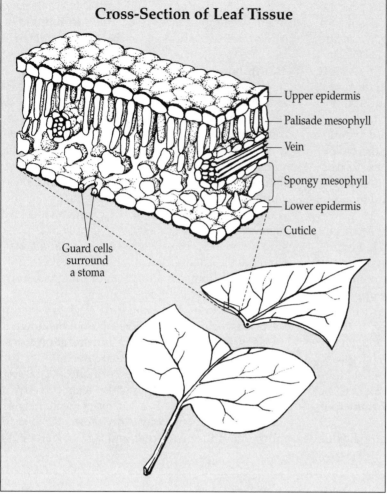

Upper epidermis

Palisade mesophyll

Vein

Spongy mesophyll

Lower epidermis

Cuticle

Guard cells surround a stoma

KIMBERLY L. DAWSON KURNIZKI

patterns: It is likely that 50 percent of rainfall in the Brazilian rain forest originates from transpired water. Plants that grow in arid conditions have developed specialized leaves to decrease the amount of water lost by transpiration. Many of these plants have leaves that are small and thick, so that surface area is reduced. The stomata may be housed in deep pits, away from wind's evaporative force. During especially dry periods, some plants even shed their leaves to reduce water loss. Others carry on an alternate form of photosynthesis that allows the stomata to remain closed during all or part of the day.

Kranz Anatomy

In certain plants known as C_4 plants, the leaves have adapted a particular way of fixing carbon; this has resulted in a ringlike arrangement of photosynthetic cells around the leaves' veins, called *Kranz anatomy*. This term (*Kranz* in German means "wreath") refers to the fact that in C_4 plants the cells that surround the water- and carbohydrate-conducting system (known as the *vascular system*) are packed very tightly together and are called *bundle sheath cells*. Surrounding the bundle sheath is a densely packed layer of mesophyll cells. The densely packed mesophyll cells are in contact with air spaces in the leaf, and because of their dense packing they keep the bundle sheath cells from contact with air. This Kranz anatomy plays a major role in C_4 photosynthesis.

In C_4 plants the initial fixation of carbon dioxide from the atmosphere takes place in the densely packed mesophyll cells. After the carbon dioxide is fixed into a four-carbon organic acid, the malate is transferred through tiny tubes from these cells to the specialized bundle sheath cells. Inside the bundle sheath cells, the malate is chemically broken down into a smaller organic molecule, and carbon dioxide is released. This carbon dioxide then enters the chloroplast of the bundle sheath cell and is fixed

sively follows. When the guard cells lose their water, they become limp and close, sealing the stomatal opening, thus greatly reducing transpiration.

Transpiration of water at the leaf surface may be affected by several factors. Wind blowing across the surface carries off water molecules, leaving room for more water molecules to take their place; an increase in temperature does the same thing. Loss of water may be slowed by opposite conditions. In rainy or foggy conditions when the air is already saturated with water, water loss from leaves is lower. Water loss also occurs slowly in cool conditions, such as those prevailing at night. An average-sized birch tree will typically lose 17,260 liters of water through transpiration in a single growing season. One acre of grass lawn may lose 102,200 liters of water in a single week.

Water transpired into the air can affect rainfall

a second time with the enzyme Rubisco and continues, as in non-C_4 plants, through the C_3 pathway.

Iona C. Baldridge, updated by Bryan Ness

See also: C_4 and CAM photosynthesis; Cacti and succulents; Gas exchange in plants; Leaf lobing and division; Leaf arrangements; Leaf abscission; Leaf margins, tips, and bases; Leaf shapes; Liquid transport systems; Photosynthesis; Photosynthetic light absorption; Photosynthetic light reactions; Plant tissues; Shoots; Stems; Water and solute movement in plants.

Sources for Further Study

Bowes, Bryan G. *A Color Atlas of Plant Structure*. Drawings by Jo Nicholson. Ames: Iowa State University Press, 2000. A fundamental guide to understanding plant structure for plant scientists, plant biologists, and horticulturists. Good color photographs and drawings.

Campbell, Neil A., and Jane B. Reece. *Biology*. 6th ed. San Francisco: Benjamin Cummings, 2002. Popular introductory general biology text that emphasizes the ideas behind concepts. The unit "Plants: Form and Function" contains five chapters devoted to plant anatomy and physiology with colorful and clear illustrations. Each chapter concludes with further readings. Includes glossary.

Mauseth, James D. *Plant Anatomy*. Menlo Park, Calif.: Benjamin/Cummings, 1988. A textbook used for undergraduate and graduate botany students but also intended as a reference book. The author emphasizes the importance of how the mechanisms of a particular structure affect the plant as a whole. Includes extensive references, glossary.

Prance, Ghillean Tolmie. *Leaves: The Formation, Characteristics, and Uses of Hundreds of Leaves Found in All Parts of the World*. New York: Crown, 1985. A wonderfully simple but quite complete book taking the reader through every aspect of the leaf. Accompanied by hundreds of beautiful photographs taken from all over the world, this volume is a beautiful addition to every botanist's library.

LEAF ARRANGEMENTS

Category: Anatomy

The study of leaf arrangements, or phyllotaxy, considers not only the descriptive classification of leaf arrangements but also theories regarding the cause of such arrangements.

The function of the arrangement of leaves (*phyllotaxy*) is to increase a plant's ability to carry on photosynthesis by positioning the leaves in such a way as to maximize the surface area available to intercept sunlight. Leaves may be either *caulescent* (on obvious stems) or *acaulescent* (with no obvious stems). Flowering plants have three basic types of arrangements: *alternate spiral*; *opposite*; and *whorled* or *verticillate*. The alternate spiral arrangement is generally considered to be the most primitive condition, with the opposite and whorled conditions being derived by suppression of internode development.

There are two major hypotheses regarding the processes governing these basic arrangements. The *field hypothesis* of phyllotaxy posits that, as leaf primordia (new leaf cells) are created by the plant, a zone that inhibits the growth of other primordia is laid down around it, and not until the shoot tip has grown beyond that zone can a new leaf primordium be laid down. The *first available space hypothesis* posits that new leaves grow as soon as the plant shoot has grown out far enough to allow space for them.

The various types of leaf arrangements are usually one of the easiest vegetative characteristics to

620 • Leaf arrangements

use in helping to identify vascular plants. This is especially true when leaf arrangement is combined with other characteristics, such as the presence or absence of petioles or the quality of being sessile or nonsessile. Other characteristics include the shape of the leaves and the appearance of the margins, bases, and apices types.

Alternate

Alternately arranged leaves produce one leaf per node. These leaves may be on alternate sides of the stem (2-ranked or *distichous*), on one side of the stem (*1-ranked* or *secund*), or in a spiral around the stem. If 2-ranked leaves overlap, as in some oncidium orchids and iris species, then they are referred to as *equitant*.

Leaves of members of the grass family (*Poaceae*) are distichous and alternate. Their leaves differ from most other vascular plant leaves in that they normally consist of a split tubular sheath that surrounds the stem and more or less linear blades held at right angles to the stem. They also have a small, tongue-like structure (*ligule*) at the junction of the sheath and blade, although in some species it may be obsolete.

Spiral

Spiral arrangements involve alternately arranged leaves in which each succeeding stem node and attached leaf is rotated slightly from the nodes below and above it. If the spiral is to the right, it is referred to as *dextrorse*; if to the left, it is referred to as *sinistrorse*.

Opposite

When two leaves occur at one node, the arrangement is called opposite. Oppositely arranged leaves may be either 2-ranked, as in Mexican heather (*Cuphea hyssopifolia*) in the henna family (*Lythraceae*), or 4-ranked or *decussate*, in which each succeeding pair of leaves is at right angles to the pairs above and below them. Decussate arrangement of leaves is characteristic of the mint family (*Lamiaceae*), the maple family (*Aceraceae*), and some members of the milkweed family (*Asclepiadaceae*), such as *Asclepias viridis*.

Whorled or Verticillate

When three or more leaves occur at one node, a whorled or verticillate arrangement is produced. The genera *Galium* and *Sherardia* in the madder family (*Rubiaceae*) are characterized by whorled leaves, as is also *Isotria* in the orchid family (*Orchidaceae*).

Rosette

Rosettes, often referred to as basal rosettes, occur in acaulescent plants, such as the common dandelion (*Taraxacum officinalis*) in the sunflower/aster family (*Asteraceae*). Acaulescent plants do have a stem, but the internodes are greatly contracted, and the leaves have a spiral alternate arrangement. Many biennial plants, such as carrots (*Daucus carota*) and poison hemlock (*Conium maculatum*) in the carrot family (*Apiaceae*), will produce a basal rosette during the first year of growth, followed by the production of a flowering stem with alternate leaves the second year.

Perfoliate

A leaf or a pair of connately fused leaves with the stem going through the center are referred to as *perfoliate*. *Montia perfoliata* and *Bupleurum rotundifolium* are examples of the perfoliate condition derived from a single leaf. *Silphium perfoliatum* is a good example of the basal connate fusion of leaves to achieve the perfoliate condition. The upper cauline leaves of henbit (*Lamium amplexicaule*) in the mint family (*Lamiaceae*) are sessile and clasping the stem but are not actually fused.

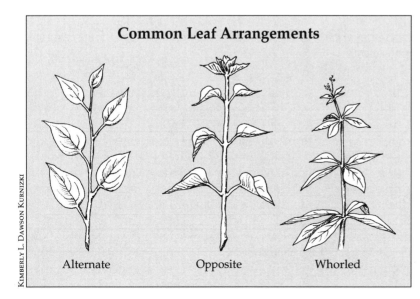

Common Leaf Arrangements

KIMBERLY L. DAWSON KURNIZKI

Alternate   Opposite   Whorled

Coniferous Leaves

The leaves of most conifers have developed with the need to minimize water loss while maximizing photosynthesis under relatively cold and dry (physiological drought) conditions where water is often not easily obtained. Needle-like leaves arranged in close, regularly spaced secund (one-sided, like a comb) divisions are referred to as *pectinate* or *comb-like*. *Acicular* leaves arranged in bundles or fascicles are typical of pines and spruces. In pines each fascicle is composed of two, three, four, five, six, seven, or eight divisions of needles, which form a more or less cylindrical shape if pushed together. The fascicles are spirally arranged on the tree branches.

In eastern red cedar (*Juniperus virginiana*), the leaves are reduced to minute *scales*, which have an opposite decussate arrangement, giving the appearance of 4-ranks. The scales are imbricate or overlapping, much like shingles on a roof. The leaves of the yew (*Taxus*) are sharp-pointed, flattened, and narrowly lance-shaped. They are spirally arranged on the branches but almost always give the appearance of being 2-ranked. This is also true for the dawn redwood (*Metasequoia*) and bald cypress (*Taxodium distichum*), both of which are deciduous in the fall, dropping entire branchlets with the attached leaves. Yew podocarpus (*Podocarpus macrophylla*) in the podocarpus family (*Podocarpaceae*), on the other hand, has a obvious spiral arrangement of the leaves.

Lawrence K. Magrath

See also: Leaf anatomy; Leaf lobing and division; Leaf margins, tips, and bases; Leaf shapes.

Sources for Further Study

Esau, Katherine. *Anatomy of Seed Plants*. 2d ed. New York: John Wiley and Sons, 1977. Discusses basic structure and development of leaves, phyllotaxy, and fibonacci series.

Jones, Samuel B., Jr., and A. E. Luchsinger. *Plant Systematics*. 2d ed. New York: McGraw-Hill, 1986. Discussion of leaf arrangement as well as structures, types, shapes, and related issues.

Prance, Ghillean Tolmie. *Leaves: The Formation, Characteristics, and Uses of Hundreds of Leaves Found in All Parts of the World*. Photographs by Kjell B. Sandved. New York: Crown, 1985. Three hundred full-color photographs make up this lavish, beautiful book. Topics include leaf arrangements. Includes index of scientific names.

LEAF LOBING AND DIVISION

Category: Anatomy

The pattern of leaf lobes (projections) or divisions, leaf arrangement, the number, and the shape of leaflets composing compound leaves are often useful characteristics for identification of plants.

Leaves, the main photosynthetic organs of plants, are usually green, flattened structures that are formed as lateral outgrowths at stem nodes. Simple leaves are composed of a single *lamina*, or blade, which may be attached to the stem via a cylindrical structure called a petiole. Leaves lacking a petiole are called *sessile*. Laminae of simple leaves may exhibit various patterns and degrees of lobing, which are often characteristic of individual species of plants and, together with reproductive features, are used in plant identification. Other species have *compound* leaves, in which the leaf laminae are subdivided into smaller leaflets. The pattern of arrangement, the number, and the shape of leaflets comprising compound leaves are often useful characteristics for identification of plants.

Some species of plants exhibit either gradual or abrupt changes in leaf lobing and division during development and are called *heterophyllous*. For example, some species exhibit a mixture of *pinnatifid* (pinnately lobed) and *pinnatisect* (pinnately compound) leaves on the same stem. Heterophylly is often observed in water plants, with one form of leaf being produced where the plant stem is submerged and another being produced where the stem is above water. Light periodicity, intensity, and quality, as well as oxygen and carbon dioxide concentrations, are known to influence leaf form in some species. In other species, different portions of a single leaf lamina can be pinnatifid or pinnatisect. The developmental mechanism for leaves of this type is not completely understood.

Lobing

Lobes typically extend greater than one-eighth of the distance from the margin to the midrib of the leaf or leaflet. The margin is the edge of the leaf lamina lying between the apex and base. The midrib is the prominent vein that subdivides the leaf or leaflet into two halves from base to apex. *Palmately lobed* margins are indented toward the base of the leaf lamina, creating a pattern like fingers extending from a hand, or a digitate pattern.

Pinnately lobed margins are indented one-quarter to one-half of the distance to the midrib, with the indentions oriented toward the midrib in a feather-like pattern. *Pinnately cleft* margins are indented a little more than half of the distance to the midrib. *Pinnately incised*, or pinnatifid, margins are deeply indented toward the midrib, extending well over half to almost completely to the midrib. The term

"pinnately lobed" is sometimes used in reference to lobed, cleft, and incised leaves, collectively.

Simple and Compound Leaves

To discriminate between *simple* and *compound* leaves, one may locate the axillary bud at the base of a leaf petiole in the node region of the stem. This area signifies the basal end of the entire leaf in both simple and compound leaves.

A simple leaf has only one blade, or lamina, associated with it. There are no leaflets. In *singly compound* leaves, the leaf is subdivided into leaflets, which attach to a central *rachis* (axis). The rachis is continuous with the petiole, which attaches to the node region of the stem, where the axillary bud will be found. In doubly, or *bipinnately, compound* leaves, the primary leaflet lamina is subdivided into smaller secondary leaflets, which attach to a secondary rachis, or *rachilla*. The secondary rachis attaches to the primary or central rachis. The primary rachis is continuous with the petiole, which attaches to the node region of the stem where the axillary bud will be found.

The number of leaflets in a compound leaf is often constant in many plant species. In *even pinnately compound* leaves, all of the leaflets are paired. There is no terminal leaflet; thus, the total number of leaf-

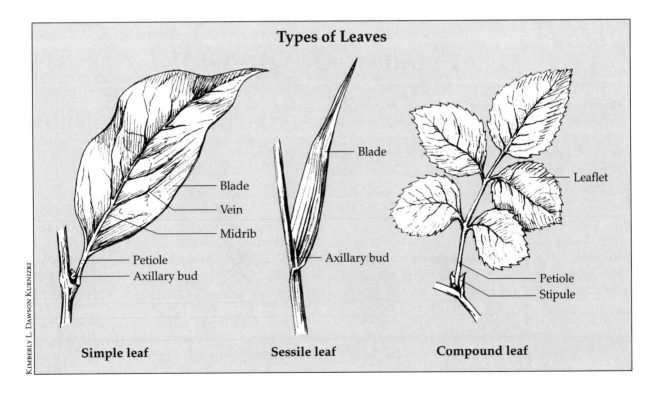

Types of Leaves

Simple leaf — Blade, Vein, Midrib, Petiole, Axillary bud

Sessile leaf — Blade, Axillary bud

Compound leaf — Leaflet, Petiole, Stipule

KIMBERLY L. DAWSON KURNIZKI

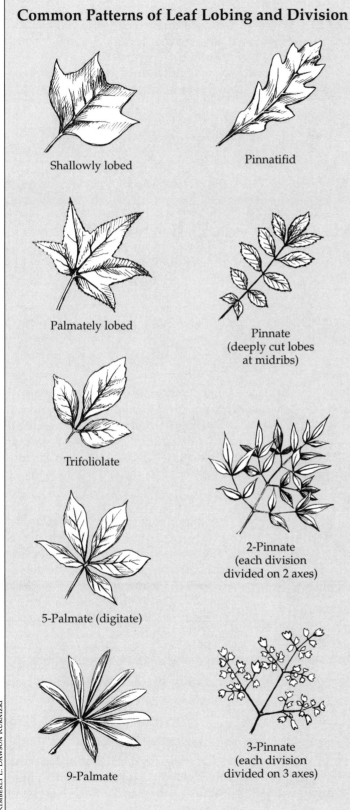

Common Patterns of Leaf Lobing and Division

Shallowly lobed

Pinnatifid

Palmately lobed

Pinnate
(deeply cut lobes
at midribs)

Trifoliolate

2-Pinnate
(each division
divided on 2 axes)

5-Palmate (digitate)

9-Palmate

3-Pinnate
(each division
divided on 3 axes)

KIMBERLY L. DAWSON KURNIZKI

lets per leaf is an even number. In *odd pinnately compound* leaves, there is one terminal, unpaired leaflet at the end of the leaf, making the total number of leaflets an odd number. *Trifoliolate* leaves have three leaflets, which may have petiolules or be sessile. The term *ternate* denotes groups of threes.

Leaf blade length is measured from where the blade joins the petiole straight to the tip of the leaf and perpendicular to the width. Width is measured at the widest part of the leaf perpendicular to the length. The petiole is the more or less round stalk that connects the leaf blade to the node region of the stem. Petiole length is measured from the point of attachment of the leaf blade to the node region of the stem. Some species lack petioles or have very short petioles.

The singly compound leaf blade is subdivided into leaflets which attach to a central rachis. The leaflet blade, or lamina, is the flat part of the leaflet. The *petiolule* is the stalk extending from the base of the leaflet lamina to the rachis. The rachis is continuous with the petiole, which attaches to the node region of the stem, where the axillary bud will be found. Leaflet length is measured from where the leaflet blade joins the petiolule straight to the tip of the leaflet perpendicular to the width. Width is measured at the widest part of the leaflet perpendicular to the length.

In the doubly or bipinnately compound leaf, the primary leaflets are themselves subdivided into still smaller secondary leaflets. The leaflet lamina is attached to a secondary rachis, or rachilla. The petiolule extends from the base of the leaflet lamina to the rachilla. The rachilla attaches to the primary or central rachis.

Roger D. Meicenheimer

See also: Leaf anatomy; Leaf arrangements; Leaf margins, tips, and bases; Leaf shapes; Shoots.

Sources for Further Study
Hardin, James W., Donald J. Leopold, and F. M. White. *Harlow and Harrar's Textbook of Dendrology*. 9th ed. Boston: McGraw-Hill, 2001. Contains an illustrated chapter on vegetative structures and associated terminology used in tree identification.
Raven, Peter H., Ray F. Evert, and Susan E. Eichhorn. *Biology of Plants*. 6th ed. New York: W. H. Freeman/Worth, 1999. Contains an illustrated chapter on leaf structure and function.
Walters, Dirk R., and David J. Keil. *Vascular Plant Taxonomy*. 4th ed. Dubuque, Iowa: Kendall/Hunt, 1996. Contains an illustrated chapter on vegetative structures and associated terminology used in plant identification.

LEAF MARGINS, TIPS, AND BASES

Category: Anatomy

The flattened part of the leaf is the leaf blade or lamina, which can be subdivided into three discrete regions: The tip or apex is the part of the lamina farthest removed from the point of attachment of the leaf to the stem. The base of the leaf is the part of the lamina that is closest to the point of attachment of the leaf to the stem. The margin is the perimeter of the leaf between the apex and base.

The form, or morphology, of leaves is often characteristic of individual species of plants and, like the reproductive features, is an important base of plant identification. Some plants have a more or less cylindrical *petiole* that joins the base of the leaf to the stem, while others lack a petiole and are called *sessile*. The *midrib* is the prominent vein that subdivides the leaf into two halves from base to apex.

Leaf Margins

The margin is the edge of the leaf lamina lying between the apex and base. *Entire* margins are smooth, without indentations or incisions. *Revolute* margins are rolled downward, toward the lower surface of the leaf. *Involute* margins are rolled upward, or toward the upper surface of the leaf. *Repand* margins are slightly and irregularly wavy, with the lamina surface undulating in a downward and upward direction. *Sinuate* leaf margins are shallowly indented and strongly wavy in the horizontal plane.

Teeth

"Teeth" are commonly seen in leaf margins and typically extend less than one-eighth of the distance from the margin to the midrib of the leaf. Various shapes and sizes of teeth are associated with leaf mar-

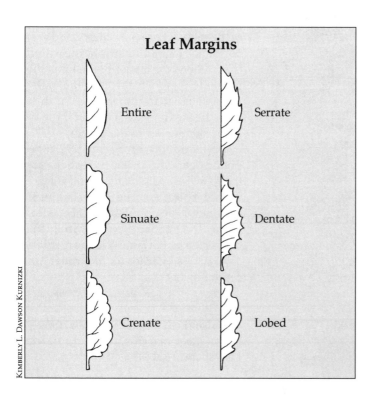

Leaf Margins

Entire

Serrate

Sinuate

Dentate

Crenate

Lobed

KIMBERLY L. DAWSON KURNIZKI

Leaf Tips

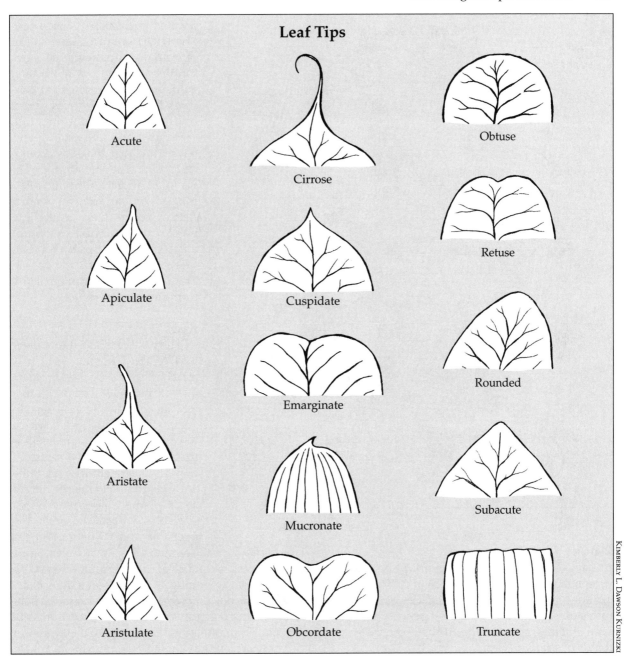

Acute

Cirrose

Obtuse

Apiculate

Cuspidate

Retuse

Aristate

Emarginate

Rounded

Aristulate

Mucronate

Subacute

Obcordate

Truncate

KIMBERLY L. DAWSON KURNIZKI

gins. *Crenate* margins have shallowly ascending round or obtuse teeth. *Crenulate* (minutely crenate) margins have minute, shallowly ascending round or obtuse teeth. *Serrate* margins have sharp, saw-toothed teeth pointing forward or toward the apex. *Serrulate* (minutely serrate) margins have very fine, sharp, saw-toothed teeth pointing forward or toward the apex. *Doubly serrate* margins have coarse, saw-toothed teeth bearing smaller teeth on the basipetal part of their edges. *Dentate* margins have sharp teeth, or indentions, pointing outward at right angles to the midrib. *Denticulate* (minutely dentate) margins have fine, sharp teeth, or indentions, pointing outward at right angles to the midrib. *Aculeate* margins have spiny or prickly projections along their edges. *Bristle tips* refer to teeth or lobes that are terminated by a sharp, flexible, elongated point.

Leaf Bases

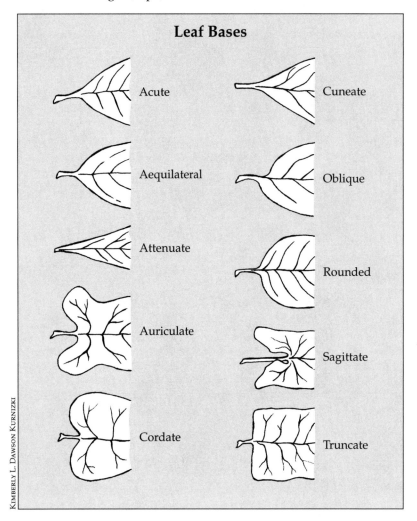

Kimberly L. Dawson Kurnizki

Acute

Aequilateral

Attenuate

Auriculate

Cordate

Cuneate

Oblique

Rounded

Sagittate

Truncate

Glands

Glands are small, protruding bumps on the leaf margin that are typically different in color from the surrounding lamina tissue. Often they occur at the apices of teeth or lobes or on leaf petioles. Use of a hand magnifying lens helps in observing the small glands of some species.

Tips or Apices

The apex is the tip of the leaf blade. *Acuminate* apices have a long, slender, sharp point, with a terminal angle less than 45 degrees, and straight to convex sides. *Acute* apices have a sharp-pointed tip, with a terminal angle between 45 and 90 degrees, and straight to convex sides. *Mucronate* apices have a tip that is terminated by a short, sharp, abrupt point. *Cuspidate* apices have a tip that is abruptly and sharply constricted into an elongated, sharp-pointed tip or cusp (a sharp, rigid point). *Obtuse* apices have a blunt or rounded tip, with the sides forming an angle of more than 90 degrees, and straight to convex sides. *Rounded* apices have a tip that is curved to form a full, sweeping arc. *Truncate* apices have a tip that looks as though it was cut off at almost a right angle to the midrib, forming a flat-topped or squared-off shape. *Retuse* apices have a shallow notch in a rounded or obtuse apex. *Emarginate* tips have a shallow and broad notch at the apex. These are only a few of the many forms leaf apices can take.

Bases

The base of a leaf is the lower part of the lamina, where it is attached to the petiole or stem. *Cuneate* bases are sharp-pointed, with an angle less than 45 degrees between opposite sides which form a wedge or triangular shape that tapers to a narrow region at the point of attachment of lamina with petiole. *Acute* bases have a sharp-pointed base, with opposite sides forming an angle between 45 and 90 degrees at the position where the lamina joins the petiole. *Obtuse* bases have a blunt or narrowly rounded base with opposite sides forming an angle greater than 90 degrees at the position where the lamina joins the petiole. *Rounded* bases are curved to form a full, sweeping arc. *Truncate* bases look as though they were cut off at nearly a right angle to the midrib, forming a flat-topped or squared-off shape. *Cordate* bases are valentine-shaped, with both right and left margins forming broad arcs that meet in the middle of the junction between lamina and petiole. *Inequilateral* bases have asymmetrical left and right sides of different sizes or shapes. *Auriculate* bases have earlike lobes where the lamina joins the petiole.

Roger D. Meicenheimer

See also: Leaf anatomy; Leaf arrangements; Leaf lobing and division; Leaf shapes; Shoots.

Sources for Further Study

Hardin, James W., Donald J. Leopold, and Fred M. White. *Harlow and Harrar's Textbook of Dendrology.* 9th ed. Boston: McGraw-Hill, 2001. Contains an illustrated chapter on vegetative structures and associated terminology used in tree identification.

Raven, Peter H., Ray F. Evert, and Susan E. Eichhorn. *Biology of Plants.* 6th ed. New York: W. H. Freeman/Worth, 1999. Contains an illustrated chapter on leaf structure and function.

Walters, Dirk R., and David J. Keil. *Vascular Plant Taxonomy.* 4th ed. Dubuque, Iowa: Kendall/Hunt, 1996. Contains an illustrated chapter on vegetative structures and associated terminology used in plant identification.

LEAF SHAPES

Category: Anatomy

The overall shapes of leaves and leaflets are often characteristic of individual species of plants and, together with reproductive features, are used in plant identification.

To assess the shape of a leaf, one examines the outline formed by the apex, margin, and base of the leaf or leaflet. If the leaf has teeth or lobes along its margin, one imagines a smooth curve interconnecting the tips of the teeth or lobes to assess the overall shape. The shapes of compound leaves can likewise be assessed by imagining a smooth curve connecting the tips of the leaflets that form the compound leaf blade. The terminology below is used to describe the shape of laminae of simple and compound leaves as well as the laminae of the leaflets of compound leaves.

Elongated Leaves

Linear leaves have a long and very narrow leaf shape, with sides that are almost parallel with one another and are usually more than four times longer than broad. *Oblong* leaves have a rectangular leaf blade two to four times longer than it is wide, with sides that are almost parallel to each other. *Ensiform* leaves resemble the shape of a broad sword. *Ligulate* leaves are straplike, resembling a tongue. *Falcate* leaves are elongated and recurved, resembling the shape of a sickle blade. *Lanceolate* leaves have a lance-shaped leaf, with the widest part of the leaf near the base and the narrowest part near the apex. The prefix *ob* means that the shape is inverted or upside down. Thus, *oblanceolate* leaves have a lance-shaped leaf, with the widest part of the leaf near the apex and the narrowest part near the base.

Circular to Elliptical Leaves

Orbicular leaves have a more or less circular leaf shape in which the width and length of the lamina are equal, or nearly so. *Elliptical* leaves have a shape that looks like an ellipse, twice as long as broad, with the widest part of the leaf near the middle. *Oval* leaves are broadly elliptical, with the blade width being more than half the length and the widest part of the leaf near the middle. Oval leaves are wider than elliptical leaves of the same length. *Ovate* leaves are egg-shaped, with the widest part of the leaf below the middle toward the base, while *obovate* leaves are egg-shaped, with the widest part of the leaf above the middle toward the apex.

Other Shapes

Reniform leaves have a shape like a kidney. *Cordate* leaves are shaped like a valentine or heart, with the lobes of the valentine at the base of the leaf and the pointed portion at the apex. *Obcordate* leaves are also valentine-shaped, but the lobes are at the apex, and the basal lamina tapers into the petiole. *Sagittate* leaves have a shape like an arrowhead. *Hastate* leaves are also shaped like an arrowhead, but the basal lobes diverge or extend away from the midrib, giving an outline that resembles a halberd. *Rhombic* leaves have a more or less diamond shape, with straight margins and with the widest part of the leaf lamina near the middle. *Spatulate* leaves have a spoon-shaped or spatula-shaped leaf where the lamina is widest near the rounded apex. *Flabellate* leaves are fan-shaped or broadly wedge-shaped, with the broadest part of the lamina at the apex. *Deltoid* leaves are delta-shaped, resembling an equiangular triangle. Often the sides of the deltoid leaves are slightly curved toward the apex.

Conifer Leaf Shapes

Needlelike, or *acicular*, leaves have a long and very narrow leaf shape, with sides that are almost parallel to each other and are usually more than ten times longer than broad. Acicular leaves are often borne on short lateral branches called *fascicles*. The number of acicular leaves per fascicle is constant within a species. *Linear* leaves have a long, narrow leaf shape, with sides that are almost parallel with each other and usually are more than four times longer than broad. Linear leaves can be flat, triangular, or square in cross section. They may also exhibit a distinct twist along their vertical axes.

Subulate leaves are short, narrow, flat, stiff, awl-shaped leaves that taper to a sharp point. *Scale* leaves are small, inconspicuous leaves that are typically appressed tightly to the stem and have overlapping margins. Scale leaves may or may not be photosynthetic (green). *Decurrent* leaves have an extension of tissue running down the stem below the point of junction of the leaf, with a stem that forms a wing or ridge of tissue. Some species have leaves that are borne on a semiwoody *peg* that extends away from the twig surface. These pegs do not abscise with leaves and remain visible on the twig for many years after leaf abscission.

Roger D. Meicenheimer

See also: Leaf anatomy; Leaf arrangements; Leaf lobing and division; Leaf margins, tips, and bases; Shoots.

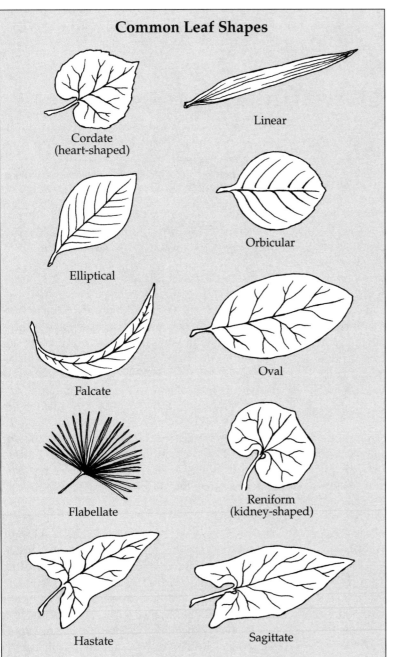

Common Leaf Shapes

Cordate (heart-shaped)

Linear

Elliptical

Orbicular

Falcate

Oval

Flabellate

Reniform (kidney-shaped)

Hastate

Sagittate

KIMBERLY L. DAWSON KURNIZKI

Sources for Further Study

Hardin, James W., Donald J. Leopold, and F. M. White. *Harlow and Harrar's Textbook of Dendrology*. 9th ed. Boston: McGraw-Hill, 2001. Contains an illustrated chapter on vegetative structures and associated terminology used in tree identification.

Raven, Peter H., Ray F. Evert, and Susan E. Eichhorn. *Biology of Plants*. 6th ed. New York: W. H. Freeman/Worth, 1999. Contains an illustrated chapter on leaf structure and function.

Walters, Dirk R., and David J. Keil. *Vascular Plant Taxonomy*. 4th ed. Dubuque, Iowa: Kendall/Hunt, 1996. Contains an illustrated chapter on vegetative structures and associated terminology used in plant identification.

LEGUMES

Categories: Agriculture; economic botany and plant uses; food

Legumes, including peas and beans, are among the most important staple food crops worldwide. Legumes are also planted among other crops because of their ability to enrich soil nitrogen content.

Legumes are plants of the pea or bean family classed in the family *Leguminosae*, which is referred to as the family *Fabaceae* in North America. With 18,000 species and 650 genera, the legumes are one of the largest families of plants in the world. The vast majority of legumes are herbaceous plants, but family members range in size from dwarf willowlike herbs of Arctic and alpine habitats to massive tropical trees. Most are flowering plants, but the family also includes a number of shrubs and trees, some quite tall and many bearing thorns or spines.

Legumes can be recognized by their distinctive fruits, which are typically elongated pods that contain and protect the seeds. The fruit is an ovary that splits along two sides at maturity. The legume family includes three subfamilies, each with a distinctive type of flower.

Subfamilies

The largest of the subfamilies is the *Papilonoidae*, named for the distinctive, butterfly-shaped wings of the flower petals. This subfamily includes the most familiar and economically important legumes, such as beans, peas, peanuts, lentils, and soybeans, the vetches and other ground covers, and animal forage crops, such as clovers, sweet clovers, lupines, and alfalfa.

Most members of the subfamily *Caesalpinioidae* are tropical and subtropical trees and shrubs, many of which are widely planted as ornamentals, including the honey-locust (*Gledtsia*) and redbud (*Cercis* species)—both very popular ornamentals—along with a number of subtropical shrubs and small trees, notably the mesquite and palo verde of the American Southwest. The subfamily also includes a number of vines, such as wisteria (*Wisteria floribunda*) and wait-a-minute vine. An extract of the wild senna (*Senna alexandrina*) is an important purgative medicine. The flowers of *Caesalpinioidae* have five uneven petals.

The subfamily *Mimosoidae* also consists mostly of tropical and subtropical trees and shrubs, such as mimosa and the Kentucky coffee tree, *Gymnocladus dioica*. Legumes of this subfamily typically have a cluster of small flowers on a single stem, the whole covered by long stamens.

Nitrogen Fertilizer from Legumes

The importance of legumes as a green fertilizer has been recognized for centuries. As early as the third century B.C.E., the Greek philosopher and scientist Theophrastus recommended that beans should be planted to enrich farm soils. Then, as now, a common agricultural practice was to rotate cereal crops, such as corn or wheat, with leguminous crops, such as clover or alfalfa (*Medicago sativa*). The legumes used as green fertilizer not only contribute nitrogen to the soil but also can be harvested as animal feed.

The ability of legumes to enrich soil nitrogen content stems from their symbiotic relationship with *Rhizobium* and other nitrogen-fixing bacteria, which enter and colonize the root hairs of certain legume seedlings. The cell walls of the root hairs respond by curling to form a nodule that houses and protects the colony of nitrogen-fixing bacteria. The bacteria fix molecular nitrogen of the atmosphere into nitrogen-containing compounds useful to the plant, which in turn provides the bacteria with water, minerals, and carbon-based products of photosynthesis.

Legumes as Crops

For centuries, people around the world have depended on a combination of grains and legumes for a healthy and sustaining diet. Legumes were among the earliest of crops cultivated in the Fertile Crescent of the Middle East (the region between the Tigris and Euphrates Rivers) some eleven thousand years ago, where farmers supplemented their wheat and barley crops with lentils and peas.

Legumes still rank among the most important of all staple food crops, especially the *pulses* (edible seeds), such as peas and beans, chickpeas, and lentils. Legumes are also important as cover plants to hold and stabilize soils, as nutrient-rich feed for livestock, as timber products, and as green fertilizer. Some species are also valuable because they can be grown in poor soils or in areas of low rainfall. Other derivatives include medicines, food flavorings, tannins, gums, resins, and dyes. An extract from *Lonchocarpus* and *Derris* called rotonene is the active ingredient in fish poisons, molluscicides, and insecticides.

Nutritionally, legumes are especially good sources of proteins, carbohydrates, minerals, vitamins, and fiber. Especially important as protein sources in areas where animal protein is scarce, legumes contribute about 18 percent of the total plant protein consumed by humans. Legume proteins contain large amounts of some essential amino acids, such as lysine, but are low in the sulfur-containing amino acids methionine and cystine. Legumes' carbohydrate content varies from 13 to 65 percent, of which half or more is starch. Many legumes are also sources of iron, calcium, phosphorus, zinc, copper, and magnesium. They are particularly high in B vitamins, such as thiamin, riboflavin, niacin, folic acid, and pantothenic acid, as well as vitamins C and E. Legumes are also valued for their low fat content.

Among the most common legumes found in the human diet are the following:

Garden pea. Native to Asia, the garden pea (*Psum sativum*) is a cool-season crop that is widely cultivated throughout the world, with China and Russia being the most important producers of dried peas and the United States and Great Britain leading the production of green peas. The garden pea also deserves mention as the plant used by Gregor Mendel in his experiments that defined the science of genetics. The garden pea is marketed as frozen, canned, dried peas, or

PHOTODISC

Legumes can be recognized by their distinctive fruits, which are typically elongated pods that contain and protect the seeds. The fruit is an ovary that is dehiscent, splitting along two sides at maturity, as do these green beans.

as snow peas and sugar peas, which represent the harvested pods.

Common bean. Another extremely important legume crop, the common bean (*Phaseolus vulgaris*) is a warm-season crop native to South America but is now one of the most widely cultivated legumes in the world, thanks to its introduction into Europe and Africa by Spanish and Portuguese during the sixteenth century. The common bean remains the most important pulse crop in tropical Africa and America, especially in Brazil. Common beans are harvested in the podded stage (snap beans or string beans) or as shell beans. Shell beans are by far the largest crop and include various types such as navy, kidney, French, string, pinto, and yellow-eye beans. Common beans are the primary ingredient in many staple and well-known foods, including Boston baked beans and chili con carne. Because of their relatively high pectin content, common beans must be soaked in water prior to consumption.

Lima bean. Also known as the butter bean or Madagascar bean, the lima bean (*Phaseolus lunatus*) is native to Central America, where it was cultivated at least seven thousand years ago. Like the common bean, the lima bean has been introduced throughout the world and is now grown in the warmer regions of Africa, Asia, and North America. The seeds are harvested and marketed canned or frozen in North America, as a bean paste in Japan, and ground into bread flour in the Philippines. Although rich in proteins and carbohydrates, lima beans contain glycosides, which can produce toxic prussic acid, and must therefore be thoroughly prepared by soaking and boiling in frequent changes of water.

Peanut. One of the most nutritious legumes, the peanut (*Arachis hypogaea*) is native to Brazil but has been widely introduced, particularly in the United States, where it has long been an important cash crop along the southern tier of states. The plant initially pushes a flowering stalk above ground, but following fertilization the stalk is pushed into the ground, where it matures, giving rise to the name ground peanut.

Soybean. Soybeans (*Glycine max*) are an important food crop that have been cultivated in the Far East for thousands of years, often in combination with rice crops. The high protein content of about 45 percent is a major reason for their importance as a food crop. An oil extracted from soybean seed is used in the manufacture of cooking oils and some margarine. Other important soybean products include tempeh, miso, tamari, and tofu.

Lentil. One of the oldest of crops, lentils (*Lens culinaris*) were domesticated as early as 8000 B.C.E. in the Fertile Crescent region of the Middle East, between the Tigris and Euphrates Rivers. Lentils are still cultivated widely in the Old World; their highest production rates are in India, but lentils also remain the most important pulse crop in Nepal and Bangladesh. Harvested lentil seeds are used in the production of flour, soups, and as a dried snack food, while the plant is used as high quality straw feed for livestock in the Middle East. Lentils are considered an excellent pulse crop because of their high protein content as well as being excellent sources of vitamins A and B, potassium, and iron. Nutritionally, they are also valuable because they lack fat content and cholesterol.

Toxic Substances

Toxic or antinutritional substances found in some legumes include alkaloids, cyanide poisons, enzyme inhibitors, saponin, and goitrogen, the last causing an enlargement of the thyroid gland. More serious still are the lectins or haemagglutinins (blood clotting agents) that may cause vomiting, diarrhea and severe abdominal pain, and lathyrism, which can produce mild to severe neurological disorders. Unpleasant but nontoxic substances that occur in certain legumes include stachyose and raffinose carbohydrates, which cause flatulence. Most of these unwanted substances can be removed by appropriate washing and cooking methods prior to consumption.

Dwight G. Smith

See also: Agriculture: traditional; Fruit crops; Fruit: structure and types; Nitrogen fixation; Organic gardening and farming; Sustainable agriculture; Vegetable crops.

Sources for Further Study

Couplan, Francois. *The Encyclopedia of Edible Plants of North America.* New Canaan, Conn.: Keats, 1998. The section on legumes details food values of fifty-five important genera that grow in North America.

Crisp, M. D., and J. J. Doyle, eds. *Phylogeny*. Vol. 7 in *Advances in Legume Systematics*. Kew, England: The Royal Botanic Gardens, 1999. This volume concentrates on various aspects of the systematics of legume biology. Each of the edited volumes in the set includes a series of papers written by experts in a particular aspect of legume taxonomy and classification.

Duke, James A. *Handbook of Legumes of World Economic Importance*. New York: Plenum Press, 1981. Includes brief summaries of all of the world's economically important legumes, with special emphasis on the pulses and fodder legumes.

National Research Council. *Tropical Legumes: Resources for the Future*. Washington, D.C.: National Academy of Sciences, 1979. A summary of the uses of all of the most important tropical legumes in the world.

Smartt, J., and N. W. Simmonds, eds. *Evolution of Crop Plants*. London: Longman Scientific and Technical Press, 1995. Includes a good history of the use of legumes as crop and fodder plants.

Vaughan, J. G., and C. Geissler. *The New Oxford Book of Food Plants*. New York: Oxford University Press, 1997. This well-written and well-illustrated book provides a good introduction to the world of legumes as food plants. Topical coverage includes sections about exotic legumes, runner beans and French beans, peas and lentils, and similar topics.

LICHENS

Categories: Algae; fungi; taxonomic groups

Lichens are composed of two distinct species, a fungus and a photosynthetic alga (or bacterium) that have coevolved to live in a symbiotic relationship with each other as a single life-form that grows on rocks, trees, and other substrates. Lichens are classified as members of the kingdom Fungi, *with most being placed under the phyla* Ascomycota *and* Basidiomycota. *It is estimated there are seventeen thousand species of lichen, representatives of which have been found nearly everywhere in the world.*

Symbiosis

Symbiosis is an extreme form of an ecological relationship known as mutualism between members of different species, in which each partner in the union derives benefits from the other. In symbiotic unions, the partners are so dependent on each other they can no longer independently survive.

In lichens, the fungal (mycobiont) symbiont provides protection, while the green-algal or bacterial (photobiont) symbiont provides sugars, created by photosynthesis. It is often suggested that the fungus in lichen species might also pass water and nutrients to the photobiont, but this function is less well documented. This special relationship allows lichens to survive in many environments, such as hot deserts and frozen Arctic tundra, that are inhospitable to most other life-forms. As a result, the lichen whole is greater than the sum of its parts. While in nature lichen partners always exist together, under laboratory conditions it is possible to take the lichen apart and grow the two partners separately.

Anatomy

Whereas in most plant species the anatomy of the organism is identified with structures associated with a single vegetative body, the "lichen body" is more aptly described as a colony of cells that share a variety of associations with one another that vary from one species of lichen to the next. In some species of lichen, fungal and algal cells merely coexist. *Coenogonium leprieurii*, for example, is a lichen that lives in low-light tropical and subtropical forests in which the filamentous green algal partner (*Trebouxia*) is dominant.

In most lichen species, however, the relationship between the symbiotic partners is more intimate, with the lichen body appearing to be a single entity.

Lichens are composed of two species which are partners: a fungus and a bacterial or algal photobiont. In most lichen species, the lichen body appears to be a single entity. Here, a yellow lichen grows on a dead tree trunk.

AP/Wide World Photos

In these species the algal symbiont has no cell walls and is penetrated by filaments from the fungal symbiont called *haustoria*, which pass sugars from the algal cell to the fungal cell and may have a role in the transportation of water and nutrients from the fungal cell. This integration is so complete that many naturalists prior to the nineteenth century mistakenly classified lichens as mosses.

In most lichen species it is nevertheless possible, with a good magnifying device, to identify several distinct regions of the thallus or lichen body. The outermost region is the *cortex*, a compacted layer composed of short, thick *hyphae* (widely dilated filaments) of the fungal symbionts that protect the lichen from abiotic factors in its environment. These hyphae extend downward into a second region, the *photobiont layer*, where they surround the algal symbionts. Below this is a third region, the *medulla*, composed of a loosely woven network of hyphae.

Underneath this is a fourth region, the *undercortex*, that is similar in appearance and structure to the cortex. The bottom of the lichen body is com-posed of *rhizines*, rootlike structures composed of bundles of hyphae that attach the lichen to its substrate (the rock, bark, or other support on which it resides). This arrangement of regions into layers serves to prevent water loss. Many species can survive complete desiccation, coming back to life when water becomes available again. The cortex also contains *pseudocyphellae*, which are pores that allow for the exchange of gases necessary for photosynthesis.

Life Cycle

Lichens typically live for ten years or more, and in some species the lichen body can survive for more than a hundred years. Reproduction in most fungal species proceeds by the development of a cup- or saucer-shaped fruiting body called an *apothecium*, which releases fungal spores to its surroundings. Procreation in lichens is more problematic in that the fungal offspring must also receive the right algal symbiont if they are to survive. The most common form of dispersion in lichen is by the accidental breaking off of small pieces of the thallus called *isidia*, which are then spread by wind to new substrates. In some species, small outgrowths of the thallus known as *soralia* arise, composed of both fungi and algae and surrounded by hyphae, to form *soredia*, which after dispersion give rise to a new thallus.

Biological and Agricultural Importance

Lichens are excellent bioindicators of air pollution, as many species are particularly sensitive to certain contaminants in their surroundings, such as sulfur dioxide. They represent a major food source for reindeer in Lapland and are used as cattle fodder there as well. One species of lichen (*Umbilicaria esculenta*) is considered a delicacy in Japan. Historically, lichens have been used as pigments for the dying of wool. The medical properties of some species of lichens for the treatment of lung disease and rabies have led to a renewed interest in them.

David W. Rudge

See also: Algae; Bacteria; Coevolution; Community-ecosystem interactions; Fungi; Green algae; Photosynthesis.

Sources for Further Study

Brodo, Irwin M., Sharnoff, Sylvia Duran, and Stephen Sharnoff. *Lichens of North America.* New Haven, Conn.: Yale University Press, 2001. A comprehensive field guide to lichens and their natural history.

Dobson, Frank. *Lichens: An Illustrated Guide.* Richmond, Surrey, England: Richmond, 1981. A massive (795-page) field guide to lichens in the British Isles. Includes more than nine hundred color photographs, as well as maps, bibliography, and index.

Hale, Mason E., Jr. *The Biology of Lichens.* 3d ed. London: Edward Arnold, 1983. A standard college-level text focused primarily on anatomy and physiology that discusses lichens as pollution monitors.

Hawksworth, D. L., and D. J. Hill. *The Lichen-Forming Fungi.* New York: Blackie & Son, 1984. A college-level text focused primarily on the life cycle of lichens, their ecology, and biogeography.

Purvis, William. *Lichens.* Washington, D.C.: Smithsonian Institution Press, 2000. Presents an introductory overview of lichen biology, ecology, and evolution, information on ecological and economic importance, how they are studied, and extensive illustrations and photographs.

LIPIDS

Categories: Cellular biology; physiology

Lipids are a diverse group of compounds sharing the common property of being hydrophobic (insoluble in water). Lipids include fatty acids, fats, oils, steroids (sterols), waxes, cutin, suberin, glycerophospholipids (phospholipids), glyceroglycolipids (glycosylglycerides), terpenes, and tochopherols.

Lipids are ubiquitous in plants, serving many important functions, including storage of metabolic energy, protection against dehydration and pathogens, the carrying of electrons, and the absorption of light. Lipids also contribute to the structure of membranes. In addition, plant lipids are agricultural commodities important to the food, medical, and manufacturing industries.

Fatty Acids

Fatty acids, the simplest of the lipids, are highly reduced compounds with a hydrophilic (water-soluble) carboxylic acid group and a hydrophobic hydrocarbon chain. Hundreds of different fatty acids have been isolated from plants. Fatty acids differ from one another in the length of the hydrocarbon chain and degree of saturation (number of carbon-carbon double bonds). The most common fatty acids have chain lengths ranging from sixteen to twenty carbon atoms, but many less common fatty acids are longer or shorter.

Saturated fatty acids have no double bonds, whereas unsaturated fatty acids have one or more double bonds. Naturally occurring *unsaturated* fatty acids have *cis* double bonds, in which the two hydrogen atoms bonded to the carbon atoms of the double bond are on the same side of the fatty acid molecule. Fatty acids with one double bond and two or more double bonds are referred to as *monounsaturated*, *diunsaturated*, and *polyunsaturated*, respectively.

Fats and Oils

Fatty acids rarely occur free in the cell. Instead they are attached by ester linkages to *glycerol*, a three-carbon sugar alcohol, to form *fats* and *oils*, the most abundant lipids. Glycerol molecules with one, two, and three fatty acids attached are referred to as *monoglycerides*, *diglycerides*, and *triglycerides* (often called *triacylglycerols*), respectively. The fatty acids attached to a triglyceride may be the same, in which case it is referred to as a simple triglyceride, or the fatty acids may be different, in which case it is referred to as a mixed triglyceride.

The degree of saturation and hydrocarbon chain length of fatty acids in triglycerides affects their melting point. Common fats, such as those from palms and coconuts, are triglycerides that contain a high proportion of saturated fatty acids and are solid at room temperature (22 degrees Celsius). Common oils, such as those from corn, peanuts, soybeans, sunflowers, and olives, are triglycerides that contain a high proportion of unsaturated fatty acids and are liquid at room temperature.

Because plants cannot control their temperatures, they contain much more oil than fat so their membranes will be fluid at ambient temperatures. The most common monounsaturated fatty acid is *oleic* acid, while the most common polyunsaturated fatty acids are *linoleic acid* and *linolenic acid*. Some plants have a high proportion of saturated fats, containing such fatty acids as *palmitic acid*, the most common saturated fatty acid found in plants. Plants also contain lesser amounts of other saturated fatty acids, such as *lauric* and *myristic acid*. *Phytanic acid*, a product of chlorophyll metabolism, is a saturated, branched chain fatty acid.

Plant oils are a mixture of triglycerides and are used as a storage form of energy in seeds. Because fats are more highly reduced than starch, they provide almost twice the energy on a per-weight basis. When fatty acids are removed from glycerol they can undergo oxidation to yield energy.

Waxes, Cutin, and Suberin

Waxes are long-chain fatty acids attached to long-chain alcohols by ester linkages. *Cutin* is a complex of hydroxylated fatty acids (fatty acids with hydroxyl groups attached to them) cross-linked to one another. Waxes and cutin are found in the cuticle, the outermost layer of plant surfaces exposed to the air, and provide protection from dehydration and pathogens. *Suberin* is a complex compound of unknown structure. It is the major component of the walls of cork cells, the outermost layer of bark. Like cutin and waxes, suberin provides protection from dehydration and pathogens.

Glycerophospholipids

Glycerophospholipids, or phosphoglycerides, contribute substantially to cellular plasma membranes. Thus, they form one of the most important classes of lipids. Glycerophospholipids are composed of glycerol phosphate to which is attached two fatty acids by ester linkages. Molecules, such as ethanolamine, choline, inositol, and serine, may be bonded to the phosphate, resulting in an even greater variety of glycerophospholipids.

Glycerglycolipids

Known as *glyceroglycolipids*, or glycosylglycerides, these lipids are primarily found in chloroplast membranes, are widespread in plants, and consist of glycerol to which is attached one or two sugar molecules and two fatty acids. The sugars attached to the glycerol are either glucose, galactose, or a digalactose unit.

Steroids

Steroids, also called *sterols*, comprise another class of lipids that are important constituents of the plant plasma membrane. They also function as plant hormones. Steroids are formed by a conjugated ring system. Side chains and groups attached to the rings result in a variety of steroids with many biological activities. Stigmasterol, beta-sitosterol, lanosterol, and ergosterol are plant steroids. Cholesterol, a common steroid in the plasma membranes of animal cells, is rarely present in plants.

Terpenes

Terpenes are lipids that are composed of two or more five-carbon isoprene units. More than twenty-two thousand terpenes have been described. The familiar flavor and aroma of many plants are due to their characteristic terpenes. Plant terpenes and terpenoid derivatives include phytol, a constituent of chlorophyll; beta-carotene, a photosynthetic pigment that is a precursor of vitamin A in animals; paclitaxel, an anticancer agent; and rubber. The blue haze often seen in the air on summer afternoons is due, in part, to terpenes emitted from leaves.

Tocopherols

Tocopherols contain an aromatic ring and a long isoprene side chain. Plant tocopherols include vitamin E, a biological antioxidant that protects unsaturated fatty acids from damage from free radical attack; vitamin K, which plays an essential blood clotting role in higher animals; and ubiquinone and plastoquinone, which are essential electron carriers in the reactions leading to the synthesis of adenosine triphosphate (ATP).

Charles L. Vigue

See also: Cell wall; Chloroplasts and other plastids; Cytosol; Membrane structure; Metabolites: primary vs. secondary; Oil bodies; Plant cells: molecular level; Plasma membranes.

Sources for Further Study

Dey, P. M., and J. B. Harborne, eds. *Plant Biochemistry*. San Diego: Academic Press, 1997. Each topic is covered by a specialist. Emphasizes plant metabolism but also includes enzymes, functions, regulation, and molecular biology. Includes references, illustrations.

Mathews, Christopher K., K. E. Van Holde, and Kevin G. Ahern. *Biochemistry*. 3d ed. New York: Addison Wesley Longman, 2000. All aspects of biochemistry covered, for the advanced student. Profusely illustrated and referenced.

Nelson, David L., and Michael M. Cox. *Lehninger Principles of Biochemistry*. 3d ed. New York: Worth, 2000. Covers general biochemistry, from structures to metabolism. Excellently illustrated and referenced. Includes a CD-ROM of art and animations.

Zubay, Geoffrey L. *Biochemistry*. 4th ed. Boston: Wm. C. Brown, 1998. General, advanced biochemistry text with especially good coverage of proteins. Includes illustrations, references. Comes with a CD-ROM of art and animations.

LIQUID TRANSPORT SYSTEMS

Categories: Anatomy; physiology; transport mechanisms

Liquid transport systems are structures that facilitate the movement of water, via the xylem, from a plant's roots to its leaves. Water then evaporates from the leaves through the stomata in the process of transpiration.

Water is the most abundant compound in plant cells. It accounts for 85-95 percent of the weight of most plants. It even makes up 5-10 percent of the weight of "dry" seeds. More than 95 percent of the water gathered by a plant, however, evaporates back into the atmosphere, often within hours after being absorbed. This evaporation of water from a living plant is called *transpiration*. Most transpiration is from leaves.

Plants transpire huge amounts of water. On a warm, dry day, an average-size maple tree transpires more than 200 liters per hour, while herbaceous plants transpire their own weight in water several times per day. A corn plant transpires almost 500 liters of water during its four-month growing season. If humans required an equivalent amount of water, a person would have to drink approximately 40 liters per day.

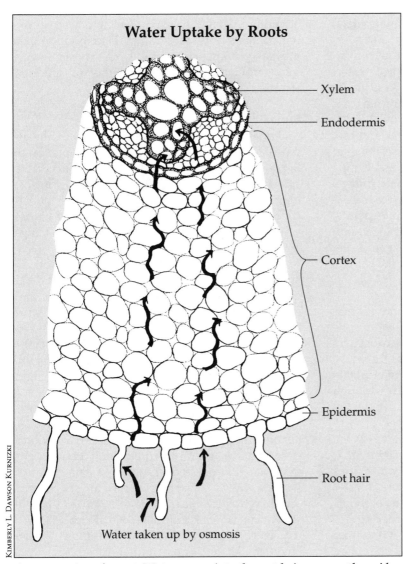

Water Uptake by Roots

Xylem

Endodermis

Cortex

Epidermis

Root hair

Water taken up by osmosis

KIMBERLY L. DAWSON KURNIZKI

A cross section of a root: Water moves into the root hairs, across the epidermis, through the cortex, across the endodermis, and into the xylem, from which it is transported up into the plant.

distant sites for storage and use. Thus, the multicellular nature of plants was largely responsible for the evolution of *xylem* and *phloem*, the two long-distance transport systems in plants.

Leaves and Roots

Leaves are the primary photosynthetic organs of most plants. The rate of gas exchange for photosynthesis depends, among other things, on the amount of available surface area. The loose internal arrangement of cells in leaves produces a large internal surface area for transpiration—an area that may be more than two hundred times greater than the leaf's external surface area.

The internal surface area of a leaf is connected with the atmosphere via an extensive system of intercellular spaces, pores called *stomata*, that occupy as much as 70 percent of a leaf's volume. Stomata are so numerous that a typical leaf of a squash plant has more than eighty million of them. Leaves also have an efficient system of *veins* (vascular tissue) for distributing water to their internal evaporative surfaces. One square centimeter of leaf may have as many as six thousand outlets of vascular tissue.

Water lost via transpiration must be replaced by water absorbed from the soil by the plant's roots. The movement of water from the roots to the leaves is very rapid. Water molecules may move as fast as 75 centimeters per minute, which is roughly equivalent to the speed of the tip of a second hand sweeping around a wall clock.

Plant survival depends on the ability to transport water and dissolved materials. Inside individual cells, *diffusion* is usually adequate for this movement. Small molecules can diffuse across a 50-micrometer-wide cell in less than one second. However, diffusion is inadequate for transport from one part of a multicellular plant to another. Multicellular plants must absorb and transport large amounts of water and dissolved minerals from the soil to their leaves, which may be many meters away from the soil. Multicellular plants must also have a system for transporting the sugars produced in leaves to

Water and its dissolved minerals move from roots to leaves in xylem. The two kinds of cells in xylem that carry water are *tracheids* and *vessels*. Both of these cell types are hollow and dead at maturity. They have thick cell walls and can therefore withstand the fluctuations in pressure associated with water flow.

Tracheids are usually long (up to 10 millimeters) and thin (10-15 micrometers in diameter) and overlap one another. Their walls have numerous thin areas that link adjacent tracheids into long, water-conducting chains. Vessels are shorter and much wider than tracheids. The walls separating adjacent vessel elements are often wholly or partially dissolved. Because of their larger diameter and dissolved walls, water moves faster in vessels than in tracheids. This increased flow rate in vessels may help explain why angiosperms dominate today's landscapes: Flowering plants, such as grasses, contain tracheids and vessels, while gymnosperms, such as pines, contain only tracheids. In woody plants, the xylem that transports water and dissolved minerals makes up the wood of the trunk.

Water Flow

Water movement through plants requires no metabolic energy; rather, water flows passively from one place to another. Although root pressure, caused by the pumping of dissolved minerals into the roots, can push water up to a few meters, in taller plants water and dissolved minerals are pulled up through the xylem. The driving force for this movement is the transpiration of water from the leaves. The hypothesis that describes the process is known as the *transpiration-cohesion hypothesis of water movement*. It states that solar-driven transpiration of water dries the walls of *mesophyll* cells of leaves; the loss of water from the cell wall then causes water from neighboring cells to enter the leaf cell. Cells bordering tracheids and vessels replace their water with water from the xylem. The loss of water from xylary elements creates a negative pressure, thereby lifting the water column up the plant. The water column does not break, because water molecules cohere strongly.

The negative pressure created in the xylem by transpiration extends all the way down to the tips of roots, even in the tallest trees. The tension in the root xylem causes water to flow passively from the soil, across the root cortex, and into the xylem of the root. This water is then pulled up the xylem to leaves to replace water lost via transpiration.

Transpiration is affected by atmospheric humidity, wind, air temperature, soil, light intensity, and the concentration of carbon dioxide in the leaf. Transpiration is greatest in plants growing in moist soil on a sunny, dry, warm, and windy day. In these conditions, transpiration often exceeds the plant's ability to absorb water. As a result, many plants wilt at midday, even if the soil in which they are growing contains abundant water. Transpiration also moves solutes in plants; for example, most minerals move from roots to shoots in the transpiration stream.

Stomata

Stomata regulate gas exchange between the atmosphere and a plant and are a key adaptation to life on land. Almost all the factors that affect transpiration do so by influencing the opening or closing of stomata. For example, decreasing the internal concentration of carbon dioxide in a leaf causes stomata to open, therefore increasing gas exchange and transpiration.

Stomata occur throughout the plant kingdom. In angiosperms, they occur on all aboveground organs, including leaves, stems, petals, stamens, and carpels. Open stomata occupy less than 2 percent of a leaf's area. In most plants the stomata are confined to the lower surfaces of leaves (away from the sun) to reduce transpiration rates. Stomata on the upper surfaces, where direct sunlight strikes, could cause serious water loss.

A stomatal complex consists of two *guard cells* and adjacent epidermal cells called *subsidiary cells*, all of which surround a *pore*. Guard cells and stomata have several distinguishing features. For example, guard cells of dicotyledons are crescent-shaped, while in most grasses they are shaped like dumbbells. Most leaves have 1,000 to 100,000 stomata per square centimeter of leaf area. Plants in dry, bright environments, such as deserts, often have smaller and more numerous stomata than plants growing in wet, shaded environments. When wide open, stomatal pores are usually 3-12 micrometers wide and 10-40 micrometers long.

Guard cells control the size of a stomatal pore by changing shape—their unusually elastic walls buckle outward when stomata open and sag inward when stomata close. Stomatal opening and closing is controlled by movement of water into and out of the guard cells. The water movement occurs by osmosis, and the direction of movement is determined by the concentrations of ions under a complex set of cellular controls. Stomatal opening is primarily caused by radial micellation of guard cells by cellulose microfibrils arranged much like the belts in a belted tire. These microfibrils are inelastic; they restrict radial expansion of guard cells while allowing increases in length. When the guard cells lengthen because of an influx of water, they bow apart and form the stomatal pore. Closing of stomata occurs when water moves out of the guard cells, causing them to become flaccid.

Sieve Tubes and Sugar

Sugars and other organic substances move in sieve tubes of the phloem. *Sieve tube members* are arranged end-to-end and are associated with files of companion cells. Companion cells and sieve elements function as a single unit. Sieve tube members are tiny cylinders approximately 40 micrometers in diameter and 1,200 micrometers long. The protoplasts of sieve tube members are connected by sievelike areas called *sieve plates*, each of which has numerous sieve pores. In woody plants, the phloem makes up the innermost layers of the bark.

Peak rates of solute transport in phloem may exceed 2 meters per hour. As a result, as much as 20 liters of sugary sap can be collected per day from severed stems of sugar palms. A sieve element 0.5 millimeter long empties and fills every two seconds, thereby delivering approximately 5 to 10 grams of sugar per hour per square centimeter of phloem area to sites of sugar storage or use.

Solutes move through the phloem via pressure flow. In 1926 plant physiologist Ernst Munch proposed the *pressure-flow hypothesis*, which states that a *turgor pressure* gradient drives the unidirectional *mass flow*, or *bulk flow*, of solutes and water through sieve tubes of the phloem. According to this model, solutes move passively through sieve tubes along a pressure gradient in a fashion analogous to the movement of water through a garden hose.

Sucrose produced at a source is actively loaded into a sieve tube. This loading causes water to enter the sieve tube via osmosis from the xylem. The influx of water into sieve tubes carries sucrose to a *sink*, which is an area where sucrose is unloaded for use or storage. Removing sucrose at the sink causes water also to move out of the sieve tube. The influx of water at the source and the efflux of water at the sink create a pressure-driven flow of fluids in the phloem.

Sugars in the cell wall are loaded into sieve tubes by companion cells. These companion cells often have numerous cell-to-cell channels called *plasmodesmata*. Their cell walls and cell membranes also have elaborate infoldings that provide a large surface area for transporting sugars from the cell wall into the sieve tube. The loading of sieve tubes requires metabolic energy in the form of *adenosine triphosphate* (ATP) and is driven indirectly by a proton gradient. Phloem transport is affected by temperature, light, and the nutritional status of the plant.

More than 90 percent of the solutes in sieve tubes are carbohydrates. In most plants, these carbohydrates are transported as sucrose. The concentration of sucrose may be as high as 30 percent, thereby giving the phloem sap a syrupy consistency. A few plant families also transport other sugars, such as raffinose and stachyose. These sugars are similar to sucrose and consist of sucrose attached to one or more molecules of D-galactose. Like sucrose, all these sugars are nonreducing sugars. This is important because nonreducing sugars such as sucrose are less reactive and less prone to enzymatic breakdown than are reducing sugars such as glucose and fructose. Sieve tubes also contain ATP and nitrogen-containing compounds. More than a dozen different amino acids occur in sieve tubes of some plants. Sieve tubes also transport hormones, alkaloids, viruses, and inorganic ions such as potassium ions.

Randy Moore

See also: Gas exchange in plants; Leaf abscission; Leaf anatomy; Nutrients; Osmosis, simple diffusion, and facilitated diffusion; Plant tissues; Roots; Water and solute movement in plants.

Sources for Further Study
Dennis, David T., et al., eds. *Plant Metabolism.* 2d ed. Harlow, England: Longman, 1997. Well-balanced overview of plant metabolism. This updated edition contains new chapters on protein synthesis and the molecular biology of plant development and on the manipulation of carbon allocations in plants.
Kaufman, Peter B., and Michael L. Evans. *Plants: Their Biology and Importance.* New York: Harper & Row, 1989. The chapters "Plant Water Relations" and "Transport of Organic Solutes and Plant Mineral Nutrition" contain good discussions of the discoveries underlying scientists' knowledge of how liquids move in plants. Includes an extensive glossary, illustrations, and references.

Kozlowski, Theodore T., and Stephen G. Pallardy. *Physiology of Woody Plants*. 2d ed. San Diego: Academic Press, 1997. Useful for graduate study in botany, forestry, or ecology.

Steward, F. C., James F. Sutcliffe, and John E. Dale, eds. *Water and Solutes in Plants*. Vol. 9 in *Plant Physiology: A Treatise*. New York: Academic Press, 1986. College-level treatment of how water and solutes move in plants. Includes representative data and many references to the landmark studies. Bibliography.

Zimmermann, Martin H. *Xylem Structure and the Ascent of Sap*. New York: Springer-Verlag, 1983. College-level text describes how the structure of xylem is ideally suited for water movement in plants. Includes illustrations and a detailed bibliography.

LIVERWORTS

Categories: Medicine and health; nonvascular plants; paleobotany; *Plantae*; taxonomic groups

Liverworts (phylum Hepatophyta*) are one of three ancient lines of bryophytes (liverworts, hornworts, and mosses): low-growing land plants that depend on free water (rain) for fertilization.*

Liverworts, with about six thousand species, generally prefer somewhat cooler, moister, shadier, and more acidic habitats than mosses. Like any bryophyte, a liverwort has a dominant (conspicuous) green gametophyte and a small, attached sporophyte, which is a single-stalked sporangium that developed from a fertilized egg. As in hornworts, liverwort gametophytes are typically dorsiventrally symmetrical (flattened). A unique feature of liverworts is the presence, in the gametophyte, of *oil bodies*, cellular organelles that produce aromatic terpenoids.

Many freshly collected liverworts have a pleasing aroma, which quickly disappears as oil bodies disintegrate. Possibly defending liverworts from herbivores, *terpenoids* (chemically diverse and found in 90 percent of liverworts) have potential medicinal value. Liverwort sporophytes mature while completely enclosed in the gametophyte. Thereby shielded from natural selection, they are far more uniform than moss sporophytes. A typical liverwort sporophyte comprises a foot, a fleshy stalk (*seta*), and a round to cylindrical capsule that splits open to release spores and *elaters*. The seta is green when young but is short-lived and grows only by cell elongation (not by meristematic cells as in other bryophytes). *Elaters*, unique to liverworts, are cells with spirally thickened walls. Their jerky, hygroscopic movements help disperse spores from the capsule.

Leafy Liverworts

Liverwort gametophytes are distinctive. They are either leafy (about two-thirds of the species) or thalloid (straplike), whereas all mosses are leafy. Liverwort leaves are often round and lobed, unlike the pointed leaves of mosses. Liverwort gametophytes are anchored by unicellular rhizoids (hairs), whereas the rhizoids of mosses are multicellular. Leafy liverworts are placed in the class *Jungermanniopsida*, with most species in the order *Jungermanniales*. The leaves are only one cell thick and lack midribs. The rounded cells have numerous chloroplasts and variable numbers of oil bodies; these resemble clusters of grapes in some species.

Stems are creeping or ascending and usually bear three rows of leaves: two rows of dorsal leaves and (in most species) one row of ventral leaves or underleaves. Leaves generally overlap and are attached to the stem at a slanted angle (a transverse angle is less common). The arrangement of the leaves in leafy liverworts can be referred to as being either *succubous* or *incubous*, based on the way the leaves overlap. In succubous species the leaves overlap, as do the shingles of a roof; the upper part of a leaf is covered by the next leaf above it (toward the apex). In incubous species, leaves overlap in the opposite way (away from the apex). Leaves of many species are divided into lobes and filaments, giving the gametophyte a delicate appearance. For example, *Frullania* has two rows of dorsal leaves, one row of bifid underleaves, and two rows of helmet-shaped ventral leaf lobes or "water sacs" (in which "wheel animals," or rotifers, may live). *Trichocolea* has leaves divided into filaments that resemble wool. The external complexity of leafy liverworts makes them well suited for capillary conduction and storage of rainwater. However, like most bryophytes, leafy liverworts have a thin cuticle (or lack one); after a rain, they soon dry out and become inactive. Upon remoistening, they quickly revive.

The archegonia (egg sacs) of leafy liverworts develop at the tips of stems and branches, whereas the antheridia (sperm sacs) are produced behind the apex. These gametangia are protected from drying out by slime hairs (which secrete mucilage) and bracts (specialized leaves). Archegonia may be concealed within an envelope of fused leaves (the *perianth*). After fertilization, the base of the archegonium swells into a calyptra that protects the embryo. The embryo may also be enclosed by a sheath of stem tissue (the *perigynium*). After sporophyte maturation (often in spring or fall) the seta elongates and forces the capsule through the protective layers. The capsule splits open, its four valves resembling a small flower, and releases the spores and elaters. A few days later, the delicate seta collapses and dies.

If a spore lands in a favorable site, it may germinate and develop into a new leafy gametophyte, which grows by means of an apical cell (initial) at the stem tip. Liverworts of moist habitats, such as *Frullania*, often show precocious spore germination, and juvenile gametophytes (rather than spores) are shed from the capsule. Juvenile gametophytes are often globular, rather than threadlike, as in mosses. Liverwort spores that are unusually tolerant of cold, dry conditions may be dispersed over great distances by wind. Many leafy liverworts also produce abundant specialized asexual propagules (gemmae), which may be dispersed by rain, wind, or the feet of animals. Asexual reproduction helps compensate for infrequent sexual reproduction; most leafy liverworts (like most bryophytes) are unisexual, and sometimes male and female plants live far apart.

Leafy liverworts flourish in humid, shaded habitats and are often pioneers on rocks, tree trunks, decaying logs, stumps, and soil by streams, ponds, footpaths, and roads. Habitats range from sunny ridges to deeply shaded gorges. A few species are aquatic, such as *Scapania undulata*, a major producer in mountain streams that is remarkably tolerant of acid mine drainage. Many species are epiphytes, festooning trees as pendent mats in temperate and

---

Classification of Liverworts

Subclass *Jungermanniae*
  Order *Calobryales*
    Family:
      Haplomitriaceae
  Order *Jungermanniales*
    Families:
      Acrobolbaceae
      Antheliaceae
      Arnelliaceae
      Calypogeiaceae
      Cephaloziaceae
      Cephaloziellaceae
      Chonecoleaceae
      Geocalycaceae
      Gymnomitriaceae
      Gyrothyraceae
      Herbertaceae
      Jubulaceae
      Jungermanniaceae
      Lejeuneaceae
      Lepidoziaceae
      Mastigophoraceae
      Mesoptychiaceae
      Plagiochilaceae
      Pleuroziaceae
      Porellaceae
      Pseudolepicoleaceae
      Ptilidiaceae
      Radulaceae
      Scapaniaceae
      Trichocoleaceae

  Order *Metzgeriales*
    Families:
      Allisoniaceae
      Aneuraceae
      Blasiaceae
      Fossombroniaceae
      Metzgeriaceae
      Pallaviciniaceae
      Pelliaceae
      Treubiaceae
Subclass *Marchantiae*
  Order *Marchantiales*
    Families:
      Aytoniaceae
      Cleveaceae
      Conocephalaceae
      Corsiniaceae
      Lunulariaceae
      Marchantiaceae
      Monosoleniaceae
      Oxymitraceae
      Ricciaceae
      Targioniaceae
  Order *Sphaerocarpales*
    Families:
      Riellaceae
      Sphaerocarpaceae

Source: Data are from U.S. Department of Agriculture, National Plant Data Center, *The PLANTS Database*, Version 3.1, http://plants.usda.gov. National Plant Data Center, Baton Rouge, LA 70874-4490 USA.

tropical rain forests. Although leafy liverworts are (like most bryophytes) typically perennial, their substrata are often "temporary" on a scale of years (fields, flood plains), decades (logs), or centuries (old-growth trees). Propagules (spores and gemmae) enable them to "shuttle" to new substrata as they become available.

Thalloid Liverworts

Thalloid liverworts typically have green cells above, scattered oil cells (one oil body per cell), ventral tissue hosting symbiotic fungi, and rhizoids arising from a central thickened area (which may include a distinct midrib). There are two kinds of thalloid liverworts, classified based on anatomy: simple and complex. Simple thalloid liverworts are placed in the class *Jungermanniopsida*, with most species in the order *Metzgeriales*. The straplike or ribbonlike thalli (bodies) grow by means of "apical initials" in marginal notches. Branching patterns vary, but thalli generally expand outward, forming circular mats. In *Pallavicinia*, a distinct midrib contains "hydroids," elongated cells specialized for water conduction, analogous to tracheids of higher plants. Archegonia and antheridia in the *Metzgeriales* are usually scattered on the thallus and are variously protected by sheaths, scales, flaps, and slime hairs. At maturity, the sporophyte bursts through the calyptra, the capsule splits open into two to four valves, and spores and elaters are released. Some species also produce gemmae. Simple thalloid liverworts grow in the same range of habitats as leafy liverworts and are often intermingled with them.

Complex thalloid liverworts, or *chamber liverworts*, are placed in the class *Marchantiopsida*. They have an upper layer of loosely packed green filaments in boxlike "air chambers" and a lower layer of compact food storage cells. Each air chamber has a pore in its "roof." The waxy epidermis of the "roof" repels excess water, while the pores permit the gas exchange necessary for photosynthesis. Although pores cannot be opened and closed (as can stomata of higher plants), the complex pores of some species can shrink under dry conditions. Although water-conducting cells occur in the midribs of some species, chamber liverworts, like all bryophytes, rely primarily on capillary water. Capillary spaces are abundant and occur among ventral scales and within a dense mat of rhizoids, smooth vertical rhizoids and horizontal rhizoids (with "pegged," wavy walls) that conduct water along the liverwort's underside.

Chamber liverworts grow by means of apical initials (protected by scales) in marginal notches. Thalli appear to branch dichotomously, repeatedly forking in two's. However, this growth is pseudo-dichotomous; in true dichotomies, each apical initial splits into two new initials. As in most bryophytes, colonies "grow ahead and die behind," so that one colony fragments into two (or more) new colonies.

Reproductive structures of chamber liverworts are highly specialized. In *Marchantia*, raindrops splash gemmae out of gemma cups. A constriction in the cup speeds up the water's movement and enhances dispersal. Also in *Marchantia*, thallus branches have been modified into umbrella-like gametangiophores; these elevate antheridia and sporangia and thereby enhance rain dispersal of sperm and wind dispersal of spores. In male plants, rain strikes the "umbrellas," with their sunken antheridia, and splashes out sperm. In female plants, sporophytes (with reduced setae) are borne on the underside of the "umbrellas"; mature capsules burst through the calyptrae, tear open irregularly, and release spores and elaters. Asexual reproduction (via gemmae) is favored by short days, whereas sexual reproduction (via sporophytes) is favored by long days.

Other chamber liverworts are less specialized than *Marchantia*. For example, *Conocephalum* has female umbrellas only and lacks gemma cups. In *Riccia* and *Ricciocarpus*, gametangia and sporangia are sunken into the thalli, and spores are liberated when the thalli decay. In both genera, spore release by disintegration is synchronized with seasonal change and is just as adaptive as the complex "umbrellas" of *Marchantia*. *Riccia fluitans* and *Ricciocarpus natans* form floating mats on ponds, but other *Riccia* species dominate the extensive "cryptogamic crusts" on arid plains in Australia. *Marchantia* is found on recently burned ground in humid areas, whereas *Conocephalum* frequently grows on shaded ledges along streams. Chamber liverworts have been extensively used as experimental subjects. Sex chromosomes in plants were first discovered in the chamber liverwort *Sphaerocarpos*.

Susan Moyle Studlar

See also: Bryophytes; Hornworts; Oil bodies; Mosses.

Sources for Further Study
Conrad, Henry Shoemaker, et al. *How to Know the Mosses and Liverworts*. 2d ed. New York: McGraw-Hill, 1979. Guide to identification achieves middle level of coverage (neither too simple nor too technical). Includes line drawings.
Greenaway, Theresa. *Mosses and Liverworts*. New York: Raintree Steck-Vaughn, 1992. Focuses on varieties and life cycles of mosses and liverworts in different climates and habitats. Includes index.
Watson, E. Vernon. *British Mosses and Liverworts*. 3d ed. New York: Cambridge University Press, 1995. Contains full descriptions and ecological details of more than two hundred species; each is illustrated. Brief notes are provided on many more species.

LOGGING AND CLEAR-CUTTING

Categories: Economic botany and plant uses; environmental issues; forests and forestry

Logging is the removal of timber from forestlands with the intention of using it for a specific purpose, such as lumber, fuelwood, or the production of pulp or chemicals. Clear-cutting is a harvesting technique in which all timber is removed from a stand at the same time.

Many people believe commercial logging is responsible for the loss of all forestland. However, acres of forestland are cleared for other purposes, particularly in tropical areas. Rain forests in Amazonia, for example, are often bulldozed to create pastureland for cattle. Rather than being harvested for timber, the wood is simply pushed into piles and burned at the site.

The Logging Process

Logging, whether of one tree or one thousand trees, involves four basic steps: selecting the timber to be harvested, felling the trees, trimming away waste material, and removing the desired portion of the tree from the woods. Equipment used in logging ranges from simple hand tools, such as axes and crosscut saws, to multifunction harvesting machines. Mechanized feller bunchers, for example, can fell the tree, trim off the branches, cut the stem into logs, and stack logs to await removal from the forest. The choice of equipment utilized in harvesting a specific stand of timber depends on factors such as the terrain, the type of timber to be logged, and whether the logger intends to harvest only selected trees or to clear-cut the site.

Soil and Atmospheric Effects

Logging and clear-cutting, if improperly done or motivated by short-term economic goals, can pose significant threats to the environment. Logging always involves some disturbance to soil and wildlife. If performed in environmentally sensitive areas, it can destroy irreplaceable habitat and contribute to erosion. Heavy equipment can compact soil, leaving ruts that may persist for many years, while clear-cutting hillsides can lead to erosion, stream siltation, and flooding. In Asia, for example, clear-cutting in the mountains of Nepal and India has caused disastrous floods in Bangladesh.

Even when logging does not inflict long-term damage on the immediate environment, the removal of trees can contribute to the threat of global warming. *Slash* (the unmarketable portions of the tree, such as tops and branches) burned at logging sites emits greenhouse gases into the atmosphere, while the loss of forest means that there are fewer trees to break those gases down into oxygen and organic compounds.

Clear-Cutting

Clear-cutting is the practice of cutting all the trees on a tract of land at the same time. At one time a standard practice in lumbering, it has become one of the most controversial harvesting techniques used in modern logging. A tract that has been clear-cut will have no trees left standing. With its windrows of slash and debris, a clear-cut tract of land may appear to the untrained eye as though a catastrophic event has devastated the landscape. Wildlife studies have indicated that certain species of birds and mammals are threatened when their habitats are clear-cut, as they either lose their nesting areas or are exposed to increased risk from predators. The northern spotted owl, for example, becomes easy prey for great horned owls when it is forced to fly across large open areas.

Clear-cutting steep hillsides can leave the land susceptible to erosion, as the removal of all trees leaves nothing to slow the flow of rainfall. The hillsides can, as a result, lose topsoil at a rapid rate, choking nearby streams with sedimentation and killing aquatic life. The large amounts of slash, or debris, left behind can pose a fire hazard. The alternative, selective harvesting, may offer better results.

Clear-cutting removes all the trees on a tract of land, leaving none standing. At one time a standard practice in logging, it has become one of the most controversial harvesting techniques used in modern logging. With its windrows of slash and debris, a clear-cut tract of land may appear as though a catastrophic event has devastated the landscape.

By contrast, however, for some species of trees selective harvesting simply does not work. Clear-cutting occurs in forests where the desired species of trees need large amounts of sunlight to regenerate. Many conifers, such as Douglas fir, are shade-intolerant. Landowners will occasionally decide to change the dominant species on a tract and so will clear-cut existing timber to allow for replanting with new, more commercially desirable trees.

Loggers are more likely to clear-cut if the timber is plantation-grown and of a uniform age and size. Clear-cutting remains an appropriate harvesting method in certain situations, as when plantation stands are harvested in rotation. Modifications in its application can help prevent damage to the environment.

Selective Harvesting

Selective harvesting, in contrast with clear-cutting, leaves trees standing on the tract. This method can be utilized with even-age plantation stands as a way of thinning them. More commonly, it is used in mixed- and uneven-age stands to harvest only trees of a desired species or size. In cutting hardwood for use as lumber, for example, 12 inches (30 centimeters) may be considered the minimum diameter of a harvestable tree. Trees smaller than that will be left in the woods to continue growing.

Selective harvesting can also be ecologically damaging, however. Logging may create stress on the residual standing timber, leading to disease and die-off of the uncut trees, while the operation of mechanized equipment can be as disrupting to nesting and foraging habits of wildlife as clear-cutting the stand would have been.

An individual, noncommercial woodcutter may fell only a few trees per year on small parcels of land. Commercial loggers, in contrast, annually harvest hundreds of thousands of trees and operate on large parcels of land. Nonetheless, significant acres of forestland are annually cleared by people who rely on noncommercial logging for wood for their own needs, such as fuel for cooking or heating their homes. Although some woodcutters may cut more than they need for their own use and then sell the surplus, fuelwood for individual households is usually gathered by members of that household. Other examples of noncommercial logging include farmers cutting trees for use as fencing or building materials on their own property.

From an environmental viewpoint, the biggest difference between commercial and noncommercial logging would seem to be one of scale, but this is not always true. An improperly logged small parcel can have more of an impact on a *watershed* or other *ecosystem* than a professionally harvested large stand. Even if no single household's logging practices pose a problem, collectively the gathering of fuelwood or other timber can be devastating. Many nations have developed programs in which professional foresters provide advice on environmentally sound harvesting practices and timberstand improvement for small property owners, but the availability of such help varies widely from country to country. With no guidance from professional foresters, trees are logged based on convenience for the woodcutter rather than principles of sustainable forestry or watershed management.

Nancy Farm Männikkö

See also: Deforestation; Forest management; Forests; Greenhouse effect; Old-growth forests; Sustainable forestry; Timber industry; Wood; Wood and charcoal as fuel resources.

Sources for Further Study

Bevis, William W. *Borneo Log: The Struggle for Sarawak's Forests*. Seattle: University of Washington Press, 1995. Readers concerned about logging in tropical forests and developing nations may find this book both disturbing and enlightening as Bevis describes the exploitation of developing nations' resources by more industrialized nations.

Chase, Alston. *In a Dark Wood: The Fight over Forests and the Rising Tyranny of Ecology*. Boston: Houghton Mifflin, 1995. Depicts the battle over the old-growth forests of the Northwest. Documents the car bombing of an environmental activist and the buyout of the Pacific Lumber Company. Explores the origins of the conflict from various points of view.

Stenzel, George, et al. *Logging and Pulpwood Production*. 2d ed. New York: Wiley, 1985. Provides a thorough description of logging practices, including discussions of environmental regulations and the various government agencies involved with commercial forestry in the United States.

Walker, Laurence C. *The Southern Forest: Geography, Ecology, and Silviculture*. Boca Raton, Fla.: CRC Press, 2000. Examines the forests of the American South, providing a comprehensive description of the region from which most wood for housing and paper in the United States will be grown and harvested in the future. Closing the old-growth forests of the Northwest and reduction of harvests in the tropics necessitate the southern states intensifying tree growth to accommodate society's requirements.

LYCOPHYTES

Categories: *Plantae*; seedless vascular plants; taxonomic groups; water-related life

The lycophytes, which compose the phylum Lycophyta, *are one of four phyla of seedless plants having vascular, or conducting, tissue. The living lycophytes are all small and herbaceous, whereas the extinct lycophytes included large trees, which were important in the formation of coal.*

There are at least twelve genera and twelve hundred species of living lycophytes. These include plants known as club mosses and spike mosses (though none are true mosses) and quillworts. The lycophytes consist of three families, each belonging to a separate order. The family *Selaginellaceae* has a single genus, *Selaginella*. Similarly, the family *Isoetaceae* has a single genus, *Isoetes*. The remaining genera belong to the family *Lycopodiaceae*. The living lycophytes are widely distributed but reach their greatest species diversity in the tropics. The lycophytes are similar to the higher vascular plants—the gymnosperms and angiosperms—in having vascular tissue and true leaves, stems, and roots.

Reproduction

The alternation of generations in lycophytes resembles, in an important way, this life cycle in the higher vascular plants: The sporophyte (the spore-bearing generation), rather than the gametophyte (the gamete-bearing generation), is the larger, more obvious generation. In contrast, in the bryophytes (mosses and their relatives, in the phylum *Bryophyta*), which are an earlier, nonvascular evolutionary line, the gametophyte is the larger, dominant generation.

Unlike the higher plants, however, the lycophytes do not produce seeds. Like all seedless plants, the lycophytes require water for the sperm to swim to the egg. In addition, the stem tip in most lycophytes forks repeatedly, resulting in branches that are of about equal length, whereas in higher plants, there is a single main axis, from which lateral branches arise.

Evolutionary Origins

Some of the features that lycophytes and the higher vascular plants have in common differ in their evolutionary origins. Such characteristics are considered convergent, or analogous, having arisen to meet similar environmental demands. The different origins are attributable to the early divergence of the lycophytes from the main lineage of vascular plants. The lycophytes first appear in the fossil record in the early Devonian period of the Paleozoic era, about 400 million years ago; they probably arose from early members of the *Zosterophyllophyta*, a phylum of seedless vascular plants that became extinct during the Devonian.

Leaves are one of the features that developed independently in lycophytes and higher plants. The leaves of lycophytes are called microphylls and are characteristic of all members of the group, extinct and living. In two of the three living lycophyte families, the microphylls are short. In all lycophytes, the microphylls are narrow and have a single, unbranched vein. In contrast, the leaves of the higher vascular plants typically have a complex system of branching veins and are called megaphylls.

Carboniferous Period

Some extinct lycophytes had, in addition to microphyllous leaves, other features analogous to, but derived independently of, those of higher vascular plants. These included tree forms and structures analogous to seeds. These do not occur in living lycophytes. Especially noteworthy among the extinct lycophytes were the scale trees. For most of the Age of Coal, in the Paleozoic's late Carboniferous period (322 million to 290 million years ago), these trees were among the dominant plants of swamp forests in what is now North America and Europe. They grew 10 to 35 meters tall and included the genera *Lepidodendron* and *Sigillaria*. The remains of these large trees contributed greatly to formation of the world's coal beds.

The trunks of most tree lycophytes branched only near the top. Stabilizing them at their bases were shallow, forking, rootlike structures, which produced spirally arranged rootlets. The stems were supported mostly by a massive bark surrounding a relatively small amount of secondary xylem, or wood. The reproductive structures were borne in cones. Like the living genera *Selaginella* and *Isoetes*, lycophyte trees were heterosporous, forming spores of two different sizes rather than a single size. Some species produced structures analogous to seeds.

The tree lycophytes vanished in the late Carboniferous period, about 296 million years ago, as the climate changed and swamps began to dry. The nearest living relative of the tree lycophytes is the

---

Classification of Lycophytes

Class *Lycopodiopsida*
    Order *Isoetales*
        Family *Isoetaceae* (quillworts)
            Genus *Isoetes*
    Order *Lycopodiales*
        Family *Lycopodiaceae* (club mosses)
            Genus *Lycopodium*
            Genus *Phylloglossum*
    Order *Selaginellales*
        Family *Selaginellaceae* (spike mosses)
            Genus *Selaginella*

Source: Data are from U.S. Department of Agriculture, National Plant Data Center, *The PLANTS Database*, Version 3.1, http://plants.usda.gov. National Plant Data Center, Baton Rouge, LA 70874-4490 USA.

The hybrid club moss Lycopedium selage x l. lucidulum *in Shenandoah National Park, Virginia.*

herbaceous plant *Isoetes* in the family *Isoetaceae*. Herbaceous members of the two other living lycophyte families, *Lycopodiaceae* and *Selaginellaceae*, also existed in the Carboniferous, and representatives of some of them survive today.

Lycopodiaceae

The family *Lycopodiaceae* includes 10 to 15 genera, with a total of 350 to 400 species, most of them tropical. The family is also represented in the temperate zone and even in the Arctic but not in arid regions. In the United States and Canada, there are seven genera, with twenty-seven species which occur in the East and Northwest but not in the dry Southwest. The most familiar of these are probably the club mosses, which belong to the family's largest genus, *Lycopodium*.

Lycopodiaceae consists mainly of trailing plants. A horizontal, branching, underground stem, or rhizome, produces roots and upright, aerial branches. The small, microphyllous leaves are generally spirally arranged. In some genera, the sporophylls— the leaves that bear the spores—look like ordinary microphylls and are interspersed with them along the aerial stems. In other genera, the sporophylls do not resemble ordinary microphylls and are grouped in cones at the ends of the aerial stems.

The spores in *Lycopodiaceae* all the same size. Each spore gives rise to a free-living gametophyte that bears both male and female reproductive structures. In some species, the gametophyte grows aboveground and produces its own food photosynthetically; in others, it grows belowground, is not photosynthetic, and is associated with a fungus. Young sporophytes may remain attached to the gametophyte for some time before becoming independent.

Selaginellaceae

Selaginella, or spike moss, is the sole genus in the family *Selaginellaceae*. With more than seven hundred species, the *Selaginellaceae* is the largest of the living lycophyte families. As with the *Lycopodiaceae*, most species of *Selaginella* are tropical, but some occur in temperate zones. In the United States and Canada, there are thirty-eight species. Many species of *Selaginella* grow in damp places, but a few occur in deserts, where they become dormant during the driest season. The resurrection plant, *Selaginella lepidophylla*, which grows in Texas, New Mexico, and Mexico, can recover from several months of complete drying.

Selaginella has a branched, prostrate stem, which produces roots and upright branches that grow a few inches tall. In some species, the horizontal and upright stems are sheathed with small leaves in four longitudinal rows. Both microphylls and sporophylls bear a small, scalelike outgrowth, called a *ligule*, at the bases of their upper surfaces.

Selaginella forms cones at the ends of the branches. The spores are of two sizes: large ones, called megaspores, and smaller ones, called microspores. Megaspores germinate to form megagametophytes, which bear egg-producing archegonia. Microspores germinate to form microgametophytes, which bear sperm-producing antheridia.

In contrast to the fully free-living gametophytes of the *Lycopodiaceae*, the gametophytes of *Selaginella* develop mostly within the spore walls. When a microgametophyte is mature, the spore wall ruptures, releasing the sperm. The megaspore wall also ruptures, allowing the megagametophyte to protrude from the wall; archegonia develop in the exposed part. Neither the microgametophyte nor the megagametophyte has chlorophyll; thus, the gametophytes, and the developing embryos in the archegonia, must obtain their nutrition from food stored within the spore.

Isoetaceae

Known as the quillwort family, the *Isoetaceae* is the smallest of the lycophyte families. Its single genus, *Isoetes*, includes about 150 species. Unlike the *Lycopodiaceae* and *Selaginellaceae*, which are predominantly tropical, *Isoetes* occurs mostly in cooler climates. In the United States and Canada, there are twenty-four species. Most Isoetes species grow partially or totally submerged in fresh water.

Unlike the *Lycopodiaceae* and *Selaginellaceae*, *Isoetes* is not a trailing plant. Instead, it has a short, thick, fleshy, bulblike underground stem, or corm. Roots grow on the lower surface of the corm, and microphyllous leaves grow on the upper surface. The corm undergoes secondary growth, which thickens it. The microphylls are not small, as they are in the *Lycopodiaceae* and *Selaginellaceae*, but instead are elongated, slender, and stiff, reaching 15 to 50 centimeters in height. They resemble quills, giving *Isoetes* the look of a young onion or a tuft of grass or rush. They are spoon-shaped at the base, where they surround the corm. As in *Selaginella*, the leaves have ligules.

Isoetes species do not produce cones. Each microphyll is a potential sporophyll. As in *Selaginella*, there are two kinds of spores: microspores and megaspores. The sporophylls that bear microspores are located nearer the center of the plant than are the sporophylls that bear megaspores. Gametophyte development takes place largely within the spore walls, as in *Selaginella*. Some species of *Isoetes* from highlands in the tropics obtain their carbon for photosynthesis from the soil rather than in the usual way, from the atmosphere. These plants have an unusual kind of photosynthesis, called CAM (crassulacean acid metabolism) photosynthesis.

Jane F. Hill

See also: Bryophytes; Coal; C_4 and CAM photosynthesis; Evolution: convergent and divergent; Evolution of plants; Fossil plants; Reproduction in plants; Seedless vascular plants; *Spermatophyta*; *Zosterophyllophyta*.

Sources for Further Study

Lellinger, David B. *A Field Manual of the Ferns and Fern-Allies of the United States and Canada.* Washington, D.C.: Smithsonian Institution Press, 1985. Covers the lycophytes, with species descriptions, geography, habitats, ecology, structure, life cycles, fossils, and growth and usage of these plants. Includes glossary, bibliography, index, and color photos.

Mabberley, D. J. *The Plant Book: A Portable Dictionary of the Vascular Plants.* 2d ed. Cambridge University Press, 1997. Reference text contains twenty thousand entries, with information on every family and genus of seed-bearing plant and ferns and other pteridophytes.

Morin, Nancy R. *Flora of North America North of Mexico.* Vol. 2. New York: Oxford University Press, 1993. Includes descriptions of North American lycophytes, their habitats, and ranges. Includes identification keys, literature reviews.

MARINE PLANTS

Categories: Algae; microorganisms; pollution; water-related life

Marine plants grow near the surface of salt water and ice, within reach of sunlight necessary for photosynthesis. Algae, the most plentiful type of marine plant, form the foundation of the food chain and crucial to a balanced ecosystem.

Water is essential to life. The earliest plants, primarily algae, formed in bodies of saline water covering prehistoric Earth. During the Silurian period, approximately 441 million to 410 million years ago, some aquatic plants began to grow on land, but many plants remained solely water-based. These marine plants have provided fundamental nourishment in the food chain. No marine animals would have evolved or been able to survive if marine plants had not existed. Marine plants support all higher saltwater life-forms. Marine sediments formed by algae often contain fossils that reveal aspects of marine plants' evolutionary history. The distribution of marine plants was affected by plate tectonics as continents moved and ocean shapes changed.

Oceans cover most of the earth's surface. Almost 99 percent of organisms, representing approximately five million species (most of them unclassified) live in oceans. As a result, oceans are significant to the well-being of life and economies. Marine plants consist of two major types, the *sea grasses* and the algae and *seaweeds*. Sea grasses represent members of some of the more complex plants, while algae and seaweeds display simple forms and are often microscopic.

Marine plants range from tiny single-celled organisms to large, intricate forms. Because all marine plants require sunlight to manufacture food, they mostly develop near water surfaces. Nutrients are also gathered from particles that currents wash up from sea floors. Marine plants can adapt to specific conditions, such as limited light and underwater caves. Some are phosphorescent, generating chemical lights.

Types

The smallest marine plants are *phytoplankton*, which are single-celled and form the basis of the marine food chain. *Diatoms* (*Bacillariophyta*) are glassy microscopic cells which frequently link together in chains. Few marine plants are angiosperms, although along tropical coasts, flowering marine plants often accumulate. *Green algae* (*Chlorophyta*) is the most common marine plant. Chlorophyll causes these algae to have bright green coloring. When algae leaves calcify, they add layers to ocean sediments. Botanists believe that 200,000 algae species exist, even though only 36,000 have been identified.

Red algae (*Rhodophyta*), tinted by the pigment phycoerythrin, are the largest type of marine plants and the most diverse. Some red algae adhere to corals, thus creating reefs. Both green and red algae species prefer warm water to cold water. In contrast, *brown algae* (*Phaeophyta*), colored with fucoxanthin pigment, are usually found in cold or temperate water, and few species live in the tropics. On reefs, brown algae frequently are the dominant organisms. Blue-green bacteria, or cyanobacteria (formerly called blue-green algae) are primarily microscopic strands which convert nitrogen from the atmosphere into forms that most marine plants can use.

Habitats

Marine plants live in diverse habitats near shores or in salt marshes and open seas worldwide. Giant kelp, a seaweed found in the South Pacific, grows in groups in warm coastal waters. In contrast, sea ice algae live on floating ice sheets. Migrating marine plants drift in a variety of water conditions.

On reefs, marine plants have several roles. Primarily, marine plants, including macroalgae and sea grasses, provide nourishment and shelter for animals. Marine plants assist corals in constructing reefs; then some plants, such as coralline algae, hold the reefs intact.

NOAA/National Undersea Research Program Collection

Coral throughout the Caribbean Sea are expelling algal marine plants as a result of unusually warm water temperatures. The coral appear bleached white because the algae remove the coral's energy and color source.

Algae live inside marine animals. Coral tissues host several million algae per square inch, and these marine plants provide 90 percent of nutrients needed by the coral. The symbiotic relationship is based on a cycle of coral enzymes which cause algae to release carbohydrates and algae to receive nitrogen from coral waste. Algae are shaded from intense sunlight by coral pigments. Algae also live in panels inside giant clams and in sponges and flatworms. In kelp bed forests, marine plants serve as food and habitats for such diverse animals as seals, eels, and octopi. Marine plants also benefit from animals; for example, some can secure nitrogen from seabird guano.

Marine plants are vulnerable to pollution. Seagrass beds and reefs have been damaged by toxins or destroyed by industrial development projects. Dredging and harvesting coral injures marine plants. Fertilizers, pesticides, oils, radioactive material, sewage, and hazardous wastes are drained into oceans. Often tropical commercial fishers use explosives to stun fish, inadvertently destroying marine plant habitats. Sea grasses have died in Maryland's polluted Chesapeake Bay.

Some scientists speculate that the growing ozone hole might place Antarctic marine plants at risk. Changing tides affect marine plant distribution because they alter water levels. Overfishing and acid spills intensify toxic sites. Toxins sicken fish, which develop cancerous tumors, and people who consume this diseased fish are often poisoned. Fungi and bacteria transported in freighters' ballast water from other regions can harm marine plants; for example, slime molds kill turtle grass. Algae frequently develop fungi because excessive nitrogen causes them to produce amino acids and deplete carbon supplies. Marine plants can be relocated by shipping vessels and can overtake native plants in distant areas.

An overabundance of algae can smother coral reefs if the supply of nitrogen is not balanced. If coral become too warm and expel algae, the coral appears bleached white because the algae remove the coral's energy and color source. When too much nitrogen floods an area, sometimes an algal bloom or toxic red tide occurs and can have devastating results. As algae multiply because of excessive nutrients, creating *algal blooms*, they usurp oxygen from other marine plants and organisms, which starve. In 1996 many Florida manatees were killed by a red algal tide. The next year, the U.S. National Aeronautical and Space Administration's Sea-Viewing Wide Field-of-View Sensor satellite began to detect concentrations of marine plants by using light wavelengths.

Uses

The oceans represent 95 percent of the earth's biosphere and affect planetary climatic conditions. Marine plants are estimated to generate approximately 70 percent of oxygen on earth and help regulate oxygen in the atmosphere. The status of marine organisms' health indicates environmental problems that humans and land organisms might encounter.

Humans have historically appropriated marine plants for medicinal uses. Because many marine plants have biotoxins, they are valuable for the development of pharmaceuticals. Using submersible technologies, oceanographers gather samples and cooperate with pharmaceutical manufacturers to seek new chemical compounds to combat disease. Because of the diversity and novelty of marine plants, scientists hope to offer new treatments for diseases resistant to existing nonmarine-plant-derived drugs. Future marine sanctuaries are envisioned to protect such potentially potent natural resources.

Marine plants have also been used as a source of nutrients. Algae with docosahexaenoic acid (DHA), a chemical usually found in human milk and vital to infants' brain development, are commercially processed. Approximately 40 percent of baby formula is made from these algae. The algae *Dunaliella bardawil* contains the orange pigment beta-carotene, which the human body converts into vitamin A. Commercial production of this algae manufactures carotene. Red algae are the chief ingredient of some seaweed drinks and are also useful as thickeners for cooking.

Other commercialization of marine plants includes harvesting seaweed for a variety of products, including foods and fertilizer. Researchers aspire to transfer proteins identified in *Dunaliella bardawil*, which resist extreme saltiness and sun exposure, to land plants that are cultivated in places with high salinity and sunlight conditions.

In an attempt to reduce crop losses, scientists study the physiological relationship of algae and water for optimum cell growth and photosynthesis to understand how such terrestrial plants as corn can manage moisture better, thereby withstanding droughts. Researchers conduct molecular examinations of marine and land plants to comprehend how water supply influences growth rate and metabolism. The cells of the alga *Chara corallina* are large enough that scientists can easily observe how dehydration affects them over a short time period.

Marine plants have a direct relationship to Earth's climate. Iron deficiencies can be detrimental when marine plants become anemic. Oceanic iron and plant absorption of carbon dioxide is connected to ice age cycles and global warming. Paleoceanographers investigated sediment samples to study the impact of a 150,000-year-period of global warming that occurred fifty-five million years ago. They hypothesize that marine plants increased in number to remove atmospheric carbon dioxide and reduce temperatures but warn that modern emissions would be too great for similar resolution.

Elizabeth D. Schafer

See also: Algae; Animal-plant interactions; Aquatic plants; Bioluminescence; Brown algae; Chrysophytes; Diatoms; Evolution of plants; Food chain; Green algae; Greenhouse effect; Medicinal plants; Nitrogen cycle; Ozone layer and ozone hole debate; Phytoplankton; Red algae.

Sources for Further Study

Dawes, Clinton J. *Marine Botany.* 2d ed. New York: John Wiley, 1998. Extensive study of marine plants, discussing physiology, habitats, ecology, biotic and abiotic factors, investigative methodology, uses, and human interaction. Includes illustrations, maps, appendices, bibliographical references, and index.

Dring, Matthew J. *The Biology of Marine Plants.* London: E. Arnold, 1982. A biological text intended for students, which provides basic information about marine plants. Includes illustrations, maps, bibliography, and index.

Lobban, Christopher S., and Michael J. Wynne, eds. *The Biology of Seaweeds.* Berkeley: University of California Press, 1981. A scholarly examination of these marine algae. Includes illustrations, bibliographies, and indexes.

Nielsen, E. Steemann. *Marine Photosynthesis: With Special Emphasis on the Ecological Aspects.* New York: Elsevier Scientific, 1975. Focuses on how marine plants convert solar energy to produce food and how this influences the ecosystem. Includes illustrations, bibliography, and index.

Phillips, Ronald C., and Ernani G. Meñez. *Seagrasses.* Washington, D.C.: Smithsonian Institution Press, 1988. Popular publication detailing these familiar marine plants that was printed for the Smithsonian Contributions to the Marine Sciences series. Includes illustrations, maps, and bibliography.

Tolbert, N. E., and C. B. Osmond, eds. *Photorespiration in Marine Plants.* Baltimore: University Park Press, 1976. Study of the physiology of marine plants found in Australia's Great Barrier Reef. Includes illustrations.

MEDICINAL PLANTS

Categories: Economic botany and plant uses; medicine and health

Because plants are so biochemically diverse, they produce thousands of substances commonly referred to as secondary metabolites. Many of these secondary metabolites have medicinal properties that have proven to be beneficial to humankind.

The use of plants for medicinal purposes predates recorded history. Primitive people's use of trial and error in their constant search for edible plants led them to discover plants containing substances that cause appetite suppression, stimulation, hallucinations, or other effects. Written records show that drugs such as opium have been in use for more than five thousand years.

From antiquity until fairly recent times, most physicians were also botanists or at least herbalists. Because modern commercial medicines are marketed in neat packages, most people do not realize that many of these drugs were first extracted from plants. Chemists have learned how to synthesize many natural products that were initially identified in a plant. However, in many cases a plant is still the only economically feasible source of the drug.

Antibacterial and Anti-inflammatory Agents

The first effective antibacterial substance was carbolic acid, but the first truly plant-derived antibacterial drug was *penicillin*, which was extracted from a very primitive plant, the fungus *Penicillium*, in 1928. The success of penicillin led to the discovery of other fungal and bacterial compounds that have antibacterial activity. The most notable of these are cephalosporin and griseofulvin.

Inflammation can be caused by mechanical or chemical damage, radiation, or foreign organisms. For centuries poultices of leaves from coriander

(*Coriandrum sativum*), thornapple (*Datura stramonium*), wintergreen (*Gaultheria procumbens*), witch hazel (*Hamamelis virginiana*), and willow (*Salix niger*) were used to treat localized inflammation. In the seventeenth and eighteenth centuries, cinchona bark was used as a source of quinine, which could be taken internally. In 1876 salicylic acid was obtained from the salicin produced by willow (*Salix*) leaves. Today, salicylic acid, also known as aspirin, and its derivatives, such as ibuprofen, are the most widely used anti-inflammatory drugs in the world.

Drugs Affecting the Reproductive System

A home remedy for preventing pregnancy was a tea made from the leaves of the Mexican plant zoapatle (*Montana tomentosa*). The drug zoapatanol and its derivatives were extracted from this plant to produce the first effective birth control substance. It has not been used in human trials, however, because of potential harmful side effects. Other plant compounds that affect the reproductive system include diosgenin, extracted from *Dioscorea* species and used as a precursor for the progesterone used in birth control pills; gossypol from cotton (*Gossypium* species), which has been shown to be an effective birth control agent for males; ergometrine, extracted from the ergot fungus (*Claviceps*) and used to control postpartum bleeding; and yohimbine, from the African tree *Corynanthe yohimbe*, which apparently has some effect as an aphrodisiac.

Circulatory, Analgesic, and Cancer-Fighting Drugs

Through the ages, dogbane (*Apocynum cannabinum*) and milkweed (*Asclepias*) have been prized for their effects on the circulatory system. These

AP/WIDE WORLD PHOTOS

Chemists have learned how to synthesize medicines from many substances that were initially identified in plants. However, in many cases a plant is still considered the best source of the drug. For example, eastern medicine uses ginseng as both a preventive and a therapeutic drug. It takes at least six years to grow ginseng before it can be harvested. The more wrinkled and gnarled the root, the better the Chinese believe it to be.

plants contain compounds called cardiac glycosides. Foxglove (*Digitalis*) has produced the most useful cardiac glycosides, digitalis and digoxin.

Opiate alkaloids such as opium, extracted from a poppy (*Papaver sonniferum*), and its derivatives, such as morphine as well as cocaine, from *Erythroxylum coca* and *Erythroxylum truxillense*, have long been known for their *analgesic* (pain-relieving) properties through their extremely dangerous and addictive effects on the central nervous system.

The primary plant-derived anticancer agents are vincristine and vinblastine, extracted from *Catheranthus roseus*, maytansinoids from *Maytentus serrata*, ellipticine and related compounds from *Ochrosia elliptica*, and paclitaxel (commonly known as taxol) from the yew tree *Taxus baccata*.

Fighting Asthma, Gastrointestinal Disorders, Parasites

The major anti-asthma drugs come from ephedrine, extracted from the ma huang plant (*Ephedra sinaica*), and its structural derivatives. Plant-derived drugs that affect the gastrointestinal tract include castor oil, senna, and aloes as laxatives, opiate alkaloids as antidiarrheals, and ipecac from *Cephaelis acuminata* as an emetic. The most useful plant-derived antiparasitic agent is quinine, derived from the bark of the chincona plant (*Chincona succirubra*). Quinine has been used to control malaria, a disease that has plagued humankind for centuries.

The Future

More plant-derived medicines await discovery, many from tropical rain-forest vegetation. Biotechnology has provided methods by which plants can be genetically modified to produce novel pharmaceuticals. Progress toward the production of specific proteins in transgenic plants provides opportunities to produce large quantities of complex pharmaceuticals and other valuable products in traditional farm environments rather than in laboratories. These novel strategies open up routes for production of a broad array of natural or nature-based products, ranging from foodstuffs with enhanced nutritive value to biopharmaceuticals.

D. R. Gossett

See also: Biotechnology; Culturally significant plants; Herbs; Metabolites: primary vs. secondary; Paclitaxel; Plants with potential.

Sources for Further Study

Cutler, Stephen J., and Horace G. Cutler, eds. *Biologically Active Natural Products: Pharmaceuticals*. Boca Raton, Fla.: CRC Press, 2000. Demonstrates the connections between agrochemicals and pharmaceuticals and explores the uses of plants and plant products in the formulation and development of pharmaceuticals.
Fetrow, Charles W. *The Complete Guide to Herbal Medicines*. Springhouse, Pa.: Springhouse, 2000. Accessible information available on more than three hundred herbal medicines.
Herrick, James W., and Dean R. Snow, eds. *Iroquois Medical Botany*. Syracuse, N.Y.: Syracuse University Press, 1997. A fascinating look at one Native American body of knowledge of herbal medicines. Important not only for those interested in herbal medicine but also for those studying American Indian cosmology as it relates to material culture. Illustrated, with references, index.
Lewis, Walter Hepworth, and M. P. F. Elvin-Lewis. *Medical Botany: Plants Affecting Man's Health*. New York: Wiley, 1990. An excellent in-depth (544-page) study of plants and the medicines they produce, examining plants' effects on human health as injurious, remedial, or psychoactive. Includes bacteria, fungi, and seaweeds as well as flowering plants.
Mann, J. *Murder, Magic, and Medicine*. Rev. ed. New York: Oxford University Press, 2000. An interesting and readable book on the use of natural plant products for medicinal purposes.
Sneader, W. *The Evolution of Modern Medicines*. New York: Wiley, 1986. Provides excellent coverage of how plants contributed to the development of many pharmaceuticals.
Stannard, Jerry, Katherine E. Stannard, and Richard Kay, eds. *Pristina Medicamenta: Ancient and Medieval Medical Botany*. Brookfield, Vt.: Ashgate, 1999. Articles on premodern texts on plants.

Stockwell, C. *Nature's Pharmacy*. London: Century, 1989. An excellent discussion of medicinal products from plants.

Sumner, Judith, and Mark J. Plotkin. *The Natural History of Medicinal Plants*. Portland, Oreg.: Timber Press, 2000. An accessible introduction to the world of medicinal plants by a Harvard University botanist, from Europe in the Middle Ages to the modern pharmacopeia.

Trease, G. E., and W. C. Evans. *Trease and Evans' Pharmacognosy*. 15th ed. W. B. Saunders, 2002. One of the most complete treatises on the production of drugs from plants. At nearly 600 pages, covers all scientific aspects of the topic, from taxonomy, cellular biology, and phytochemistry through genetics. Drugs are examined in chapters that group them by chemical class. The scope is broad, including vitamins and hormones and even alternative therapies such as homeopathic medicine and aromatherapy. Professionals will appreciate the chapters on investigative methodologies. Appendices, index.

Walter, Lynne Paige, and Ellen Hodgson Brown. *Nature's Pharmacy: Break the Drug Cycle with Safe, Natural Treatments for Two Hundred Everyday Ailments*. Upper Saddle River, N.J.: Prentice Hall, 1999. Typical of a wave of similar publications that began to appear after the 1996 law that removed "food supplements" from FDA regulatory responsibility, this 400-plus-page reference catalogs common ailments, from acne to whooping cough, offering signs, symptoms, and suggestions for alternative treatments in addition to traditional Western medicine.

MEDITERRANEAN SCRUB

Category: Biomes

Mediterranean scrub vegetation is dominated by fire-adapted shrubs. The biome fringes the Mediterranean Sea, for which it is named, but is also found along western coasts of continents in areas with warm, dry summers and moist, cool winters.

Regions with mediterranean vegetation are coastal regions between 30 and 45 degrees north latitude or between 30 and 45 degrees south latitude. The air circulating around high-pressure zones over adjacent oceans guides storms away from the coast in the warm season but changes position in concert with the tilt of the earth on its axis and brings storms onto the coast in the cool season. As a result, the warm season is dry, and the cool season is moist. Fire is an important component of mediterranean environments, especially after the warm, dry summer.

North America's representative of mediterranean scrub is the *chaparral* of the Pacific Coast of Southern California and northern Baja California, Mexico. In chaparral and some other mediterranean regions, winds blowing from continental high-pressure regions toward the coast help push storm tracks offshore during the warm season. In California these winds are called Santa Ana winds and are best known for driving chaparral fires. Lightning started such fires before human settlement, but they are often started by careless people today. With the lower temperatures of autumn and winter the continental pressure wanes, and the Santa Ana winds decrease. At the same time, the oceanic high-pressure region shifts, and winter storms track onto the coast, bringing the cool season rains.

Character and Components

Mediterranean scrub is found in small, scattered areas around the world. The plant species that occur in this biome on one continent are unrelated to

AP/WIDE WORLD PHOTOS

Chaparral is an ecological community composed of shrubby plants adapted to dry summers and moist winters that occur especially in Southern California. Shrubs well adapted to fire are widespread throughout chaparral and all mediterranean scrub ecosystems.

those that occur in the same biome on other continents. As a result, mediterranean scrub presents a classical example of convergent evolution, the environmentally driven development of similar characteristics in unrelated species. Under the influence of mediterranean climate, entire communities of unrelated species become similar to one another. Many mediterranean areas also contain a large number of endemic plant species, species that grow nowhere else.

Mediterranean scrub is dominated by shrubs well adapted to fire. Some species have specialized underground structures that are undamaged by the fire and send up new growth shortly after the fire passes. Other species have specialized, long-lived seeds that require intense heat to stimulate germi-

nation. Still other species combine the two strategies. In communities that burn regularly, such species have a great advantage over their competitors.

Mediterranean shrubs are not just adapted to recover after a fire; they are actually adapted to carry the fire once it is started. These species synthesize and store highly flammable chemicals in their leaves and stems. The flammable vegetation ensures that most fires will burn large areas.

The most widespread shrub in North American chaparral is chamise (*Adenostoma fasciculatum*), which sprouts from underground structures and produces large numbers of seedlings after a fire. Various species of manzanita (*Arctostaphylos*) and wild lilac (*Ceanothus*) are also widespread throughout chaparral. Some species in each genus both sprout and produce large numbers of seedlings after fires. Other species in each genus depend entirely on heat-stimulated seeds to reestablish their presence in a burned area.

Mediterranean vegetation also occurs on western coasts in southern Australia, where it is called *mallee*; the Cape region of South Africa (*fynbos*); the central coast of Chile (*matorral*); and around the Mediterranean Sea (*maquis*). In all these areas, the vegetation has the same adaptive characteristics and appearance, but the species are not related to those of other areas. Although there are differences among the regions besides the species that occur in each, the similar physical and vegetational characteristics lend a continuity that is widely recognized as the mediterranean scrub biome.

Concerns

As people moved into Mediterranean scrub regions, two major and related concerns surfaced. First, the fires, which are such an important part of scrub ecology, were destructive and dangerous, leading to *fire suppression*. Second, fire suppression may actually increase fire damage and may threaten the mediterranean scrub biome's very existence when combined with other human activities. A comparison of the fire history in the chaparral of California and that of Baja California lends credibility to the idea that fire suppression increases fire damage. Fire suppression has long been practiced in Southern California. In contrast, much less fire suppression has gone on in Baja. Fewer, larger, and more destructive fires burn in Southern California chaparral than in Baja chaparral. The simplest explanation is that fire suppression allows fuel to build up, so that when a fire starts it is essen-

tially unstoppable, as often occurs in California chaparral. With less fire suppression and less fuel accumulation, Baja fires burn more frequently but are smaller and less destructive. The small fires remove the fuel periodically, thus decreasing the danger of large, destructive fires.

There are other differences between California and Baja chaparral that may account for the differences in the fire regimes, but the foregoing hypothesis is interesting from the perspective of human impact on chaparral as well as that of fire's impact on humans. Population growth and its attendant activities threaten the very existence of the chaparral. Humans destroy chaparral to build home sites, suppress fires, and plant grass in burned areas to stabilize the soil and to mitigate future fires. The grasses compete with chaparral plants and retard chaparral recovery. The impact of these and other activities on the native chaparral ecosystem is not well understood but is almost certainly negative. Other mediterranean scrub areas suffer similar fates. Although mediterranean scrub is still well represented in comparison to some biomes, its response to human impact should be carefully studied and monitored, both to protect human investment in mediterranean ecosystems and to preserve the intriguing mediterranean scrub and its many unique plant species.

Carl W. Hoagstrom

See also: African flora; Australian flora; Biomes: types; Central American flora; Community-ecosystem interactions; European flora; Evolution: convergent and divergent; Forest fires; North American flora; South American flora.

Sources for Further Study

Barbour, Michael G., and William Dwight Billings, eds. *North American Terrestrial Vegetation.* 2d ed. New York: Cambridge University Press, 2000. One chapter covers chaparral as one of North America's shrublands.

Dallman, Peter R. *Plant Life in the World's Mediterranean Climates.* Berkeley: University of California Press, 1998. Covers all mediterranean climates, some of which support woodlands and grasslands as well as mediterranean scrub.

Vankat, John L. *The Natural Vegetation of North America: An Introduction.* Melbourne, Fla.: Krieger, 1992. One chapter introduces North American chaparral in the context of shrublands and woodlands.

MEMBRANE STRUCTURE

Categories: Anatomy; cellular biology; physiology

All cells, whether prokaryotic or eukaryotic, are surrounded by a membrane called the plasma membrane, an essential barrier between the external environment and the cytoplasm inside the cell. In addition, eukaryotic cells contain other membranes that are part of a variety of organelles, such as nuclei, plastids, mitochondria, vacuoles, Golgi bodies, and the endomembrane system.

Compartmentalization by membranes allows the function of competing processes, such as respiration and photosynthesis, in separate areas of the same cell. In addition, membranes control which molecules enter or leave the cell and the various organelles. Finally, proteins associated with membranes are responsible for extracellular interactions and the energy transactions involved in photosynthesis and respiration.

Fluid Mosaic Model

In the 1960's it was believed that all cellular membranes were structured as two outer layers of protein surrounding a lipid layer. In 1972 Jonathan Singer and Garth Nicolson proposed the now-accepted *fluid mosaic model*. The lipid component of the membrane forms the basic structure, while the proteins act as enzymes, receptors, and transporters. The lipid molecules, most of which are phospholipids, each have a hydrophilic ("water-loving") end and a hydrophobic ("water-fearing") end and associate together such that they form a lipid bilayer. The hydrophobic ends of the lipids from one layer point toward the hydrophobic ends of the other layer and, by associating only with each other, avoid all contact with water. The hydrophilic ends then form the two water-exposed surfaces.

The major lipids in plant cell membranes are phospholipids (lipids with a phosphorus atom bonded to the hydrophilic end) and sterols. In addition, sugar-containing lipids (glycolipids) and sulfur-containing lipids (sulfolipids) are found to different degrees, depending on the particular membrane. By having different hydrophilic ends, the two surfaces can have a different chemical composition and, therefore, different *membrane properties*.

Proteins are associated with membranes in one of two ways. Those that are loosely bound to the surface are called peripheral proteins, while those tightly bound to the interior through hydrophobic interactions are referred to as integral proteins. Integral proteins may also have large hydrophilic portions extending from the surface on either or both sides of the membrane. Some of the membrane-associated proteins are able to diffuse sideways within the plane of the membrane, giving the membrane a certain fluidity. Membrane proteins with sugar groups attached to the hydrophilic ends are termed glycoproteins and are very common on the membrane surface facing the outside of the cell.

Membrane Properties

The lipid and protein makeup of membranes provides them with several important properties. Due to the hydrophobic nature of the membrane interior, water, some gases, and a few small, non-charged molecules are the only compounds that can cross freely. All other molecules need the help of a transport protein. Membranes have a high electrical resistance and therefore are capable of maintaining a difference in voltage (called a *membrane potential*) from one surface to the other. Membrane potentials are used to drive the transport of charged ions and are also involved in sensing of the environment. In addition, membranes have a low surface tension (that is, they are very "wettable") and a net negative surface charge, so they are capable of binding a variety of water-soluble minerals, ions, and proteins (such as the peripheral proteins).

Transport Across Membranes

Large, hydrophilic, or electrically charged solutes cannot pass directly through a cell membrane, yet plants need to be able to move a variety of molecules among the various organelles in the cell. Therefore, the transport of these solute molecules across membranes is made possible by specific integral, membrane-associated proteins called transport proteins, of which there are two basic types: *channel proteins* (usually just called channels) and *carrier proteins* (also called carriers, transporters, or porters).

Several channel proteins together form a pore through the membrane that is filled with water and lined with electrical charges. The size of the pore and the types and number of charges inside it make each individual channel specific for a particular water-soluble ion. Because the charged solutes merely diffuse through the channel in response to a concentration gradient, no metabolic energy is expended during this type of transport,

which is therefore called passive transport.

For carrier proteins, the solute to be transported binds to a portion of the carrier protein, which induces a change in its shape, and that causes the solute to be moved to the other side of the membrane, where it is released. The carrier protein assumes its previous shape and is then available to bind and transport another solute molecule. Frequently, energy (in the form of adenosine triphosphate, or ATP) is consumed during the operation of carrier proteins in what is known as active transport.

Robert R. Wise

See also: Active transport; Cell wall; Cytosol; Lipids; Liquid transport systems; Osmosis, simple diffusion, and facilitated diffusion; Plasma membranes; Proteins and amino acids; Vesicle-mediated transport.

Sources for Further Study

Buchanan, Bob B., Wilhelm Gruissem, and Russell L. Jones. *Biochemistry and Molecular Biology of Plants*. Rockville, Md.: American Society of Plant Physiologists, 2000. Contains a thorough and detailed treatment of plant membrane structure and function. Includes tables, photographs, and colored drawings.

Hopkins, William G. *Introduction to Plant Physiology*. New York: John Wiley & Sons, 1999. College text for an upper-level plant physiology course. Includes tables, photographs.

Raven, Peter H., Ray F. Evert, and Susan E. Eichhorn. *Biology of Plants*. 6th ed. New York: W. H. Freeman/Worth, 1999. This introductory college textbook has a chapter on membrane structure and function with a strong focus on the role membranes play in water transport. Includes tables, photographs, and colored drawings.

Singer, S. J., and G. L. Nicholson. "The Fluid Mosaic Model of the Structure of Cell Membranes." *Science* 175 (1975): 720-731. The original article on the fluid mosaic model of membrane structure.

METABOLITES: PRIMARY VS. SECONDARY

Categories: Cellular biology; physiology; poisonous, toxic, and invasive plants

Metabolites are compounds synthesized by plants for both essential functions, such as growth and development (primary metabolites), and specific functions, such as pollinator attraction or defense against herbivory (secondary metabolites).

Metabolites are organic compounds synthesized by organisms using enzyme-mediated chemical reactions called *metabolic pathways*. Primary metabolites have functions that are essential to growth and development and are therefore present in all plants. In contrast, secondary metabolites are variously distributed in the plant kingdom, and their functions are specific to the plants in which they are found. Secondary metabolites are often colored, fragrant, or flavorful compounds, and they typically mediate the interaction of plants with other organisms. Such interactions include those of plant-pollinator, plant-pathogen, and plant-herbivore.

Primary Metabolites

Primary metabolites comprise many different types of organic compounds, including, but not limited to, carbohydrates, lipids, proteins, and nucleic acids. They are found universally in the plant kingdom because they are the components or products of fundamental metabolic pathways or cycles such as glycolysis, the Krebs cycle, and the Calvin cycle. Because of the importance of these and other primary pathways in enabling a plant to synthe-

size, assimilate, and degrade organic compounds, primary metabolites are essential.

Examples of primary metabolites include energy-rich fuel molecules, such as sucrose and starch, structural components such as cellulose, informational molecules such as DNA (deoxyribonucleic acid) and RNA (ribonucleic acid), and pigments, such as chlorophyll. In addition to having fundamental roles in plant growth and development, some primary metabolites are precursors (starting materials) for the synthesis of secondary metabolites.

Secondary Metabolites

Secondary metabolites largely fall into three classes of compounds: alkaloids, terpenoids, and phenolics. However, these classes of compounds

Many plant-derived metabolites classed as alkaloids have medicinal properties, including cocaine from the coca plant. These alkaloids affect the nervous system and can be used as painkillers, as long as the dose is carefully administered. In Bolivia, a government minister is shown cutting coca plants.

also include primary metabolites, so whether a compound is a primary or secondary metabolite is a distinction based not only on its chemical structure but also on its function and distribution within the plant kingdom.

Many thousands of secondary metabolites have been isolated from plants, and many of them have powerful physiological effects in humans and are used as medicines. It is only since the late twentieth century that secondary metabolites have been clearly recognized as having important functions in plants. Research has focused on the role of secondary metabolites in plant defense. This is discussed below with reference to alkaloids, though it is relevant to many types of secondary metabolites.

Alkaloids

Alkaloids are a large group of nitrogen-containing compounds, examples of which are known to occur in approximately 20 percent of all flowering plants. Closely related plant species often contain alkaloids of related chemical structure. The primary metabolites from which they are derived include amino acids such as tryptophan, tyrosine, and lysine. Alkaloid biosynthetic pathways can be long, and many alkaloids have correspondingly complex chemical structures. Alkaloids accumulate in plant organs such as leaves or fruits and are ingested by animals that consume those plant parts. Many alkaloids are extremely toxic, especially to mammals, and act as potent nerve poisons, enzyme inhibitors, or membrane transport inhibitors. In addition to being toxic, many alkaloids are also bitter or otherwise bad-tasting. Therefore, the presence of alkaloids and other toxic secondary metabolites can serve as a deterrent to animals, which learn to avoid eating such plants.

Sometimes domesticated animals that have not previously been exposed to alkaloid-containing plants do not have acquired avoidance mechanisms, and they become poisoned. For example, groundsel contains the alkaloid senecionine, which has resulted in many recorded cases of livestock fatalities due to liver failure. More frequently, over time, natural selection has resulted in animals developing biochemical mechanisms or behavioral traits that lead to avoidance of alkaloid-containing plants.

In other, more unusual cases, animals may evolve a mechanism for sequestering (storing) or breaking down a potentially toxic compound, thus "disarming" the plant. For instance, caterpillars of the cinnabar moth can devour groundsel plants and sequester senecionine without suffering any ill effects. Moreover, the caterpillars thereby acquire their own weapon against predators: the plant-derived alkaloid stored within their bodies. Over time, plants acquire new capabilities to synthesize additional defense compounds to combat animals that have developed "resistance" to the original chemicals. This type of an "arms race" is a form of coevolution and may help to account for the incredible abundance of secondary metabolites in flowering plants.

Medicinal Alkaloids

Many potentially toxic plant-derived alkaloids have medicinal properties, as long as they are administered in carefully regulated doses. *Alkaloids* with important medicinal uses include morphine and codeine from the opium poppy and cocaine from the coca plant. These alkaloids act on the nervous system and are used as painkillers. Atropine, from the deadly nightshade plant, also acts on the nervous system and is used in anesthesia and ophthalmology. Vincristine and vinblastine from the periwinkle plant are inhibitors of cell division and are used to treat cancers of the blood and lymphatic systems. Quinine from the bark of the cinchona tree is toxic to the *Plasmodium* parasite, which causes malaria, and has long been used in tropical and subtropical regions of the world. Other alkaloids are used as stimulants, including caffeine, present in coffee, tea, and cola plants (and the drinks derived from these plants), and nicotine, which is present in tobacco. Nicotine preparations are, paradoxically, also used as an aid in smoking cessation. Nicotine is also a very potent insecticide. For many years ground-up tobacco leaves were used for insect control, but this practice was superseded by the use of special formulations of nicotine. More recently the use of nicotine as an insecticide has been discouraged because of its toxicity to humans.

Terpenoids

Terpenoids are derived from acetyl coenzyme A or from intermediates in glycolysis. They are classified by the number of five-carbon isoprenoid units they contain. Monoterpenes (containing two C_5-units) are exemplified by the aromatic oils (such as menthol) contained in the leaves of members of the mint family. In addition to giving these plants their characteristic taste and fragrance, these aromatic oils have insect-repellent qualities. The pyrethroids, which are monoterpene esters from the flowers of chrysanthemum and related species, are used commercially as insecticides. They fatally affect the nervous systems of insects while being biodegradable and nontoxic to mammals, including humans.

Diterpenes are formed from four C_5-units. Paclitaxel (commonly known by the brand name Taxol), a diterpene found in bark of the Pacific yew tree, is a potent inhibitor of cell division in animals. At the end of the twentieth century, paclitaxel was developed as a powerful new chemotherapeutic treatment for people with solid tumors, such as ovarian cancer patients.

Triterpenoids (formed from six C_5-units) comprise the plant steroids, some of which act as plant hormones. These also can protect plants from insect attack, though their mode of action is quite different from that of the pyrethroids. For example, the phytoecdysones are a group of plant sterols that resemble insect molting hormones. When ingested in excess, phytoecdysones can disrupt the normal molting cycle with often lethal consequences to the insect.

Tetraterpenoids (eight C_5-units) include important pigments such as beta-carotene, which is a precursor of vitamin A, and lycopene, which gives tomatoes their red color. Rather than functioning in plant defense, the colored pigments that accumulate in ripening fruits can serve as attractants to animals, which actually aid the plant in seed dispersal.

The polyterpenes are polymers that may contain several thousand isoprenoid units. Rubber, a polyterpene in the latex of rubber trees that probably aids in wound healing in the plant, is also very important for the manufacture of tires and other products.

Phenolic Compounds

Phenolic compounds are defined by the presence of one or more aromatic rings bearing a hydroxyl functional group. Many are synthesized from the amino acid phenylalanine. Simple phenolic compounds, such as salicylic acid, can be important in defense against fungal pathogens. Salicylic acid concentration increases in the leaves of certain plants in response to fungal attack and enables the plant to mount a complex defense response. Interestingly, aspirin, a derivative of salicylic acid, is routinely used in humans to reduce inflammation, pain, and fever. Other phenolic compounds, called *isoflavones*, are synthesized rapidly in plants of the legume family when they are attacked by bacterial or fungal pathogens, and they have strong antimicrobial activity.

Lignin, a complex phenolic macromolecule, is laid down in plant secondary cell walls and is the main component of wood. It is a very important structural molecule in all woody plants, allowing them to achieve height, girth, and longevity. Lignin is also valuable for plant defense: Plant parts containing cells with lignified walls are much less palatable to insects and other animals than are nonwoody plants and are much less easily digested by fungal enzymes than plant parts that contain only cells with primary cellulose walls.

Other phenolics function as attractants. Anthocyanins and anthocyanidins are phenolic pigments that impart pink and purple colors to flowers and fruits. This pigmentation attracts insects and other animals that move between individual plants and accomplish pollination and fruit dispersal. Often the plant pigment and the pollinator's visual systems are well matched: Plants with red flowers attract birds and mammals because these animals possess the correct photoreceptors to see red pigments.

Valerie M. Sponsel

See also: Angiosperm evolution; Animal-plant interactions; Biochemical coevolution in angiosperms; Calvin cycle; Coevolution; Estrogens from plants; Glycolysis and fermentation; Hormones; Krebs cycle; Medicinal plants; Paclitaxel; Pheromones; Pigments in plants; Pollination; Resistance to plant diseases; Rubber.

Sources for Further Study

Levetin, Estelle, and Karen McMahon. *Plants and Society.* Boston: WCB/McGraw-Hill, 1999. Units on commercial products derived from plants and human health describe the important uses of plant secondary metabolites.

Moore, Randy, et al. *Botany.* 2d ed. Boston: WCB/McGraw-Hill, 1998. Chapter 2 introduces the structures of secondary metabolites, and the epilogue describes nature's botanical medicine cabinet.